Study and Solutions Guide

CALCULUS
SEVENTH EDITION
Larson/Hostetler/Edwards

Volume II
Chapters 10-14
and
Appendix A

Bruce H. Edwards
University of Florida

HOUGHTON MIFFLIN COMPANY   Boston   New York

Editor-in-Chief: Jack Shira
Managing Editor: Cathy Cantin
Development Manager: Maureen Ross
Development Editor: Laura Wheel
Assistant Editor: Rosalind Horn
Supervising Editor: Karen Carter
Project Editor: Patty Bergin
Editorial Assistant: Lindsey Gulden
Production Technology Supervisor: Gary Crespo
Marketing Manager: Michael Busnach
Marketing Assistant: Nicole Mollica
Senior Manufacturing Coordinator: Jane Spelman

Printed in the United States of America

ISBN 0-618-14923-6

9-CRS-06 05

Preface

This *Study and Solutions Guide* is designed as a supplement to *Calculus*, Seventh Edition, by Ron Larson, Robert P. Hostetler, and Bruce H. Edwards. All references to chapters, theorems, and exercises relate to the main text. Solutions to every odd-numbered exercise in the text are given with all essential algebraic steps included. Although this supplement is not a substitute for good study habits, it can be valuable when incorporated into a well-planned course of study. For suggestions that may assist you in the use of this text, your lecture notes, and this *Guide*, please refer to the student web site for your text at *college.hmco.com*.

I have made every effort to see that the solutions are correct. However, I would appreciate hearing about any errors or other suggestions for improvement. Good luck with your study of calculus.

Bruce H. Edwards
University of Florida
Gainesville, Florida 32611
(be@math.ufl.edu)

CONTENTS

CONTENTS

C H A P T E R   1 0
Vectors and the Geometry of Space

Section 10.1   Vectors in the Plane

Solutions to Odd-Numbered Exercises

1. (a) $\mathbf{v} = \langle 5 - 1, 3 - 1 \rangle = \langle 4, 2 \rangle$

(b)
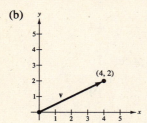

3. (a) $\mathbf{v} = \langle -4 - 3, -2 - (-2) \rangle = \langle -7, 0 \rangle$

(b)
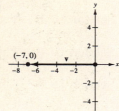

5. $\mathbf{u} = \langle 5 - 3, 6 - 2 \rangle = \langle 2, 4 \rangle$

$\mathbf{v} = \langle 1 - (-1), 8 - 4 \rangle = \langle 2, 4 \rangle$

$\mathbf{u} = \mathbf{v}$

7. $\mathbf{u} = \langle 6 - 0, -2 - 3 \rangle = \langle 6, -5 \rangle$

$\mathbf{v} = \langle 9 - 3, 5 - 10 \rangle = \langle 6, -5 \rangle$

$\mathbf{u} = \mathbf{v}$

9. (b) $\mathbf{v} = \langle 5 - 1, 5 - 2 \rangle = \langle 4, 3 \rangle$

(a) and (c).
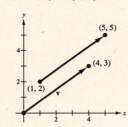

11. (b) $\mathbf{v} = \langle 6 - 10, -1 - 2 \rangle = \langle -4, -3 \rangle$

(a) and (c).
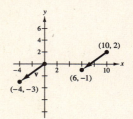

13. (b) $\mathbf{v} = \langle 6 - 6, 6 - 2 \rangle = \langle 0, 4 \rangle$

(a) and (c).
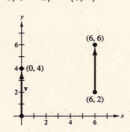

15. (b) $\mathbf{v} = \left\langle \frac{1}{2} - \frac{3}{2}, 3 - \frac{4}{3} \right\rangle = \left\langle -1, \frac{5}{3} \right\rangle$

(a) and (c).
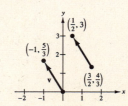

17. (a) $2\mathbf{v} = \langle 4, 6 \rangle$

(b) $-3\mathbf{v} = \langle -6, -9 \rangle$

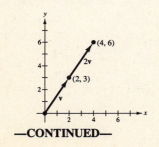

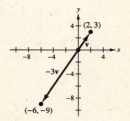

—CONTINUED—

17. —CONTINUED—

(c) $\frac{7}{2}\mathbf{v} = \left\langle 7, \frac{21}{2} \right\rangle$

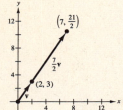

(d) $\frac{2}{3}\mathbf{v} = \left\langle \frac{4}{3}, 2 \right\rangle$

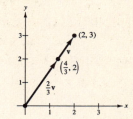

19.

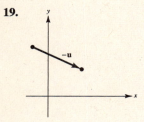

21.

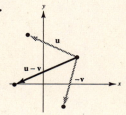

23. (a) $\frac{2}{3}\mathbf{u} = \frac{2}{3}\langle 4, 9 \rangle = \left\langle \frac{8}{3}, 6 \right\rangle$

(b) $\mathbf{v} - \mathbf{u} = \langle 2, -5 \rangle - \langle 4, 9 \rangle = \langle -2, -14 \rangle$

(c) $2\mathbf{u} + 5\mathbf{v} = 2\langle 4, 9 \rangle + 5\langle 2, -5 \rangle = \langle 18, -7 \rangle$

25. $\mathbf{v} = \frac{3}{2}(2\mathbf{i} - \mathbf{j}) = 3\mathbf{i} - \frac{3}{2}\mathbf{j}$

$\qquad = \left\langle 3, -\frac{3}{2} \right\rangle$

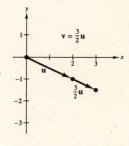

27. $\mathbf{v} = (2\mathbf{i} - \mathbf{j}) + 2(\mathbf{i} + 2\mathbf{j})$

$\qquad = 4\mathbf{i} + 3\mathbf{j} = \langle 4, 3 \rangle$

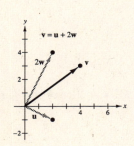

29. $u_1 - 4 = -1$

$\qquad u_2 - 2 = 3$

$\qquad u_1 = 3$

$\qquad u_2 = 5$

$\qquad Q = (3, 5)$

31. $\|\mathbf{v}\| = \sqrt{16 + 9} = 5$

33. $\|\mathbf{v}\| = \sqrt{36 + 25} = \sqrt{61}$

35. $\|\mathbf{v}\| = \sqrt{0 + 16} = 4$

37. $\|\mathbf{u}\| = \sqrt{3^2 + 12^2} = \sqrt{153}$

$\mathbf{v} = \dfrac{\mathbf{u}}{\|\mathbf{u}\|} = \dfrac{\langle 3, 12 \rangle}{\sqrt{153}} = \left\langle \dfrac{3}{\sqrt{153}}, \dfrac{12}{\sqrt{153}} \right\rangle$

$\qquad = \left\langle \dfrac{\sqrt{17}}{17}, \dfrac{4\sqrt{17}}{17} \right\rangle$ unit vector

39. $\|\mathbf{u}\| = \sqrt{\left(\dfrac{3}{2}\right)^2 + \left(\dfrac{5}{2}\right)^2} = \dfrac{\sqrt{34}}{2}$

$\mathbf{v} = \dfrac{\mathbf{u}}{\|\mathbf{u}\|} = \dfrac{\langle (3/2), (5/2) \rangle}{\sqrt{34}/2} = \left\langle \dfrac{3}{\sqrt{34}}, \dfrac{5}{\sqrt{34}} \right\rangle$

$\qquad = \left\langle \dfrac{3\sqrt{34}}{34}, \dfrac{5\sqrt{34}}{34} \right\rangle$ unit vector

41. $\|\mathbf{u}\| = \langle 1, -1 \rangle, \mathbf{v} = \langle -1, 2 \rangle$

   (a) $\|\mathbf{u}\| = \sqrt{1+1} = \sqrt{2}$

   (b) $\|\mathbf{v}\| = \sqrt{1+4} = \sqrt{5}$

   (c) $\mathbf{u} + \mathbf{v} = \langle 0, 1 \rangle$

      $\|\mathbf{u} + \mathbf{v}\| = \sqrt{0+1} = 1$

   (d) $\dfrac{\mathbf{u}}{\|\mathbf{u}\|} = \dfrac{1}{\sqrt{2}} \langle 1, -1 \rangle$

      $\left\| \dfrac{\mathbf{u}}{\|\mathbf{u}\|} \right\| = 1$

   (e) $\dfrac{\mathbf{v}}{\|\mathbf{v}\|} = \dfrac{1}{\sqrt{5}} \langle -1, 2 \rangle$

      $\left\| \dfrac{\mathbf{v}}{\|\mathbf{v}\|} \right\| = 1$

   (f) $\dfrac{\mathbf{u} + \mathbf{v}}{\|\mathbf{u} + \mathbf{v}\|} = \langle 0, 1 \rangle$

      $\left\| \dfrac{\mathbf{u} + \mathbf{v}}{\|\mathbf{u} + \mathbf{v}\|} \right\| = 1$

43. $\mathbf{u} = \left\langle 1, \dfrac{1}{2} \right\rangle, \mathbf{v} = \langle 2, 3 \rangle$

   (a) $\|\mathbf{u}\| = \sqrt{1 + \dfrac{1}{4}} = \dfrac{\sqrt{5}}{2}$

   (b) $\|\mathbf{v}\| = \sqrt{4+9} = \sqrt{13}$

   (c) $\mathbf{u} + \mathbf{v} = \left\langle 3, \dfrac{7}{2} \right\rangle$

      $\|\mathbf{u} + \mathbf{v}\| = \sqrt{9 + \dfrac{49}{4}} = \dfrac{\sqrt{85}}{2}$

   (d) $\dfrac{\mathbf{u}}{\|\mathbf{u}\|} = \dfrac{2}{\sqrt{5}} \left\langle 1, \dfrac{1}{2} \right\rangle$

      $\left\| \dfrac{\mathbf{u}}{\|\mathbf{u}\|} \right\| = 1$

   (e) $\dfrac{\mathbf{v}}{\|\mathbf{v}\|} = \dfrac{1}{\sqrt{13}} \langle 2, 3 \rangle$

      $\left\| \dfrac{\mathbf{v}}{\|\mathbf{v}\|} \right\| = 1$

   (f) $\dfrac{\mathbf{u} + \mathbf{v}}{\|\mathbf{u} + \mathbf{v}\|} = \dfrac{2}{\sqrt{85}} \left\langle 3, \dfrac{7}{2} \right\rangle$

      $\left\| \dfrac{\mathbf{u} + \mathbf{v}}{\|\mathbf{u} + \mathbf{v}\|} \right\| = 1$

45. $\mathbf{u} = \langle 2, 1 \rangle$

    $\|\mathbf{u}\| = \sqrt{5} \approx 2.236$

    $\mathbf{v} = \langle 5, 4 \rangle$

    $\|\mathbf{v}\| = \sqrt{41} \approx 6.403$

    $\mathbf{u} + \mathbf{v} = \langle 7, 5 \rangle$

    $\|\mathbf{u} + \mathbf{v}\| = \sqrt{74} \approx 8.602$

    $\|\mathbf{u} + \mathbf{v}\| \le \|\mathbf{u}\| + \|\mathbf{v}\|$

47. $\dfrac{\mathbf{u}}{\|\mathbf{u}\|} = \dfrac{1}{\sqrt{2}} \langle 1, 1 \rangle$

    $4 \left(\dfrac{\mathbf{u}}{\|\mathbf{u}\|} \right) = 2\sqrt{2} \langle 1, 1 \rangle$

    $\mathbf{v} = \langle 2\sqrt{2}, 2\sqrt{2} \rangle$

49. $\dfrac{\mathbf{u}}{\|\mathbf{u}\|} = \dfrac{1}{2\sqrt{3}} \langle \sqrt{3}, 3 \rangle$

    $2 \left(\dfrac{\mathbf{u}}{\|\mathbf{u}\|} \right) = \dfrac{1}{\sqrt{3}} \langle \sqrt{3}, 3 \rangle$

    $\mathbf{v} = \langle 1, \sqrt{3} \rangle$

51. $\mathbf{v} = 3[(\cos 0°)\mathbf{i} + (\sin 0°)\mathbf{j}] = 3\mathbf{i} = \langle 3, 0 \rangle$

53. $\mathbf{v} = 2[(\cos 150°)\mathbf{i} + (\sin 150°)\mathbf{j}]$

    $= -\sqrt{3}\mathbf{i} + \mathbf{j} = \langle -\sqrt{3}, 1 \rangle$

55. $\mathbf{u} = \mathbf{i}$

    $\mathbf{v} = \dfrac{3\sqrt{2}}{2}\mathbf{i} + \dfrac{3\sqrt{2}}{2}\mathbf{j}$

    $\mathbf{u} + \mathbf{v} = \left(\dfrac{2 + 3\sqrt{2}}{2} \right)\mathbf{i} + \dfrac{3\sqrt{2}}{2}\mathbf{j}$

57. $\mathbf{u} = 2(\cos 4)\mathbf{i} + 2(\sin 4)\mathbf{j}$

$\mathbf{v} = (\cos 2)\mathbf{i} + (\sin 2)\mathbf{j}$

$\mathbf{u} + \mathbf{v} = (2\cos 4 + \cos 2)\mathbf{i} + (2\sin 4 + \sin 2)\mathbf{j}$

59. A scalar is a real number. A vector is represented by a directed line segment. A vector has both length and direction.

61. To normalize $\mathbf{v}$, when $\mathbf{v} \neq 0$, you find a unit vector $\mathbf{u}$ in the direction of $\mathbf{v}$:

$$\mathbf{u} = \frac{\mathbf{v}}{\|\mathbf{v}\|}.$$

For Exercises 63–67, $a\mathbf{u} + b\mathbf{w} = a(\mathbf{i} + 2\mathbf{j}) + b(\mathbf{i} - \mathbf{j}) = (a + b)\mathbf{i} + (2a - b)\mathbf{j}.$

63. $\mathbf{v} = 2\mathbf{i} + \mathbf{j}$. Therefore, $a + b = 2$, $2a - b = 1$. Solving simultaneously, we have $a = 1$, $b = 1$.

65. $\mathbf{v} = 3\mathbf{i}$. Therefore, $a + b = 3$, $2a - b = 0$. Solving simultaneously, we have $a = 1$, $b = 2$.

67. $\mathbf{v} = \mathbf{i} + \mathbf{j}$. Therefore, $a + b = 1$, $2a - b = 1$. Solving simultaneously, we have $a = \frac{2}{3}$, $b = \frac{1}{3}$.

69. $y = x^3$, $y' = 3x^2 = 3$ at $x = 1$.

  (a) $m = 3$. Let $\mathbf{w} = \langle 1, 3 \rangle$, then

  $$\frac{\mathbf{w}}{\|\mathbf{w}\|} = \pm\frac{1}{\sqrt{10}}\langle 1, 3 \rangle.$$

  (b) $m = -\frac{1}{3}$. Let $\mathbf{w} = \langle 3, -1 \rangle$, then

  $$\frac{\mathbf{w}}{\|\mathbf{w}\|} = \pm\frac{1}{\sqrt{10}}\langle 3, -1 \rangle.$$

71. $f(x) = \sqrt{25 - x^2}$

  $$f'(x) = \frac{-x}{\sqrt{25 - x^2}} = \frac{-3}{4} \text{ at } x = 3.$$

  (a) $m = -\frac{3}{4}$. Let $\mathbf{w} = \langle -4, 3 \rangle$, then

  $$\frac{\mathbf{w}}{\|\mathbf{w}\|} = \pm\frac{1}{5}\langle -4, 3 \rangle.$$

  (b) $m = \frac{4}{3}$. Let $\mathbf{w} = \langle 3, 4 \rangle$, then

  $$\frac{\mathbf{w}}{\|\mathbf{w}\|} = \pm\frac{1}{5}\langle 3, 4 \rangle$$

73. $\mathbf{u} = \frac{\sqrt{2}}{2}\mathbf{i} + \frac{\sqrt{2}}{2}\mathbf{j}$

$\mathbf{u} + \mathbf{v} = \sqrt{2}\mathbf{j}$

$\mathbf{v} = (\mathbf{u} + \mathbf{v}) - \mathbf{u} = -\frac{\sqrt{2}}{2}\mathbf{i} + \frac{\sqrt{2}}{2}\mathbf{j}$

75. Programs will vary.

77. $\|\mathbf{F}_1\| = 2$, $\theta_{\mathbf{F}_1} = 33°$

$\|\mathbf{F}_2\| = 3$, $\theta_{\mathbf{F}_2} = -125°$

$\|\mathbf{F}_3\| = 2.5$, $\theta_{\mathbf{F}_3} = 110°$

$\|\mathbf{R}\| = \|\mathbf{F}_1 + \mathbf{F}_2 + \mathbf{F}_3\| \approx 1.33$

$\theta_{\mathbf{R}} = \theta_{\mathbf{F}_1 + \mathbf{F}_2 + \mathbf{F}_3} \approx 132.5°$

79. (a) $180(\cos 30°\,\mathbf{i} + \sin 30°\,\mathbf{j}) + 275\mathbf{i} \approx 430.88\mathbf{i} + 90\mathbf{j}$

  Direction: $\alpha \approx \arctan\left(\dfrac{90}{430.88}\right) \approx 0.206(\approx 11.8°)$

  Magnitude: $\sqrt{430.88^2 + 90^2} \approx 440.18$ newtons

  (b) $M = \sqrt{(275 + 180\cos\theta)^2 + (180\sin\theta)^2}$

  $\alpha = \arctan\left[\dfrac{180\sin\theta}{275 + 180\cos\theta}\right]$

—CONTINUED—

79. —CONTINUED—

(c)

θ	0°	30°	60°	90°	120°	150°	180°
M	455	440.2	396.9	328.7	241.9	149.3	95
α	0°	11.8°	23.1°	33.2°	40.1°	37.1°	0

(d)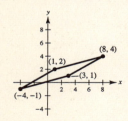

(e) M decreases because the forces change from acting in the same direction to acting in the opposite direction as θ increases from 0° to 180°.

81. $\mathbf{F}_1 + \mathbf{F}_2 + \mathbf{F}_3 = (75 \cos 30°\mathbf{i} + 75 \sin 30°\mathbf{j}) + (100 \cos 45°\mathbf{i} + 100 \sin 45°\mathbf{j}) + (125 \cos 120°\mathbf{i} + 125 \sin 120°\mathbf{j})$

$$= \left(\frac{75}{2}\sqrt{3} + 50\sqrt{2} - \frac{125}{2}\right)\mathbf{i} + \left(\frac{75}{2} + 50\sqrt{2} + \frac{125}{2}\sqrt{3}\right)\mathbf{j}$$

$$\|\mathbf{R}\| = \|\mathbf{F}_1 + \mathbf{F}_2 + \mathbf{F}_3\| \approx 228.5 \text{ lb}$$

$$\theta_\mathbf{R} = \theta_{\mathbf{F}_1 + \mathbf{F}_2 + \mathbf{F}_3} \approx 71.3°$$

83. (a) The forces act along the same direction. $\theta = 0°$.

(b) The forces cancel out each other. $\theta = 180°$.

(c) No, the magnitude of the resultant can not be greater than the sum.

85. $(-4, -1), (6, 5), (10, 3)$

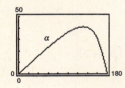

87. $\mathbf{u} = \overrightarrow{CB} = \|\mathbf{u}\|(\cos 30° \, \mathbf{i} + \sin 30° \, \mathbf{j})$

$\mathbf{v} = \overrightarrow{CA} = \|\mathbf{v}\|(\cos 130° \, \mathbf{i} + \sin 130° \, \mathbf{j})$

Vertical components: $\|\mathbf{u}\| \sin 30° + \|\mathbf{v}\| \sin 130° = 2000$

Horizontal components: $\|\mathbf{u}\| \cos 30° + \|\mathbf{v}\| \cos 130° = 0$

Solving this system, you obtain

$\|\mathbf{u}\| \approx 1305.5$ pounds and $\|\mathbf{v}\| \approx 1758.8$ pounds.

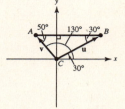

89. Horizontal component $= \|\mathbf{v}\| \cos \theta = 1200 \cos 6° \approx 1193.43$ ft/sec

Vertical component $= \|\mathbf{v}\| \sin \theta = 1200 \sin 6° \approx 125.43$ ft/sec

91. $\mathbf{u} = 900[\cos 148° \, \mathbf{i} + \sin 148° \, \mathbf{j}]$

$\mathbf{v} = 100[\cos 45° \, \mathbf{i} + \sin 45° \, \mathbf{j}]$

$\mathbf{u} + \mathbf{v} = [900 \cos 148° + 100 \cos 45°]\mathbf{i} + [900 \sin 148° + 100 \sin 45°]\mathbf{j}$

$\approx -692.53 \, \mathbf{i} + 547.64 \, \mathbf{j}$

$\theta \approx \arctan\left(\dfrac{547.64}{-692.53}\right) \approx -38.34°. \quad 38.34° \text{ North of West.}$

$\|\mathbf{u} + \mathbf{v}\| \approx \sqrt{(-692.53)^2 + (547.64)^2} \approx 882.9 \text{ km/hr.}$

93. $\mathbf{F}_1 + \mathbf{F}_2 + \mathbf{F}_3 = \mathbf{0}$

$-3600\mathbf{j} + T_2(\cos 35°\mathbf{i} - \sin 35° \, \mathbf{j}) + T_3(\cos 92°\mathbf{i} + \sin 92°\mathbf{j}) = \mathbf{0}$

$T_2 \cos 35° + T_3 \cos 92° = 0$

$-T_2 \sin 35° + T_3 \sin 92° = 3600$

$T_2 = \dfrac{-T_3 \cos 92°}{\cos 35°} \Longrightarrow \dfrac{T_3 \cos 92°}{\cos 35°} \sin 35° + T_3 \sin 92° = 3600 \text{ and } T_3(0.97495) \approx 3600 \Longrightarrow T_3 \approx 3692.48$

Finally, $T_2 \approx 157.32$

95.   Let the triangle have vertices at $(0, 0)$, $(a, 0)$, and (b, c).
Let $\mathbf{u}$ be the vector joining $(0, 0)$ and (b, c), as indicated
in the figure. Then $\mathbf{v}$, the vector joining the midpoints, is

$\mathbf{v} = \left(\dfrac{a + b}{2} - \dfrac{a}{2}\right)\mathbf{i} + \dfrac{c}{2}\mathbf{j}$

$= \dfrac{b}{2}\mathbf{i} + \dfrac{c}{2}\mathbf{j} = \dfrac{1}{2}(b\mathbf{i} + c\mathbf{j}) = \dfrac{1}{2}\mathbf{u}$

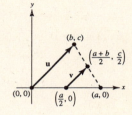

97. $\mathbf{w} = \|\mathbf{u}\|\mathbf{v} + \|\mathbf{v}\|\mathbf{u}$

$= \|\mathbf{u}\|[\|\mathbf{v}\| \cos \theta_v\mathbf{i} + \|\mathbf{v}\| \sin \theta_v\mathbf{j}] + \|\mathbf{v}\|[\|\mathbf{u}\| \cos \theta_u\mathbf{i} + \|\mathbf{u}\| \sin \theta_u\mathbf{j}] = \|\mathbf{u}\| \, \|\mathbf{v}\|[(\cos \theta_u + \cos \theta_v)\mathbf{i} + (\sin \theta_u + \sin \theta_v)\mathbf{j}]$

$= 2\|\mathbf{u}\| \, \|\mathbf{v}\|\left[\cos\left(\dfrac{\theta_u + \theta_v}{2}\right)\cos\left(\dfrac{\theta_u - \theta_v}{2}\right)\mathbf{i} + \sin\left(\dfrac{\theta_u + \theta_v}{2}\right)\cos\left(\dfrac{\theta_u - \theta_v}{2}\right)\mathbf{j}\right]$

$\tan \theta_w = \dfrac{\sin\left(\dfrac{\theta_u + \theta_v}{2}\right)\cos\left(\dfrac{\theta_u - \theta_v}{2}\right)}{\cos\left(\dfrac{\theta_u + \theta_v}{2}\right)\cos\left(\dfrac{\theta_u - \theta_v}{2}\right)} = \tan\left(\dfrac{\theta_u + \theta_v}{2}\right)$

Thus, $\theta_w = (\theta_u + \theta_v)/2$ and $\mathbf{w}$ bisects the angle between $\mathbf{u}$ and $\mathbf{v}$.

99. True   **101.** True   **103.** False

$\|a\mathbf{i} + b\mathbf{j}\| = \sqrt{2}\,|a|$

Section 10.2   Space Coordinates and Vectors in Space

1.

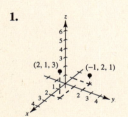

3.

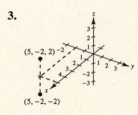

5. $A(2, 3, 4)$

$B(-1, -2, 2)$

7. $x = -3, y = 4, z = 5$: $(-3, 4, 5)$

9. $y = z = 0, x = 10$: $(10, 0, 0)$

11. The z-coordinate is 0.

13. The point is 6 units above the xy-plane.

15. The point is on the plane parallel to the yz-plane that passes through $x = 4$.

17. The point is to the left of the xz-plane.

19. The point is on or between the planes $y = 3$ and $y = -3$.

21. The point (x, y, z) is 3 units below the xy-plane, and below either quadrant I or III.

23. The point could be above the xy-plane and thus above quadrants II or IV, or below the xy-plane, and thus below quadrants I or III.

25. $d = \sqrt{(5 - 0)^2 + (2 - 0)^2 + (6 - 0)^2}$

$= \sqrt{25 + 4 + 36} = \sqrt{65}$

27. $d = \sqrt{(6 - 1)^2 + (-2 - (-2))^2 + (-2 - 4)^2}$

$= \sqrt{25 + 0 + 36} = \sqrt{61}$

29. $A(0, 0, 0), B(2, 2, 1), C(2, -4, 4)$

$|AB| = \sqrt{4 + 4 + 1} = 3$

$|AC| = \sqrt{4 + 16 + 16} = 6$

$|BC| = \sqrt{0 + 36 + 9} = 3\sqrt{5}$

$|BC|^2 = |AB|^2 + |AC|^2$

Right triangle, not isosceles

31. $A(1, -3, -2), B(5, -1, 2), C(-1, 1, 2)$

$|AB| = \sqrt{16 + 4 + 16} = 6$

$|AC| = \sqrt{4 + 16 + 16} = 6$

$|BC| = \sqrt{36 + 4 + 0} = 2\sqrt{10}$

Since $|AB| = |AC|$, the triangle is isosceles, not a right triangle.

33. The z-coordinate is changed by 5 units:

$(0, 0, 5), (2, 2, 6), (2, -4, 9)$

35. $\left(\dfrac{5 + (-2)}{2}, \dfrac{-9 + 3}{2}, \dfrac{7 + 3}{2}\right) = \left(\dfrac{3}{2}, -3, 5\right)$

37. Center: $(0, 2, 5)$

Radius: 2

$(x - 0)^2 + (y - 2)^2 + (z - 5)^2 = 4$

$x^2 + y^2 + z^2 - 4y - 10z + 25 = 0$

39. Center: $\dfrac{(2, 0, 0) + (0, 6, 0)}{2} = (1, 3, 0)$

Radius: $\sqrt{10}$

$(x - 1)^2 + (y - 3)^2 + (z - 0)^2 = 10$

$x^2 + y^2 + z^2 - 2x - 6y = 0$

41. $x^2 + y^2 + z^2 - 2x + 6y + 8z + 1 = 0$

$(x^2 - 2x + 1) + (y^2 + 6y + 9) + (z^2 + 8z + 16) = -1 + 1 + 9 + 16$

$(x - 1)^2 + (y + 3)^2 + (z + 4)^2 = 25$

Center: $(1, -3, -4)$

Radius: 5

43. $9x^2 + 9y^2 + 9z^2 - 6x + 18y + 1 = 0$

$$x^2 + y^2 + z^2 - \frac{2}{3}x + 2y + \frac{1}{9} = 0$$

$$\left(x^2 - \frac{2}{3}x + \frac{1}{9}\right) + (y^2 + 2y + 1) + z^2 = -\frac{1}{9} + \frac{1}{9} + 1$$

$$\left(x - \frac{1}{3}\right)^2 + (y + 1)^2 + (z - 0)^2 = 1$$

Center: $\left(\frac{1}{3}, -1, 0\right)$

Radius: 1

45. $x^2 + y^2 + z^2 \le 36$

Solid ball of radius 6 centered at origin.

47. (a) $\mathbf{v} = (2 - 4)\mathbf{i} + (4 - 2)\mathbf{j} + (3 - 1)\mathbf{k}$

$$= -2\mathbf{i} + 2\mathbf{j} + 2\mathbf{k} = \langle -2, 2, 2 \rangle$$

(b)

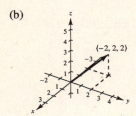

49. (a) $\mathbf{v} = (0 - 3)\mathbf{i} + (3 - 3)\mathbf{j} + (3 - 0)\mathbf{k}$

$$= -3\mathbf{i} + 3\mathbf{k} = \langle -3, 0, 3 \rangle$$

(b)

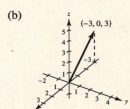

51. $\langle 4 - 3, 1 - 2, 6 - 0 \rangle = \langle 1, -1, 6 \rangle$

$\|\langle 1, -1, 6 \rangle\| = \sqrt{1 + 1 + 36} = \sqrt{38}$

Unit vector: $\dfrac{\langle 1, -1, 6 \rangle}{\sqrt{38}} = \left\langle \dfrac{1}{\sqrt{38}}, \dfrac{-1}{\sqrt{38}}, \dfrac{6}{\sqrt{38}} \right\rangle$

53. $\langle -5 - (-4), 3 - 3, 0 - 1 \rangle = \langle -1, 0, -1 \rangle$

$\|\langle -1, 0, -1 \rangle\| = \sqrt{1 + 1} = \sqrt{2}$

Unit vector: $\left\langle \dfrac{-1}{\sqrt{2}}, 0, \dfrac{-1}{\sqrt{2}} \right\rangle$

55. (b) $\mathbf{v} = (3 + 1)\mathbf{i} + (3 - 2)\mathbf{j} + (4 - 3)\mathbf{k}$

$$= 4\mathbf{i} + \mathbf{j} + \mathbf{k} = \langle 4, 1, 1 \rangle$$

(a) and (c).

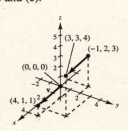

57. $(q_1, q_2, q_3) - (0, 6, 2) = (3, -5, 6)$

$Q = (3, 1, 8)$

59. (a) $2\mathbf{v} = \langle 2, 4, 4 \rangle$

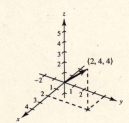

(b) $-\mathbf{v} = \langle -1, -2, -2 \rangle$

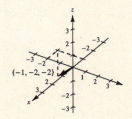

—CONTINUED—

59. **—CONTINUED—**

(c) $\frac{3}{2}\mathbf{v} = \left\langle \frac{3}{2}, 3, 3 \right\rangle$

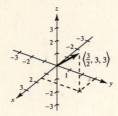

(d) $0\mathbf{v} = \langle 0, 0, 0 \rangle$

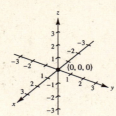

61. $\mathbf{z} = \mathbf{u} - \mathbf{v} = \langle 1, 2, 3 \rangle - \langle 2, 2, -1 \rangle = \langle -1, 0, 4 \rangle$

63. $\mathbf{z} = 2\mathbf{u} + 4\mathbf{v} - \mathbf{w} = \langle 2, 4, 6 \rangle + \langle 8, 8, -4 \rangle - \langle 4, 0, -4 \rangle = \langle 6, 12, 6 \rangle$

65. $2\mathbf{z} - 3\mathbf{u} = 2\langle z_1, z_2, z_3 \rangle - 3\langle 1, 2, 3 \rangle = \langle 4, 0, -4 \rangle$

$2z_1 - 3 = 4 \implies z_1 = \frac{7}{2}$

$2z_2 - 6 = 0 \implies z_2 = 3$

$2z_3 - 9 = -4 \implies z_3 = \frac{5}{2}$

$\mathbf{z} = \left\langle \frac{7}{2}, 3, \frac{5}{2} \right\rangle$

67. (a) and (b) are parallel since $\langle -6, -4, 10 \rangle = -2\langle 3, 2, -5 \rangle$ and $\left\langle 2, \frac{4}{3}, -\frac{10}{3} \right\rangle = \frac{2}{3}\langle 3, 2, -5 \rangle$.

69. $\mathbf{z} = -3\mathbf{i} + 4\mathbf{j} + 2\mathbf{k}$

(a) is parallel since $-6\mathbf{i} + 8\mathbf{j} + 4\mathbf{k} = 2\mathbf{z}$.

71. $P(0, -2, -5)$, $Q(3, 4, 4)$, $R(2, 2, 1)$

$\overrightarrow{PQ} = \langle 3, 6, 9 \rangle$

$\overrightarrow{PR} = \langle 2, 4, 6 \rangle$

$\langle 3, 6, 9 \rangle = \frac{3}{2}\langle 2, 4, 6 \rangle$

Therefore, $\overrightarrow{PQ}$ and $\overrightarrow{PR}$ are parallel. The points are collinear.

73. $P(1, 2, 4)$, $Q(2, 5, 0)$, $R(0, 1, 5)$

$\overrightarrow{PQ} = \langle 1, 3, -4 \rangle$

$\overrightarrow{PR} = \langle -1, -1, 1 \rangle$

Since $\overrightarrow{PQ}$ and $\overrightarrow{PR}$ are not parallel, the points are not collinear.

75. $A(2, 9, 1)$, $B(3, 11, 4)$, $C(0, 10, 2)$, $D(1, 12, 5)$

$\overrightarrow{AB} = \langle 1, 2, 3 \rangle$

$\overrightarrow{CD} = \langle 1, 2, 3 \rangle$

$\overrightarrow{AC} = \langle -2, 1, 1 \rangle$

$\overrightarrow{BD} = \langle -2, 1, 1 \rangle$

Since $\overrightarrow{AB} = \overrightarrow{CD}$ and $\overrightarrow{AC} = \overrightarrow{BD}$, the given points form the vertices of a parallelogram.

77. $\|\mathbf{v}\| = 0$

79. $\mathbf{v} = \langle 1, -2, -3 \rangle$

$\|\mathbf{v}\| = \sqrt{1 + 4 + 9} = \sqrt{14}$

81. $\mathbf{v} = \langle 0, 3, -5 \rangle$

$\|\mathbf{v}\| = \sqrt{0 + 9 + 25} = \sqrt{34}$

83. $\mathbf{u} = \langle 2, -1, 2 \rangle$

$\|\mathbf{u}\| = \sqrt{4 + 1 + 4} = 3$

(a) $\dfrac{\mathbf{u}}{\|\mathbf{u}\|} = \frac{1}{3}\langle 2, -1, 2 \rangle$

(b) $-\dfrac{\mathbf{u}}{\|\mathbf{u}\|} = -\frac{1}{3}\langle 2, -1, 2 \rangle$

85. $\mathbf{u} = \langle 3, 2, -5 \rangle$

$\|\mathbf{u}\| = \sqrt{9 + 4 + 25} = \sqrt{38}$

(a) $\dfrac{\mathbf{u}}{\|\mathbf{u}\|} = \frac{1}{\sqrt{38}}\langle 3, 2, -5 \rangle$

(b) $-\dfrac{\mathbf{u}}{\|\mathbf{u}\|} = -\frac{1}{\sqrt{38}}\langle 3, 2, -5 \rangle$

87. Programs will vary.

89. $c\mathbf{v} = \langle 2c, 2c, -c \rangle$

$\|c\mathbf{v}\| = \sqrt{4c^2 + 4c^2 + c^2} = 5$

$9c^2 = 25$

$c = \pm\dfrac{5}{3}$

91. $\mathbf{v} = 10\dfrac{\mathbf{u}}{\|\mathbf{u}\|} = 10\left\langle 0, \dfrac{1}{\sqrt{2}}, \dfrac{1}{\sqrt{2}} \right\rangle$

$= \left\langle 0, \dfrac{10}{\sqrt{2}}, \dfrac{10}{\sqrt{2}} \right\rangle$

93. $\mathbf{v} = \dfrac{3}{2}\dfrac{\mathbf{u}}{\|\mathbf{u}\|} = \dfrac{3}{2}\left\langle \dfrac{2}{3}, \dfrac{-2}{3}, \dfrac{1}{3} \right\rangle = \left\langle 1, -1, \dfrac{1}{2} \right\rangle$

95. $\mathbf{v} = 2[\cos(\pm 30°)\mathbf{j} + \sin(\pm 30°)\mathbf{k}]$

$= \sqrt{3}\,\mathbf{j} \pm \mathbf{k} = \langle 0, \sqrt{3}, \pm 1 \rangle$

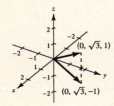

97. $\mathbf{v} = \langle -3, -6, 3 \rangle$

$\dfrac{2}{3}\mathbf{v} = \langle -2, -4, 2 \rangle$

$(4, 3, 0) + (-2, -4, 2) = (2, -1, 2)$

99. (a)

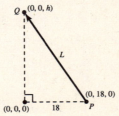

(b) $\mathbf{w} = a\mathbf{u} + b\mathbf{v} = a\mathbf{i} + (a + b)\mathbf{j} + b\mathbf{k} = \mathbf{0}$

$a = 0, a + b = 0, b = 0$

Thus, a and b are both zero.

(c) $a\mathbf{i} + (a + b)\mathbf{j} + b\mathbf{k} = \mathbf{i} + 2\mathbf{j} + \mathbf{k}$

$a = 1, b = 1$

$\mathbf{w} = \mathbf{u} + \mathbf{v}$

(d) $a\mathbf{i} + (a + b)\mathbf{j} + b\mathbf{k} = \mathbf{i} + 2\mathbf{j} + 3\mathbf{k}$

$a = 1, a + b = 2, b = 3$

Not possible

101. $d = \sqrt{(x_2 - x_1)^2 + (y_2 - y_1)^2 + (z_2 - z_1)^2}$

103.   Two nonzero vectors **u** and **v** are parallel if $\mathbf{u} = c\mathbf{v}$ for some scalar c.

105. (a)   The height of the right triangle is $h = \sqrt{L^2 - 18^2}$. The vector $\overrightarrow{PQ}$ is given by

$\overrightarrow{PQ} = \langle 0, -18, h \rangle.$

The tension vector **T** in each wire is

$\mathbf{T} = c\langle 0, -18, h \rangle$ where $ch = \dfrac{24}{3} = 8.$

Hence, $\mathbf{T} = \dfrac{8}{h}\langle 0, -18, h \rangle$ and

$T = \|\mathbf{T}\| = \dfrac{8}{h}\sqrt{18^2 + h^2} = \dfrac{8}{\sqrt{L^2 - 18^2}}\sqrt{18^2 + (L^2 - 18^2)} = \dfrac{8L}{\sqrt{L^2 - 18^2}}$

(b)

L	20	25	30	35	40	45	50
T	18.4	11.5	10	9.3	9.0	8.7	8.6

—**CONTINUED**—

105. —CONTINUED—

(c)

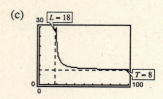

$x = 18$ is a vertical asymptote and $y = 8$ is a horizontal asymptote.

(d) $\displaystyle\lim_{L \to 18^+} \frac{8L}{\sqrt{L^2 - 18^2}} = \infty$

$\displaystyle\lim_{L \to \infty} \frac{8L}{\sqrt{L^2 - 18^2}} = \lim_{L \to \infty} \frac{8}{\sqrt{1 - (18/L)^2}} = 8$

(e) From the table, $T = 10$ implies $L = 30$ inches.

107. Let α be the angle between **v** and the coordinate axes.

$$\mathbf{v} = (\cos \alpha)\mathbf{i} + (\cos \alpha)\mathbf{j} + (\cos \alpha)\mathbf{k}$$

$$\|\mathbf{v}\| = \sqrt{3} \cos \alpha = 1$$

$$\cos \alpha = \frac{1}{\sqrt{3}} = \frac{\sqrt{3}}{3}$$

$$\mathbf{v} = \frac{\sqrt{3}}{3}(\mathbf{i} + \mathbf{j} + \mathbf{k}) = \frac{\sqrt{3}}{3}\langle 1, 1, 1\rangle$$

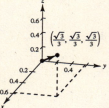

109. $\overrightarrow{AB} = \langle 0, 70, 115\rangle, \mathbf{F}_1 = C_1\langle 0, 70, 115\rangle$

$\overrightarrow{AC} = \langle -60, 0, 115\rangle, \mathbf{F}_2 = C_2\langle -60, 0, 115\rangle$

$\overrightarrow{AD} = \langle 45, -65, 115\rangle, \mathbf{F}_3 = C_3\langle 45, -65, 115\rangle$

$\mathbf{F} = \mathbf{F}_1 + \mathbf{F}_2 + \mathbf{F}_3 = \langle 0, 0, 500\rangle$

Thus:

$$-60C_2 + 45C_3 = 0$$
$$70C_1 \qquad -65C_3 = 0$$
$$115(C_1 + C_2 + C_3) = 500$$

Solving this system yields $C_1 = \frac{104}{69}$, $C_2 = \frac{28}{23}$, and $C_3 = \frac{112}{69}$. Thus:

$\|\mathbf{F}_1\| \approx 202.919N$

$\|\mathbf{F}_2\| \approx 157.909N$

$\|\mathbf{F}_3\| \approx 226.521N$

111. $d(AP) = 2d(BP)$

$$\sqrt{x^2 + (y + 1)^2 + (z - 1)^2} = 2\sqrt{(x - 1)^2 + (y - 2)^2 + z^2}$$

$$x^2 + y^2 + z^2 + 2y - 2z + 2 = 4(x^2 + y^2 + z^2 - 2x - 4y + 5)$$

$$0 = 3x^2 + 3y^2 + 3z^2 - 8x - 18y + 2z + 18$$

$$-6 + \frac{16}{9} + 9 + \frac{1}{9} = \left(x^2 - \frac{8}{3}x + \frac{16}{9}\right) + (y^2 - 6y + 9) + \left(z^2 + \frac{2}{3}z + \frac{1}{9}\right)$$

$$\frac{44}{9} = \left(x - \frac{4}{3}\right)^2 + (y - 3)^2 + \left(z + \frac{1}{3}\right)^2$$

Sphere; center: $\left(\frac{4}{3}, 3, -\frac{1}{3}\right)$, radius: $\frac{2\sqrt{11}}{3}$

Section 10.3   The Dot Product of Two Vectors

1. $\mathbf{u} = \langle 3, 4 \rangle, \mathbf{v} = \langle 2, -3 \rangle$

   (a) $\mathbf{u} \cdot \mathbf{v} = 3(2) + 4(-3) = -6$

   (b) $\mathbf{u} \cdot \mathbf{u} = 3(3) + 4(4) = 25$

   (c) $\|\mathbf{u}\|^2 = 25$

   (d) $(\mathbf{u} \cdot \mathbf{v})\mathbf{v} = -6\langle 2, -3 \rangle = \langle -12, 18 \rangle$

   (e) $\mathbf{u} \cdot (2\mathbf{v}) = 2(\mathbf{u} \cdot \mathbf{v}) = 2(-6) = -12$

3. $\mathbf{u} = \langle 2, -3, 4 \rangle, \mathbf{v} = \langle 0, 6, 5 \rangle$

   (a) $\mathbf{u} \cdot \mathbf{v} = 2(0) + (-3)(6) + (4)(5) = 2$

   (b) $\mathbf{u} \cdot \mathbf{u} = 2(2) + (-3)(-3) + 4(4) = 29$

   (c) $\|\mathbf{u}\|^2 = 29$

   (d) $(\mathbf{u} \cdot \mathbf{v})\mathbf{v} = 2\langle 0, 6, 5 \rangle = \langle 0, 12, 10 \rangle$

   (e) $\mathbf{u} \cdot (2\mathbf{v}) = 2(\mathbf{u} \cdot \mathbf{v}) = 2(2) = 4$

5. $\mathbf{u} = 2\mathbf{i} - \mathbf{j} + \mathbf{k}, \mathbf{v} = \mathbf{i} - \mathbf{k}$

   (a) $\mathbf{u} \cdot \mathbf{v} = 2(1) + (-1)(0) + 1(-1) = 1$

   (b) $\mathbf{u} \cdot \mathbf{u} = 2(2) + (-1)(-1) + (1)(1) = 6$

   (c) $\|\mathbf{u}\|^2 = 6$

   (d) $(\mathbf{u} \cdot \mathbf{v})\mathbf{v} = \mathbf{v} = \mathbf{i} - \mathbf{k}$

   (e) $\mathbf{u} \cdot (2\mathbf{v}) = 2(\mathbf{u} \cdot \mathbf{v}) = 2$

7. $\mathbf{u} = \langle 3240, 1450, 2235 \rangle$

     $\mathbf{v} = \langle 2.22, 1.85, 3.25 \rangle$

  $\mathbf{u} \cdot \mathbf{v} = \$17,139.05$

This gives the total amount that the person earned on his products.

9. $\dfrac{\mathbf{u} \cdot \mathbf{v}}{\|\mathbf{u}\| \, \|\mathbf{v}\|} = \cos \theta$

   $\mathbf{u} \cdot \mathbf{v} = (8)(5) \cos \dfrac{\pi}{3} = 20$

11. $\mathbf{u} = \langle 1, 1 \rangle, \mathbf{v} = \langle 2, -2 \rangle$

   $\cos \theta = \dfrac{\mathbf{u} \cdot \mathbf{v}}{\|\mathbf{u}\| \, \|\mathbf{v}\|} = \dfrac{0}{\sqrt{2}\sqrt{8}} = 0$

   $\theta = \dfrac{\pi}{2}$

13. $\mathbf{u} = 3\mathbf{i} + \mathbf{j}, \mathbf{v} = -2\mathbf{i} + 4\mathbf{j}$

   $\cos \theta = \dfrac{\mathbf{u} \cdot \mathbf{v}}{\|\mathbf{u}\| \, \|\mathbf{v}\|} = \dfrac{-2}{\sqrt{10}\sqrt{20}} = \dfrac{-1}{5\sqrt{2}}$

   $\theta = \arccos\left(-\dfrac{1}{5\sqrt{2}}\right) \approx 98.1°$

15. $\mathbf{u} = \langle 1, 1, 1 \rangle, \mathbf{v} = \langle 2, 1, -1 \rangle$

   $\cos \theta = \dfrac{\mathbf{u} \cdot \mathbf{v}}{\|\mathbf{u}\| \, \|\mathbf{v}\|} = \dfrac{2}{\sqrt{3}\sqrt{6}} = \dfrac{\sqrt{2}}{3}$

   $\theta = \arccos \dfrac{\sqrt{2}}{3} \approx 61.9°$

17. $\mathbf{u} = 3\mathbf{i} + 4\mathbf{j}, \mathbf{v} = -2\mathbf{j} + 3\mathbf{k}$

   $\cos \theta = \dfrac{\mathbf{u} \cdot \mathbf{v}}{\|\mathbf{u}\| \, \|\mathbf{v}\|} = \dfrac{-8}{5\sqrt{13}} = \dfrac{-8\sqrt{13}}{65}$

   $\theta = \arccos\left(-\dfrac{8\sqrt{13}}{65}\right) \approx 116.3°$

19. $\mathbf{u} = \langle 4, 0 \rangle, \mathbf{v} = \langle 1, 1 \rangle$

   $\mathbf{u} \ne c\mathbf{v} \implies$ not parallel

   $\mathbf{u} \cdot \mathbf{v} = 4 \ne 0 \implies$ not orthogonal

   Neither

21. $\mathbf{u} = \langle 4, 3 \rangle, \mathbf{v} = \left\langle \dfrac{1}{2}, -\dfrac{2}{3} \right\rangle$

   $\mathbf{u} \ne c\mathbf{v} \implies$ not parallel

   $\mathbf{u} \cdot \mathbf{v} = 0 \implies$ orthogonal

23. $\mathbf{u} = \mathbf{j} + 6\mathbf{k}, \mathbf{v} = \mathbf{i} - 2\mathbf{j} - \mathbf{k}$

   $\mathbf{u} \ne c\mathbf{v} \implies$ not parallel

   $\mathbf{u} \cdot \mathbf{v} = -8 \ne 0 \implies$ not orthogonal

   Neither

25. $\mathbf{u} = \langle 2, -3, 1 \rangle, \mathbf{v} = \langle -1, -1, -1 \rangle$

   $\mathbf{u} \ne c\mathbf{v} \implies$ not parallel

   $\mathbf{u} \cdot \mathbf{v} = 0 \implies$ orthogonal

27. $\mathbf{u} = \mathbf{i} + 2\mathbf{j} + 2\mathbf{k}, \|\mathbf{u}\| = 3$

$\cos \alpha = \dfrac{1}{3}$

$\cos \beta = \dfrac{2}{3}$

$\cos \gamma = \dfrac{2}{3}$

$\cos^2 \alpha + \cos^2 \beta + \cos^2 \gamma = \dfrac{1}{9} + \dfrac{4}{9} + \dfrac{4}{9} = 1$

29. $\mathbf{u} = \langle 0, 6, -4 \rangle, \|\mathbf{u}\| = \sqrt{52} = 2\sqrt{13}$

$\cos \alpha = 0$

$\cos \beta = \dfrac{3}{\sqrt{13}}$

$\cos \gamma = -\dfrac{2}{\sqrt{13}}$

$\cos^2 \alpha + \cos^2 \beta + \cos^2 \gamma = 0 + \dfrac{9}{13} + \dfrac{4}{13} = 1$

31. $\mathbf{u} = \langle 3, 2, -2 \rangle \quad \|\mathbf{u}\| = \sqrt{17}$

$\cos \alpha = \dfrac{3}{\sqrt{17}} \implies \alpha \approx 0.7560 \text{ or } 43.3°$

$\cos \beta = \dfrac{2}{\sqrt{17}} \implies \beta \approx 1.0644 \text{ or } 61.0°$

$\cos \gamma = \dfrac{-2}{\sqrt{17}} \implies \gamma \approx 2.0772 \text{ or } 119.0°$

33. $\mathbf{u} = \langle -1, 5, 2 \rangle \quad \|\mathbf{u}\| = \sqrt{30}$

$\cos \alpha = \dfrac{-1}{\sqrt{30}} \implies \alpha \approx 1.7544 \text{ or } 100.5°$

$\cos \beta = \dfrac{5}{\sqrt{30}} \implies \beta \approx 0.4205 \text{ or } 24.1°$

$\cos \gamma = \dfrac{2}{\sqrt{30}} \implies \gamma \approx 1.1970 \text{ or } 68.6°$

35. $\mathbf{F}_1: C_1 = \dfrac{50}{\|\mathbf{F}_1\|} \approx 4.3193$

$\mathbf{F}_2: C_2 = \dfrac{80}{\|\mathbf{F}_2\|} \approx 5.4183$

$\mathbf{F} = \mathbf{F}_1 + \mathbf{F}_2$

$\quad \approx 4.3193 \langle 10, 5, 3 \rangle + 5.4183 \langle 12, 7, -5 \rangle$

$\quad = \langle 108.2126, 59.5246, -14.1336 \rangle$

$\|\mathbf{F}\| \approx 124.310 \text{ lb}$

$\cos \alpha \approx \dfrac{108.2126}{\|\mathbf{F}\|} \implies \alpha \approx 29.48°$

$\cos \beta \approx \dfrac{59.5246}{\|\mathbf{F}\|} \implies \beta \approx 61.39°$

$\cos \gamma \approx \dfrac{-14.1336}{\|\mathbf{F}\|} \implies \gamma \approx 96.53°$

37. Let s = length of a side.

$\mathbf{v} = \langle s, s, s \rangle$

$\|\mathbf{v}\| = s\sqrt{3}$

$\cos \alpha = \cos \beta = \cos \gamma = \dfrac{s}{s\sqrt{3}} = \dfrac{1}{\sqrt{3}}$

$\alpha = \beta = \gamma = \arccos\left(\dfrac{1}{\sqrt{3}}\right) \approx 54.7°$

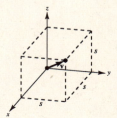

39. $\overrightarrow{OA} = \langle 0, 10, 10 \rangle$

$\cos \alpha = \dfrac{0}{\sqrt{0^2 + 10^2 + 10^2}} = 0 \implies \alpha = 90°$

$\cos \beta = \cos \gamma = \dfrac{10}{\sqrt{0^2 + 10^2 + 10^2}}$

$\quad = \dfrac{1}{\sqrt{2}} \implies \beta = \gamma = 45°$

41. $\mathbf{w}_2 = \mathbf{u} - \mathbf{w}_1 = \langle 6, 7 \rangle - \langle 2, 8 \rangle = \langle 4, -1 \rangle$

43. $\mathbf{w}_2 = \mathbf{u} - \mathbf{w}_1 = \langle 0, 3, 3 \rangle - \langle -2, 2, 2 \rangle = \langle 2, 1, 1 \rangle$

45. $\mathbf{u} = \langle 2, 3 \rangle, \mathbf{v} = \langle 5, 1 \rangle$

(a) $\mathbf{w}_1 = \left(\dfrac{\mathbf{u} \cdot \mathbf{v}}{\|\mathbf{v}\|^2}\right)\mathbf{v} = \dfrac{13}{26}\langle 5, 1 \rangle = \left\langle \dfrac{5}{2}, \dfrac{1}{2} \right\rangle$

(b) $\mathbf{w}_2 = \mathbf{u} - \mathbf{w}_1 = \left\langle -\dfrac{1}{2}, \dfrac{5}{2} \right\rangle$

47. $\mathbf{u} = \langle 2, 1, 2 \rangle$, $\mathbf{v} = \langle 0, 3, 4 \rangle$

   (a) $\mathbf{w}_1 = \left(\dfrac{\mathbf{u} \cdot \mathbf{v}}{\|\mathbf{v}\|^2} \right) \mathbf{v}$

       $= \dfrac{11}{25} \langle 0, 3, 4 \rangle = \left\langle 0, \dfrac{33}{25}, \dfrac{44}{25} \right\rangle$

   (b) $\mathbf{w}_2 = \mathbf{u} - \mathbf{w}_1 = \left\langle 2, -\dfrac{8}{25}, \dfrac{6}{25} \right\rangle$

49. $\mathbf{u} \cdot \mathbf{v} = \langle u_1, u_2, u_3 \rangle \cdot \langle v_1, v_2, v_3 \rangle = u_1 v_1 + u_2 v_2 + u_3 v_3$

51. (a) Orthogonal, $\theta = \dfrac{\pi}{2}$ (b) Acute, $0 < \theta < \dfrac{\pi}{2}$ (c) Obtuse, $\dfrac{\pi}{2} < \theta < \pi$

53. See page 738. Direction cosines of $\mathbf{v} = \langle v_1, v_2, v_3 \rangle$ are

   $\cos \alpha = \dfrac{v_1}{\|\mathbf{v}\|}$, $\cos \beta = \dfrac{v_2}{\|\mathbf{v}\|}$, $\cos \gamma = \dfrac{v_3}{\|\mathbf{v}\|}$.

   α, β, and γ are the direction angles. See Figure 10.26.

55. (a) $\left(\dfrac{\mathbf{u} \cdot \mathbf{v}}{\|\mathbf{v}\|^2} \right) \mathbf{v} = \mathbf{u} \implies \mathbf{u} = c\mathbf{v} \implies \mathbf{u}$ and $\mathbf{v}$ are parallel.

   (b) $\left(\dfrac{\mathbf{u} \cdot \mathbf{v}}{\|\mathbf{v}\|^2} \right) \mathbf{v} = \mathbf{0} \implies \mathbf{u} \cdot \mathbf{v} = 0 \implies \mathbf{u}$ and $\mathbf{v}$
     are orthogonal.

57. Programs will vary.

59. Programs will vary.

61. Because $\mathbf{u}$ appears to be perpendicular to $\mathbf{v}$, the projection of $\mathbf{u}$ onto $\mathbf{v}$ is $\mathbf{0}$. Analytically,

   $\text{proj}_{\mathbf{v}}\,\mathbf{u} = \dfrac{\mathbf{u} \cdot \mathbf{v}}{\|\mathbf{v}\|^2} \mathbf{v} = \dfrac{\langle 2, -3 \rangle \cdot \langle 6, 4 \rangle}{\|\langle 6, 4 \rangle\|^2} \langle 6, 4 \rangle = 0 \langle 6, 4 \rangle = \mathbf{0}$.

63. $\mathbf{u} = \dfrac{1}{2}\mathbf{i} - \dfrac{2}{3}\mathbf{j}$. Want $\mathbf{u} \cdot \mathbf{v} = 0$.

   $\mathbf{v} = 8\mathbf{i} + 6\mathbf{j}$ and $-\mathbf{v} = -8\mathbf{i} - 6\mathbf{j}$ are orthogonal to $\mathbf{u}$.

65. $\mathbf{u} = \langle 3, 1, -2 \rangle$. Want $\mathbf{u} \cdot \mathbf{v} = 0$.

   $\mathbf{v} = \langle 0, 2, 1 \rangle$ and $-\mathbf{v} = \langle 0, -2, -1 \rangle$ are orthogonal to $\mathbf{u}$.

67. (a) Gravitational Force $\mathbf{F} = -48,000\,\mathbf{j}$

      $\mathbf{v} = \cos 10° \,\mathbf{i} + \sin 10° \,\mathbf{j}$

      $\mathbf{w}_1 = \dfrac{\mathbf{F} \cdot \mathbf{v}}{\|\mathbf{v}\|^2} \mathbf{v} = (\mathbf{F} \cdot \mathbf{v})\mathbf{v} = (-48,000)(\sin 10°)\mathbf{v}$

                $\approx -8335.1(\cos 10° \,\mathbf{i} + \sin 10° \,\mathbf{j})$

      $\|\mathbf{w}_1\| \approx 8335.1$ lb

   (b) $\mathbf{w}_2 = \mathbf{F} \cdot \mathbf{w}_1 = -48,000\,\mathbf{j} + 8335.1(\cos 10° \,\mathbf{i} + \sin 10° \,\mathbf{j})$

         $= 8208.5\,\mathbf{i} - 46,552.6\,\mathbf{j}$

      $\|\mathbf{w}_2\| \approx 47,270.8$ lb

69. $\mathbf{F} = 85 \left(\dfrac{1}{2}\mathbf{i} + \dfrac{\sqrt{3}}{2}\mathbf{j} \right)$

   $\mathbf{v} = 10\mathbf{i}$

   $W = \mathbf{F} \cdot \mathbf{v} = 425$ ft $\cdot$ lb

71. $\overrightarrow{PQ} = \langle 4, 7, 5 \rangle$

   $\mathbf{v} = \langle 1, 4, 8 \rangle$

   $W = \overrightarrow{PQ} \cdot \mathbf{v} = 72$

73. False. Let $\mathbf{u} = \langle 2, 4 \rangle$, $\mathbf{v} = \langle 1, 7 \rangle$ and $\mathbf{w} = \langle 5, 5 \rangle$. Then $\mathbf{u} \cdot \mathbf{v} = 2 + 28 = 30$ and $\mathbf{u} \cdot \mathbf{w} = 10 + 20 = 30$.

75. In a rhombus, $\|\mathbf{u}\| = \|\mathbf{v}\|$. The diagonals are $\mathbf{u} + \mathbf{v}$ and $\mathbf{u} - \mathbf{v}$.

   $(\mathbf{u} + \mathbf{v}) \cdot (\mathbf{u} - \mathbf{v}) = (\mathbf{u} + \mathbf{v}) \cdot \mathbf{u} - (\mathbf{u} + \mathbf{v}) \cdot \mathbf{v}$

                         $= \mathbf{u} \cdot \mathbf{u} + \mathbf{v} \cdot \mathbf{u} - \mathbf{u} \cdot \mathbf{v} - \mathbf{v} \cdot \mathbf{v}$

                         $= \|\mathbf{u}\|^2 - \|\mathbf{v}\|^2 = 0$

   Therefore, the diagonals are orthogonal.

77. $\mathbf{u} = \langle \cos \alpha, \sin \alpha, 0 \rangle, \mathbf{v} = \langle \cos \beta, \sin \beta, 0 \rangle$

The angle between $\mathbf{u}$ and $\mathbf{v}$ is $\alpha - \beta$. (Assuming that $\alpha > \beta$). Also,

$$\cos(\alpha - \beta) = \frac{\mathbf{u} \cdot \mathbf{v}}{\|\mathbf{u}\| \|\mathbf{v}\|} = \frac{\cos \alpha \cos \beta + \sin \alpha \sin \beta}{(1)(1)} = \cos \alpha \cos \beta + \sin \alpha \sin \beta.$$

79. $\|\mathbf{u} - \mathbf{v}\|^2 = (\mathbf{u} - \mathbf{v}) \cdot (\mathbf{u} - \mathbf{v})$

$\qquad = (\mathbf{u} - \mathbf{v}) \cdot \mathbf{u} - (\mathbf{u} - \mathbf{v}) \cdot \mathbf{v}$

$\qquad = \mathbf{u} \cdot \mathbf{u} - \mathbf{v} \cdot \mathbf{u} - \mathbf{u} \cdot \mathbf{v} + \mathbf{v} \cdot \mathbf{v}$

$\qquad = \|\mathbf{u}\|^2 - \mathbf{u} \cdot \mathbf{v} - \mathbf{u} \cdot \mathbf{v} + \|\mathbf{v}\|^2$

$\qquad = \|\mathbf{u}\|^2 + \|\mathbf{v}\|^2 - 2\mathbf{u} \cdot \mathbf{v}$

81. $\|\mathbf{u} + \mathbf{v}\|^2 = (\mathbf{u} + \mathbf{v}) \cdot (\mathbf{u} + \mathbf{v})$

$\qquad = (\mathbf{u} + \mathbf{v}) \cdot \mathbf{u} + (\mathbf{u} + \mathbf{v}) \cdot \mathbf{v}$

$\qquad = \mathbf{u} \cdot \mathbf{u} + \mathbf{v} \cdot \mathbf{u} + \mathbf{u} \cdot \mathbf{v} + \mathbf{v} \cdot \mathbf{v}$

$\qquad = \|\mathbf{u}\|^2 + 2\mathbf{u} \cdot \mathbf{v} + \|\mathbf{v}\|^2$

$\qquad \leq \|\mathbf{u}\|^2 + 2\|\mathbf{u}\| \|\mathbf{v}\| + \|\mathbf{v}\|^2$ from Exercise 80

$\qquad \leq (\|\mathbf{u}\| + \|\mathbf{v}\|)^2$

Therefore, $\|\mathbf{u} + \mathbf{v}\| \leq \|\mathbf{u}\| + \|\mathbf{v}\|$.

Section 10.4   The Cross Product of Two Vectors in Space

1. $\mathbf{j} \times \mathbf{i} = \begin{vmatrix} \mathbf{i} & \mathbf{j} & \mathbf{k} \\ 0 & 1 & 0 \\ 1 & 0 & 0 \end{vmatrix} = -\mathbf{k}$

3. $\mathbf{j} \times \mathbf{k} = \begin{vmatrix} \mathbf{i} & \mathbf{j} & \mathbf{k} \\ 0 & 1 & 0 \\ 0 & 0 & 1 \end{vmatrix} = \mathbf{i}$

5. $\mathbf{i} \times \mathbf{k} = \begin{vmatrix} \mathbf{i} & \mathbf{j} & \mathbf{k} \\ 1 & 0 & 0 \\ 0 & 0 & 1 \end{vmatrix} = -\mathbf{j}$

7. (a) $\mathbf{u} \times \mathbf{v} = \begin{vmatrix} \mathbf{i} & \mathbf{j} & \mathbf{k} \\ -2 & 3 & 4 \\ 3 & 7 & 2 \end{vmatrix} = \langle -22, 16, -23 \rangle$

(b) $\mathbf{v} \times \mathbf{u} = -(\mathbf{u} \times \mathbf{v}) = \langle 22, -16, 23 \rangle$

(c) $\mathbf{v} \times \mathbf{v} = \begin{vmatrix} \mathbf{i} & \mathbf{j} & \mathbf{k} \\ 3 & 7 & 2 \\ 3 & 7 & 2 \end{vmatrix} = \mathbf{0}$

9. (a) $\mathbf{u} \times \mathbf{v} = \begin{vmatrix} \mathbf{i} & \mathbf{j} & \mathbf{k} \\ 7 & 3 & 2 \\ 1 & -1 & 5 \end{vmatrix} = \langle 17, -33, -10 \rangle$

(b) $\mathbf{v} \times \mathbf{u} = -(\mathbf{u} \times \mathbf{v}) = \langle -17, 33, 10 \rangle$

(c) $\mathbf{v} \times \mathbf{v} = \mathbf{0}$

11. $\mathbf{u} = \langle 2, -3, 1 \rangle, \mathbf{v} = \langle 1, -2, 1 \rangle$

$\mathbf{u} \times \mathbf{v} = \begin{vmatrix} \mathbf{i} & \mathbf{j} & \mathbf{k} \\ 2 & -3 & 1 \\ 1 & -2 & 1 \end{vmatrix} = -\mathbf{i} - \mathbf{j} - \mathbf{k} = \langle -1, -1, -1 \rangle$

$\mathbf{u} \cdot (\mathbf{u} \times \mathbf{v}) = 2(-1) + (-3)(-1) + (1)(-1) = 0 \implies \mathbf{u} \perp \mathbf{u} \times \mathbf{v}$

$\mathbf{v} \cdot (\mathbf{u} \times \mathbf{v}) = 1(-1) + (-2)(-1) + (1)(-1) = 0 \implies \mathbf{v} \perp \mathbf{u} \times \mathbf{v}$

13. $u = \langle 12, -3, 0 \rangle, v = \langle -2, 5, 0 \rangle$

$$u \times v = \begin{vmatrix} i & j & k \\ 12 & -3 & 0 \\ -2 & 5 & 0 \end{vmatrix} = 54k = \langle 0, 0, 54 \rangle$$

$$u \cdot (u \times v) = 12(0) + (-3)(0) + 0(54)$$
$$= 0 \implies u \perp u \times v$$

$$v \cdot (u \times v) = -2(0) + 5(0) + 0(54)$$
$$= 0 \implies v \perp u \times v$$

15. $u = i + j + k, v = 2i + j - k$

$$u \times v = \begin{vmatrix} i & j & k \\ 1 & 1 & 1 \\ 2 & 1 & -1 \end{vmatrix} = -2i + 3j - k = \langle -2, 3, -1 \rangle$$

$$u \cdot (u \times v) = 1(-2) + 1(3) + 1(-1)$$
$$= 0 \implies u \perp u \times v$$

$$v \cdot (u \times v) = 2(-2) + 1(3) + (-1)(-1)$$
$$= 0 \implies v \perp u \times v$$

17.

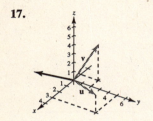

19. $(-v) \times u = -(v \times u) = u \times v$

21. $u = \langle 4, -3.5, 7 \rangle$

$v = \langle -1, 8, 4 \rangle$

$u \times v = \left\langle -70, -23, \dfrac{57}{2} \right\rangle$

$\dfrac{u \times v}{\|u \times v\|} = \left\langle \dfrac{-140}{\sqrt{24{,}965}}, \dfrac{-46}{\sqrt{24{,}965}}, \dfrac{57}{\sqrt{24{,}965}} \right\rangle$

23. $u = -3i + 2j - 5k$

$v = \dfrac{1}{2}i - \dfrac{3}{4}j + \dfrac{1}{10}k$

$u \times v = \left\langle -\dfrac{71}{20}, -\dfrac{11}{5}, \dfrac{5}{4} \right\rangle$

$\dfrac{u \times v}{\|u \times v\|} = \dfrac{20}{\sqrt{7602}} \left\langle -\dfrac{71}{20}, -\dfrac{11}{5}, \dfrac{5}{4} \right\rangle$

$= \left\langle -\dfrac{71}{\sqrt{7602}}, -\dfrac{44}{\sqrt{7602}}, \dfrac{25}{\sqrt{7602}} \right\rangle$

25. Programs will vary.

27. $u = j$

$v = j + k$

$$u \times v = \begin{vmatrix} i & j & k \\ 0 & 1 & 0 \\ 0 & 1 & 1 \end{vmatrix} = i$$

$A = \|u \times v\| = \|i\| = 1$

29. $u = \langle 3, 2, -1 \rangle$

$v = \langle 1, 2, 3 \rangle$

$$u \times v = \begin{vmatrix} i & j & k \\ 3 & 2 & -1 \\ 1 & 2 & 3 \end{vmatrix} = \langle 8, -10, 4 \rangle$$

$A = \|u \times v\| = \|\langle 8, -10, 4 \rangle\| = \sqrt{180} = 6\sqrt{5}$

31. $A(1, 1, 1,), B(2, 3, 4), C(6, 5, 2), D(7, 7, 5)$

$\overrightarrow{AB} = \langle 1, 2, 3 \rangle, \overrightarrow{AC} = \langle 5, 4, 1 \rangle, \overrightarrow{CD} = \langle 1, 2, 3 \rangle,$
$\overrightarrow{BD} = \langle 5, 4, 1 \rangle$

Since $\overrightarrow{AB} = \overrightarrow{CD}$ and $\overrightarrow{AC} = \overrightarrow{BD}$, the figure is a parallelogram. $\overrightarrow{AB}$ and $\overrightarrow{AC}$ are adjacent sides and

$$\overrightarrow{AB} \times \overrightarrow{AC} = \begin{vmatrix} i & j & k \\ 1 & 2 & 3 \\ 5 & 4 & 1 \end{vmatrix} = -10i + 14j - 6k.$$

$A = \|\overrightarrow{AB} \times \overrightarrow{AC}\| = \sqrt{332} = 2\sqrt{83}$

33. $A(0, 0, 0), B(1, 2, 3), C(-3, 0, 0)$

$\overrightarrow{AB} = \langle 1, 2, 3 \rangle, \overrightarrow{AC} = \langle -3, 0, 0 \rangle$

$$\overrightarrow{AB} \times \overrightarrow{AC} = \begin{vmatrix} i & j & k \\ 1 & 2 & 3 \\ -3 & 0 & 0 \end{vmatrix} = -9j + 6k$$

$A = \dfrac{1}{2}\|\overrightarrow{AB} \times \overrightarrow{AC}\| = \dfrac{1}{2}\sqrt{117} = \dfrac{3}{2}\sqrt{13}$

35. $A(2, -7, 3), B(-1, 5, 8), C(4, 6, -1)$

$\overrightarrow{AB} = \langle -3, 12, 5 \rangle, \overrightarrow{AC} = \langle 2, 13, -4 \rangle$

$\overrightarrow{AB} \times \overrightarrow{AC} = \begin{vmatrix} \mathbf{i} & \mathbf{j} & \mathbf{k} \\ -3 & 12 & 5 \\ 2 & 13 & -4 \end{vmatrix} = \langle -113, -2, -63 \rangle$

Area $= \frac{1}{2}\|\overrightarrow{AB} \times \overrightarrow{AC}\| = \frac{1}{2}\sqrt{16{,}742}$

37. $\mathbf{F} = -20\mathbf{k}$

$\overrightarrow{PQ} = \frac{1}{2}(\cos 40°\mathbf{j} + \sin 40°\mathbf{k})$

$\overrightarrow{PQ} \times \mathbf{F} = \begin{vmatrix} \mathbf{i} & \mathbf{j} & \mathbf{k} \\ 0 & \cos 40°/2 & \sin 40°/2 \\ 0 & 0 & -20 \end{vmatrix} = -10\cos 40°\mathbf{i}$

$\|\overrightarrow{PQ} \times \mathbf{F}\| = 10\cos 40° \approx 7.66 \text{ ft} \cdot \text{lb}$

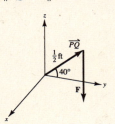

39. (a) $\overrightarrow{OA} = \frac{3}{2}\mathbf{k}$

$\mathbf{F} = -60(\sin\theta\mathbf{j} + \cos\theta\mathbf{k})$

$\overrightarrow{OA} \times \mathbf{F} = \begin{vmatrix} \mathbf{i} & \mathbf{j} & \mathbf{k} \\ 0 & 0 & 3/2 \\ 0 & -60\sin\theta & -60\cos\theta \end{vmatrix} = 90\sin\theta\mathbf{i}$

$\|\overrightarrow{OA} \times \mathbf{F}\| = 90\sin\theta$

(b) When $\theta = 45°$: $\|\overrightarrow{OA} \times \mathbf{F}\| = 90\left(\frac{\sqrt{2}}{2}\right) = 45\sqrt{2} \approx 63.64$.

(c) Let $T = 90\sin\theta$.

$\frac{dT}{d\theta} = 90\cos\theta = 0$ when $\theta = 90°$.

This is what we expected. When $\theta = 90°$ the pipe wrench is horizontal.

41. $\mathbf{u} \cdot (\mathbf{v} \times \mathbf{w}) = \begin{vmatrix} 1 & 0 & 0 \\ 0 & 1 & 0 \\ 0 & 0 & 1 \end{vmatrix} = 1$

43. $\mathbf{u} \cdot (\mathbf{v} \times \mathbf{w}) = \begin{vmatrix} 2 & 0 & 1 \\ 0 & 3 & 0 \\ 0 & 0 & 1 \end{vmatrix} = 6$

45. $\mathbf{u} \cdot (\mathbf{v} \times \mathbf{w}) = \begin{vmatrix} 1 & 1 & 0 \\ 0 & 1 & 1 \\ 1 & 0 & 1 \end{vmatrix} = 2$

$V = |\mathbf{u} \cdot (\mathbf{v} \times \mathbf{w})| = 2$

47. $\mathbf{u} = \langle 3, 0, 0 \rangle$

$\mathbf{v} = \langle 0, 5, 1 \rangle$

$\mathbf{w} = \langle 2, 0, 5 \rangle$

$\mathbf{u} \cdot (\mathbf{v} \times \mathbf{w}) = \begin{vmatrix} 3 & 0 & 0 \\ 0 & 5 & 1 \\ 2 & 0 & 5 \end{vmatrix} = 75$

$V = |\mathbf{u} \cdot (\mathbf{v} \times \mathbf{w})| = 75$

49. $\mathbf{u} \times \mathbf{v} = \langle u_1, u_2, u_3 \rangle \cdot \langle v_1, v_2, v_3 \rangle = (u_2v_3 - u_3v_2)\mathbf{i} - (u_1v_3 - u_3v_1)\mathbf{j} + (u_1v_2 - u_2v_1)\mathbf{k}$

51. The magnitude of the cross product will increase by a factor of 4.

53. If the vectors are ordered pairs, then the cross product does not exist. False.

55. True

57. $\mathbf{u} = \langle u_1, u_2, u_3 \rangle$, $\mathbf{v} = \langle v_1, v_2, v_3 \rangle$, $\mathbf{w} = \langle w_1, w_2, w_3 \rangle$

$$\mathbf{u} \times (\mathbf{v} + \mathbf{w}) = \begin{vmatrix} \mathbf{i} & \mathbf{j} & \mathbf{k} \\ u_1 & u_2 & u_3 \\ v_1 + w_1 & v_2 + w_2 & v_3 + w_3 \end{vmatrix}$$

$$= [u_2(v_3 + w_3) - u_3(v_2 + w_2)]\mathbf{i} - [u_1(v_3 + w_3) - u_3(v_1 + w_1)]\mathbf{j} + [u_1(v_2 + w_2) - u_2(v_1 + w_1)]\mathbf{k}$$

$$= (u_2v_3 - u_3v_2)\mathbf{i} - (u_1v_3 - u_3v_1)\mathbf{j} + (u_1v_2 - u_2v_1)\mathbf{k} + (u_2w_3 - u_3w_2)\mathbf{i} - $$

$$(u_1w_3 - u_3w_1)\mathbf{j} + (u_1w_2 - u_2w_1)\mathbf{k}$$

$$= (\mathbf{u} \times \mathbf{v}) + (\mathbf{u} \times \mathbf{w})$$

59. $\mathbf{u} = \langle u_1, u_2, u_3 \rangle$

$$\mathbf{u} \times \mathbf{u} = \begin{vmatrix} \mathbf{i} & \mathbf{j} & \mathbf{k} \\ u_1 & u_2 & u_3 \\ u_1 & u_2 & u_3 \end{vmatrix} = (u_2u_3 - u_3u_2)\mathbf{i} - (u_1u_3 - u_3u_1)\mathbf{j} + (u_1u_2 - u_2u_1)\mathbf{k} = \mathbf{0}$$

61. $\qquad \mathbf{u} \times \mathbf{v} = (u_2v_3 - u_3v_2)\mathbf{i} - (u_1v_3 - u_3v_1)\mathbf{j} + (u_1v_2 - u_2v_1)\mathbf{k}$

$(\mathbf{u} \times \mathbf{v}) \cdot \mathbf{u} = (u_2v_3 - u_3v_2)u_1 + (u_3v_1 - u_1v_3)u_2 + (u_1v_2 - u_2v_1)u_3 = 0$

$(\mathbf{u} \times \mathbf{v}) \cdot \mathbf{v} = (u_2v_3 - u_3v_2)v_1 + (u_3v_1 - u_1v_3)v_2 + (u_1v_2 - u_2v_1)v_3 = 0$

Thus, $\mathbf{u} \times \mathbf{v} \perp \mathbf{u}$ and $\mathbf{u} \times \mathbf{v} \perp \mathbf{v}$.

63. $\|\mathbf{u} \times \mathbf{v}\| = \|\mathbf{u}\|\,\|\mathbf{v}\| \sin \theta$

If $\mathbf{u}$ and $\mathbf{v}$ are orthogonal, $\theta = \pi/2$ and $\sin \theta = 1$. Therefore, $\|\mathbf{u} \times \mathbf{v}\| = \|\mathbf{u}\|\,\|\mathbf{v}\|$.

Section 10.5   Lines and Planes in Space

1. $x = 1 + 3t, y = 2 - t, z = 2 + 5t$

(a)

(b) When $t = 0$ we have $P = (1, 2, 2)$. When $t = 3$ we have $Q = (10, -1, 17)$.

$$\vec{PQ} = \langle 9, -3, 15 \rangle$$

The components of the vector and the coefficients of t are proportional since the line is parallel to $\vec{PQ}$.

(c) $y = 0$ when $t = 2$. Thus, $x = 7$ and $z = 12$.
Point: $(7, 0, 12)$

$x = 0$ when $t = -\dfrac{1}{3}$. Point: $\left(0, \dfrac{7}{3}, \dfrac{1}{3}\right)$

$z = 0$ when $t = -\dfrac{2}{5}$. Point: $\left(-\dfrac{1}{5}, \dfrac{12}{5}, 0\right)$

3. Point: $(0, 0, 0)$

Direction vector: $\mathbf{v} = \langle 1, 2, 3 \rangle$

Direction numbers: $1, 2, 3$

(a) Parametric: $x = t, y = 2t, z = 3t$

(b) Symmetric: $x = \dfrac{y}{2} = \dfrac{z}{3}$

5. Point: $(-2, 0, 3)$

Direction vector: $\mathbf{v} = \langle 2, 4, -2 \rangle$

Direction numbers: $2, 4, -2$

(a) Parametric: $x = -2 + 2t, y = 4t, z = 3 - 2t$

(b) Symmetric: $\dfrac{x + 2}{2} = \dfrac{y}{4} = \dfrac{z - 3}{-2}$

7. Point: $(1, 0, 1)$

Direction vector: $\mathbf{v} = 3\mathbf{i} - 2\mathbf{j} + \mathbf{k}$

Direction numbers: $3, -2, 1$

(a) Parametric: $x = 1 + 3t, y = -2t, z = 1 + t$

(b) Symmetric: $\dfrac{x - 1}{3} = \dfrac{y}{-2} = \dfrac{z - 1}{1}$

9. Points: $(5, -3, -2), \left(\dfrac{-2}{3}, \dfrac{2}{3}, 1\right)$

Direction vector: $\mathbf{v} = \dfrac{17}{3}\mathbf{i} - \dfrac{11}{3}\mathbf{j} - 3\mathbf{k}$

Direction numbers: $17, -11, -9$

(a) Parametric: $x = 5 + 17t, y = -3 - 11t, z = -2 - 9t$

(b) Symmetric: $\dfrac{x - 5}{17} = \dfrac{y + 3}{-11} = \dfrac{z + 2}{-9}$

11. Points: $(2, 3, 0), (10, 8, 12)$

Direction vector: $\langle 8, 5, 12 \rangle$

Direction numbers: $8, 5, 12$

(a) Parametric: $x = 2 + 8t, y = 3 + 5t, z = 12t$

(b) Symmetric: $\dfrac{x - 2}{8} = \dfrac{y - 3}{5} = \dfrac{z}{12}$

13. Point: $(2, 3, 4)$

Direction vector: $\mathbf{v} = \mathbf{k}$

Direction numbers: $0, 0, 1$

Parametric: $x = 2, y = 3, z = 4 + t$

15. Point: $(-2, 3, 1)$

Direction vector: $\mathbf{v} = 4\mathbf{i} - \mathbf{k}$

Direction numbers: $4, 0, -1$

Parametric: $x = -2 + 4t, y = 3, z = 1 - t$

Symmetric: $\dfrac{x + 2}{4} = \dfrac{z - 1}{-1}, y = 3$

(a) On line

(b) On line

(c) Not on line $(y \neq 3)$

(d) Not on line $\left(\dfrac{6 + 2}{4} \neq \dfrac{-2 - 1}{-1}\right)$

17. $L_1\colon \mathbf{v} = \langle -3, 2, 4 \rangle$ $(6, -2, 5)$ on line

$L_2\colon \mathbf{v} = \langle 6, -4, -8 \rangle$ $(6, -2, 5)$ on line

$L_3\colon \mathbf{v} = \langle -6, 4, 8 \rangle$ $(6, -2, 5)$ not on line

$L_4\colon \mathbf{v} = \langle 6, 4, -6 \rangle$ not parallel to $L_1, L_2,$ nor L_3

Hence, L_1 and L_2 are identical.

$L_1 = L_2$ and L_3 are parallel.

19. At the point of intersection, the coordinates for one line equal the corresponding coordinates for the other line. Thus,

    (i) $4t + 2 = 2s + 2$, (ii) $3 = 2s + 3$, and (iii) $-t + 1 = s + 1$.

From (ii), we find that $s = 0$ and consequently, from (iii), $t = 0$. Letting $s = t = 0$, we see that equation (i) is satisfied and therefore the two lines intersect. Substituting zero for s or for t, we obtain the point $(2, 3, 1)$.

    $\mathbf{u} = 4\mathbf{i} - \mathbf{k}$ (First line)

    $\mathbf{v} = 2\mathbf{i} + 2\mathbf{j} + \mathbf{k}$ (Second line)

$$\cos \theta = \frac{|\mathbf{u} \cdot \mathbf{v}|}{\|\mathbf{u}\| \, \|\mathbf{v}\|} = \frac{8 - 1}{\sqrt{17}\sqrt{9}} = \frac{7}{3\sqrt{17}} = \frac{7\sqrt{17}}{51}$$

21. Writing the equations of the lines in parametric form we have

    $x = 3t$ $y = 2 - t$ $z = -1 + t$

    $x = 1 + 4s$ $y = -2 + s$ $z = -3 - 3s.$

For the coordinates to be equal, $3t = 1 + 4s$ and $2 - t = -2 + s$. Solving this system yields $t = \frac{17}{7}$ and $s = \frac{11}{7}$. When using these values for s and t, the z coordinates are not equal. The lines do not intersect.

23. $x = 2t + 3$ $x = -2s + 7$

    $y = 5t - 2$ $y = s + 8$

    $z = -t + 1$ $z = 2s - 1$

Point of intersection: $(7, 8, -1)$

Note: $t = 2$ and $s = 0$

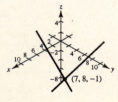

25. $4x - 3y - 6z = 6$

(a) $P = (0, 0, -1), Q = (0, -2, 0), R = (3, 4, -1)$

$\overrightarrow{PQ} = \langle 0, -2, 1 \rangle, \overrightarrow{PR} = \langle 3, 4, 0 \rangle$

(b) $\overrightarrow{PQ} \times \overrightarrow{PR} = \begin{vmatrix} \mathbf{i} & \mathbf{j} & \mathbf{k} \\ 0 & -2 & 1 \\ 3 & 4 & 0 \end{vmatrix} = \langle -4, 3, 6 \rangle$

The components of the cross product are proportional to the coefficients of the variables in the equation. The cross product is parallel to the normal vector.

27. Point: $(2, 1, 2)$

$\mathbf{n} = \mathbf{i} = \langle 1, 0, 0 \rangle$

$1(x - 2) + 0(y - 1) + 0(z - 2) = 0$

$x - 2 = 0$

29. Point: $(3, 2, 2)$

Normal vector: $\mathbf{n} = 2\mathbf{i} + 3\mathbf{j} - \mathbf{k}$

$2(x - 3) + 3(y - 2) - 1(z - 2) = 0$

$2x + 3y - z = 10$

31. Point: $(0, 0, 6)$

Normal vector: $\mathbf{n} = -\mathbf{i} + \mathbf{j} - 2\mathbf{k}$

$-1(x - 0) + 1(y - 0) - 2(z - 6) = 0$

$-x + y - 2z + 12 = 0$

$x - y + 2z = 12$

33. Let $\mathbf{u}$ be the vector from $(0, 0, 0)$ to $(1, 2, 3)$:

$\mathbf{u} = \mathbf{i} + 2\mathbf{j} + 3\mathbf{k}$

Let $\mathbf{v}$ be the vector from $(0, 0, 0)$ to $(-2, 3, 3)$:

$\mathbf{v} = -2\mathbf{i} + 3\mathbf{j} + 3\mathbf{k}$

Normal vector: $\mathbf{u} \times \mathbf{v} = \begin{vmatrix} \mathbf{i} & \mathbf{j} & \mathbf{k} \\ 1 & 2 & 3 \\ -2 & 3 & 3 \end{vmatrix}$

$= -3\mathbf{i} + (-9)\mathbf{j} + 7\mathbf{k}$

$-3(x - 0) - 9(y - 0) + 7(z - 0) = 0$

$3x + 9y - 7z = 0$

35. Let $\mathbf{u}$ be the vector from $(1, 2, 3)$ to $(3, 2, 1)$: $\mathbf{u} = 2\mathbf{i} - 2\mathbf{k}$

Let $\mathbf{v}$ be the vector from $(1, 2, 3)$ to $(-1, -2, 2)$: $\mathbf{v} = -2\mathbf{i} - 4\mathbf{j} - \mathbf{k}$

Normal vector: $\left(\tfrac{1}{2}\mathbf{u}\right) \times (-\mathbf{v}) = \begin{vmatrix} \mathbf{i} & \mathbf{j} & \mathbf{k} \\ 1 & 0 & -1 \\ 2 & 4 & 1 \end{vmatrix} = 4\mathbf{i} - 3\mathbf{j} + 4\mathbf{k}$

$4(x - 1) - 3(y - 2) + 4(z - 3) = 0$

$4x - 3y + 4z = 10$

37. $(1, 2, 3)$, Normal vector: $\mathbf{v} = \mathbf{k}, 1(z - 3) = 0, z = 3$

39. The direction vectors for the lines are $\mathbf{u} = -2\mathbf{i} + \mathbf{j} + \mathbf{k}$, $\mathbf{v} = -3\mathbf{i} + 4\mathbf{j} - \mathbf{k}$.

Normal vector: $\mathbf{u} \times \mathbf{v} = \begin{vmatrix} \mathbf{i} & \mathbf{j} & \mathbf{k} \\ -2 & 1 & 1 \\ -3 & 4 & -1 \end{vmatrix} = -5(\mathbf{i} + \mathbf{j} + \mathbf{k})$

Point of intersection of the lines: $(-1, 5, 1)$

$(x + 1) + (y - 5) + (z - 1) = 0$

$x + y + z = 5$

41. Let $\mathbf{v}$ be the vector from $(-1, 1, -1)$ to $(2, 2, 1)$: $\mathbf{v} = 3\mathbf{i} + \mathbf{j} + 2\mathbf{k}$

Let $\mathbf{n}$ be a vector normal to the plane $2x - 3y + z = 3$: $\mathbf{n} = 2\mathbf{i} - 3\mathbf{j} + \mathbf{k}$

Since $\mathbf{v}$ and $\mathbf{n}$ both lie in the plane P, the normal vector to P is

$\mathbf{v} \times \mathbf{n} = \begin{vmatrix} \mathbf{i} & \mathbf{j} & \mathbf{k} \\ 3 & 1 & 2 \\ 2 & -3 & 1 \end{vmatrix} = 7\mathbf{i} + \mathbf{j} - 11\mathbf{k}$

$7(x - 2) + 1(y - 2) - 11(z - 1) = 0$

$7x + y - 11z = 5$

43. Let $\mathbf{u} = \mathbf{i}$ and let $\mathbf{v}$ be the vector from $(1, -2, -1)$ to $(2, 5, 6)$: $\mathbf{v} = \mathbf{i} + 7\mathbf{j} + 7\mathbf{k}$

Since $\mathbf{u}$ and $\mathbf{v}$ both lie in the plane P, the normal vector to P is:

$$\mathbf{u} \times \mathbf{v} = \begin{vmatrix} \mathbf{i} & \mathbf{j} & \mathbf{k} \\ 1 & 0 & 0 \\ 1 & 7 & 7 \end{vmatrix} = -7\mathbf{j} + 7\mathbf{k} = -7(\mathbf{j} - \mathbf{k})$$

$$[y - (-2)] - [z - (-1)] = 0$$

$$y - z = -1$$

45. The normal vectors to the planes are

$$\mathbf{n}_1 = \langle 5, -3, 1 \rangle, \mathbf{n}_2 = \langle 1, 4, 7 \rangle, \cos\theta = \frac{|\mathbf{n}_1 \cdot \mathbf{n}_2|}{\|\mathbf{n}_1\| \|\mathbf{n}_2\|} = 0.$$

Thus, $\theta = \pi/2$ and the planes are orthogonal.

47. The normal vectors to the planes are

$$\mathbf{n}_1 = \mathbf{i} - 3\mathbf{j} + 6\mathbf{k}, \ \mathbf{n}_2 = 5\mathbf{i} + \mathbf{j} - \mathbf{k},$$

$$\cos\theta = \frac{|\mathbf{n}_1 \cdot \mathbf{n}_2|}{\|\mathbf{n}_1\| \|\mathbf{n}_2\|} = \frac{|5 - 3 - 6|}{\sqrt{46}\sqrt{27}} = \frac{4\sqrt{138}}{414}.$$

Therefore, $\theta = \arccos\left(\frac{4\sqrt{138}}{414}\right) \approx 83.5°$.

49. The normal vectors to the planes are $\mathbf{n}_1 = \langle 1, -5, -1 \rangle$ and $\mathbf{n}_2 = \langle 5, -25, -5 \rangle$. Since $\mathbf{n}_2 = 5\mathbf{n}_1$, the planes are parallel, but not equal.

51. $4x + 2y + 6z = 12$

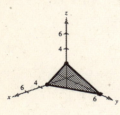

53. $2x - y + 3z = 4$

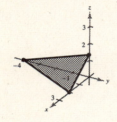

55. $y + z = 5$

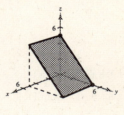

57. $x = 5$

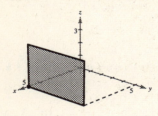

59. $2x + y - z = 6$

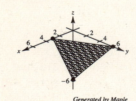

Generated by Maple

61. $-5x + 4y - 6z + 8 = 0$

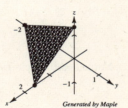

Generated by Maple

63. P_1: $\mathbf{n} = \langle 3, -2, 5 \rangle$ $(1, -1, 1)$ on plane

   P_2: $\mathbf{n} = \langle -6, 4, -10 \rangle$ $(1, -1, 1)$ not on plane

   P_3: $\mathbf{n} = \langle -3, 2, 5 \rangle$

   P_4: $\mathbf{n} = \langle 75, -50, 125 \rangle$ $(1, -1, 1)$ on plane

   P_1 and P_4 are identical.

   $P_1 = P_4$ is parallel to P_2.

65. Each plane passes through the points

   $(c, 0, 0), (0, c, 0),$ and $(0, 0, c)$.

67. The normals to the planes are $\mathbf{n}_1 = 3\mathbf{i} + 2\mathbf{j} - \mathbf{k}$ and $\mathbf{n}_2 = \mathbf{i} - 4\mathbf{j} + 2\mathbf{k}$. The direction vector for the line is

$$\mathbf{n}_2 \times \mathbf{n}_1 = \begin{vmatrix} \mathbf{i} & \mathbf{j} & \mathbf{k} \\ 1 & -4 & 2 \\ 3 & 2 & -1 \end{vmatrix} = 7(\mathbf{j} + 2\mathbf{k}).$$

Now find a point of intersection of the planes.

$$6x + 4y - 2z = 14$$
$$x - 4y + 2z = 0$$
$$7x \qquad = 14$$
$$x = 2$$

Substituting 2 for x in the second equation, we have $-4y + 2z = -2$ or $z = 2y - 1$. Letting $y = 1$, a point of intersection is $(2, 1, 1)$.

$$x = 2, y = 1 + t, z = 1 + 2t$$

69. Writing the equation of the line in parametric form and substituting into the equation of the plane we have:

$$x = \frac{1}{2} + t, \ y = \frac{-3}{2} - t, \ z = -1 + 2t$$

$$2\left(\frac{1}{2} + t\right) - 2\left(\frac{-3}{2} - t\right) + (-1 + 2t) = 12, \ t = \frac{3}{2}$$

Substituting $t = 3/2$ into the parametric equations for the line we have the point of intersection $(2, -3, 2)$. The line does not lie in the plane.

71. Writing the equation of the line in parametric form and substituting into the equation of the plane we have:

$$x = 1 + 3t, \ y = -1 - 2t, \ z = 3 + t$$
$$2(1 + 3t) + 3(-1 - 2t) = 10, \ -1 = 10, \text{ contradiction}$$

Therefore, the line does not intersect the plane.

73. Point: $Q(0, 0, 0)$

Plane: $2x + 3y + z - 12 = 0$

Normal to plane: $\mathbf{n} = \langle 2, 3, 1 \rangle$

Point in plane: $P(6, 0, 0)$

Vector $\overrightarrow{PQ} = \langle -6, 0\ 0 \rangle$

$$D = \frac{|\overrightarrow{PQ} \cdot \mathbf{n}|}{\|\mathbf{n}\|} = \frac{|-12|}{\sqrt{14}} = \frac{6\sqrt{14}}{7}$$

75. Point: $Q(2, 8, 4)$

Plane: $2x + y + z = 5$

Normal to plane: $\mathbf{n} = \langle 2, 1, 1 \rangle$

Point in plane: $P\langle 0, 0, 5 \rangle$

Vector: $\overrightarrow{PQ} = \langle 2, 8, -1 \rangle$

$$D = \frac{|\overrightarrow{PQ} \cdot \mathbf{n}|}{\|\mathbf{n}\|} = \frac{11}{\sqrt{6}} = \frac{11\sqrt{6}}{6}$$

77. The normal vectors to the planes are $\mathbf{n}_1 = \langle 1, -3, 4 \rangle$ and $\mathbf{n}_2 = \langle 1, -3, 4 \rangle$. Since $\mathbf{n}_1 = \mathbf{n}_2$, the planes are parallel. Choose a point in each plane.

$P = (10, 0, 0)$ is a point in $x - 3y + 4z = 10$.
$Q = (6, 0, 0)$ is a point in $x - 3y + 4z = 6$.

$$\overrightarrow{PQ} = \langle -4, 0, 0 \rangle, D = \frac{|\overrightarrow{PQ} \cdot \mathbf{n}_1|}{\|\mathbf{n}_1\|} = \frac{4}{\sqrt{26}} = \frac{2\sqrt{26}}{13}$$

79. The normal vectors to the planes are $\mathbf{n}_1 = \langle -3, 6, 7 \rangle$ and $\mathbf{n}_2 = \langle 6, -12, -14 \rangle$. Since $\mathbf{n}_2 = -2\mathbf{n}_1$, the planes are parallel. Choose a point in each plane.

$P = (0, -1, 1)$ is a point in $-3x + 6y + 7z = 1$.

$Q = \left(\frac{25}{6}, 0, 0\right)$ is a point in $6x - 12y - 14z = 25$.

$$\overrightarrow{PQ} = \left\langle \frac{25}{6}, 1, -1 \right\rangle$$

$$D = \frac{|\overrightarrow{PQ} \cdot \mathbf{n}_1|}{\|\mathbf{n}_1\|} = \frac{|-27/2|}{\sqrt{94}} = \frac{27}{2\sqrt{94}} = \frac{27\sqrt{94}}{188}$$

81. $\mathbf{u} = \langle 4, 0, -1 \rangle$ is the direction vector for the line. $Q(1, 5, -2)$ is the given point, and $P(-2, 3, 1)$ is on the line. Hence, $\overrightarrow{PQ} = \langle 3, 2, -3 \rangle$ and

$$\overrightarrow{PQ} \times \mathbf{u} = \begin{vmatrix} \mathbf{i} & \mathbf{j} & \mathbf{k} \\ 3 & 2 & -3 \\ 4 & 0 & -1 \end{vmatrix} = \langle -2, -9, -8 \rangle$$

$$D = \frac{\|\overrightarrow{PQ} \times \mathbf{u}\|}{\|\mathbf{u}\|} = \frac{\sqrt{149}}{\sqrt{17}} = \frac{\sqrt{2533}}{17}$$

83. The parametric equations of a line L parallel to $\mathbf{v} = \langle a, b, c, \rangle$ and passing through the point $P(x_1, y_1, z_1)$ are

$$x = x_1 + at, y = y_1 + bt, z = z_1 + ct.$$

The symmetric equations are

$$\frac{x - x_1}{a} = \frac{y - y_1}{b} = \frac{z - z_1}{c}.$$

85. Solve the two linear equations representing the planes to find two points of intersection. Then find the line determined by the two points.

87. (a) Sphere

$$(x - 3)^2 + (y + 2)^2 + (z - 5)^2 = 16$$

$$x^2 + y^2 + z^2 - 6x + 4y - 10z + 22 = 0$$

(b) Parallel planes

$$4x - 3y + z = 10 \pm 4\|\mathbf{n}\| = 10 \pm 4\sqrt{26}$$

89. (a) $z = 28.7 - 1.83x - 1.09y$

Year	1980	1985	1990	1994	1995	1996	1997
z (approx.)	16.16	14.23	9.81	8.60	8.42	8.27	8.23

(b) An increase in x or y will cause a decrease in z. In fact, any increase in two variables will cause a decrease in the third.

(c)

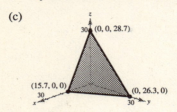

91. True

Section 10.6   Surfaces in Space

1. Ellipsoid

Matches graph (c)

3. Hyperboloid of one sheet

Matches graph (f)

5. Elliptic paraboloid

Matches graph (d)

7. $z = 3$

Plane parallel to the xy-coordinate plane

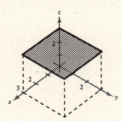

9. $y^2 + z^2 = 9$

The x-coordinate is missing so we have a cylindrical surface with rulings parallel to the x-axis. The generating curve is a circle.

11. $y = x^2$

The z-coordinate is missing so we have a cylindrical surface with rulings parallel to the z-axis. The generating curve is a parabola.

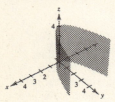

13. $4x^2 + y^2 = 4$

$$\frac{x^2}{1} + \frac{y^2}{4} = 1$$

The z-coordinate is missing so we have a cylindrical surface with rulings parallel to the z-axis. The generating curve is an ellipse.

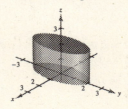

15. $z = \sin y$

The x-coordinate is missing so we have a cylindrical surface with rulings parallel to the x-axis. The generating curve is the sine curve.

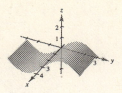

17. $z = x^2 + y^2$

(a) You are viewing the paraboloid from the x-axis: $(20, 0, 0)$

(b) You are viewing the paraboloid from above, but not on the z-axis: $(10, 10, 20)$

(c) You are viewing the paraboloid from the z-axis: $(0, 0, 20)$

(d) You are viewing the paraboloid from the y-axis: $(0, 20, 0)$

19. $\dfrac{x^2}{1} + \dfrac{y^2}{4} + \dfrac{z^2}{1} = 1$

Ellipsoid

xy-trace: $\dfrac{x^2}{1} + \dfrac{y^2}{4} = 1$ ellipse

xz-trace: $x^2 + z^2 = 1$ circle

yz-trace: $\dfrac{y^2}{4} + \dfrac{z^2}{1} = 1$ ellipse

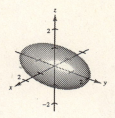

21. $16x^2 - y^2 + 16z^2 = 4$

$4x^2 - \dfrac{y^2}{4} + 4z^2 = 1$

Hyperboloid on one sheet

xy-trace: $4x^2 - \dfrac{y^2}{4} = 1$ hyperbola

xz-trace: $4(x^2 + z^2) = 1$ circle

yz-trace: $\dfrac{-y^2}{4} + 4z^2 = 1$ hyperbola

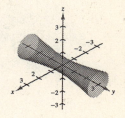

23. $x^2 - y + z^2 = 0$

Elliptic paraboloid

xy-trace: $y = x^2$

xz-trace: $x^2 + z^2 = 0$,
        point $(0, 0, 0)$

yz-trace: $y = z^2$

$y = 1: x^2 + z^2 = 1$

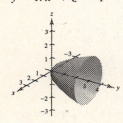

25. $x^2 - y^2 + z = 0$

Hyperbolic paraboloid

xy-trace: $y = \pm x$

xz-trace: $z = -x^2$

yz-trace: $z = y^2$

$y = \pm 1: z = 1 - x^2$

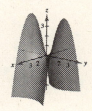

27. $z^2 = x^2 + \dfrac{y^2}{4}$

Elliptic Cone

xy-trace: point $(0, 0, 0)$

xz-trace: $z = \pm x$

yz-trace: $z = \dfrac{\pm 1}{2} y$

$z = \pm 1: x^2 + \dfrac{y^2}{4} = 1$

29. $16x^2 + 9y^2 + 16z^2 - 32x - 36y + 36 = 0$

$16(x^2 - 2x + 1) + 9(y^2 - 4y + 4) + 16z^2 = -36 + 16 + 36$

$16(x - 1)^2 + 9(y - 2)^2 + 16z^2 = 16$

$\dfrac{(x - 1)^2}{1} + \dfrac{(y - 2)^2}{16/9} + \dfrac{z^2}{1} = 1$

Ellipsoid with center $(1, 2, 0)$.

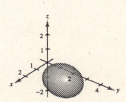

31. $z = 2 \sin x$

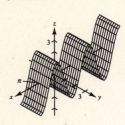

33. $z^2 = x^2 + 4y^2$

$z = \pm\sqrt{x^2 + 4y^2}$

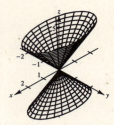

35. $x^2 + y^2 = \left(\dfrac{2}{z}\right)^2$

$y = \pm\sqrt{\dfrac{4}{z^2} - x^2}$

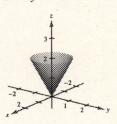

37. $z = 4 - \sqrt{|xy|}$

39. $4x^2 - y^2 + 4z^2 = -16$

$z = \pm\sqrt{\dfrac{y^2}{4} - x^2 - 4}$

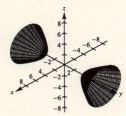

41. $z = 2\sqrt{x^2 + y^2}$

$z = 2$

$2\sqrt{x^2 + y^2} = 2$

$x^2 + y^2 = 1$

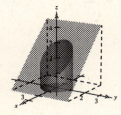

43. $x^2 + y^2 = 1$

$x + z = 2$

$z = 0$

45. $x^2 + z^2 = [r(y)]^2$ and $z = r(y) = \pm 2\sqrt{y}$; therefore,

$x^2 + z^2 = 4y.$

47. $x^2 + y^2 = [r(z)]^2$ and $y = r(z) = \dfrac{z}{2}$; therefore,

$x^2 + y^2 = \dfrac{z^2}{4}, \ 4x^2 + 4y^2 = z^2.$

49. $y^2 + z^2 = [r(x)]^2$ and $y = r(x) = \dfrac{2}{x}$; therefore,

$y^2 + z^2 = \left(\dfrac{2}{x}\right)^2, \ y^2 + z^2 = \dfrac{4}{x^2}.$

51. $x^2 + y^2 - 2z = 0$

$x^2 + y^2 = \left(\sqrt{2z}\right)^2$

Equation of generating curve: $y = \sqrt{2z}$ or $x = \sqrt{2z}$

53. Let C be a curve in a plane and let L be a line not in a parallel plane. The set of all lines parallel to L and intersecting C is called a cylinder.

55. See pages 765 and 766.

57. $V = 2\pi \displaystyle\int_0^4 x(4x - x^2)\, dx$

$= 2\pi\left[\dfrac{4x^3}{3} - \dfrac{x^4}{4}\right]_0^4 = \dfrac{128\pi}{3}$

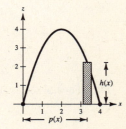

59. $z = \dfrac{x^2}{2} + \dfrac{y^2}{4}$

(a) When $z = 2$ we have $2 = \dfrac{x^2}{2} + \dfrac{y^2}{4}$, or $1 = \dfrac{x^2}{4} + \dfrac{y^2}{8}$.

Major axis: $2\sqrt{8} = 4\sqrt{2}$

Minor axis: $2\sqrt{4} = 4$

$c^2 = a^2 - b^2,\ c^2 = 4,\ c = 2$

Foci: $(0, \pm 2, 2)$

(b) When $z = 8$ we have $8 = \dfrac{x^2}{2} + \dfrac{y^2}{4}$, or $1 = \dfrac{x^2}{16} + \dfrac{y^2}{32}$.

Major axis: $2\sqrt{32} = 8\sqrt{2}$

Minor axis: $2\sqrt{16} = 8$

$c^2 = 32 - 16 = 16,\ c = 4$

Foci: $(0, \pm 4, 8)$

61. If (x, y, z) is on the surface, then

$$(y + 2)^2 = x^2 + (y - 2)^2 + z^2$$
$$y^2 + 4y + 4 = x^2 + y^2 - 4y + 4 + z^2$$
$$x^2 + z^2 = 8y$$

Elliptic paraboloid

Traces parallel to xz-plane are circles.

63. $\dfrac{x^2}{3963^2} + \dfrac{y^2}{3963^2} + \dfrac{z^2}{3942^2} = 1$

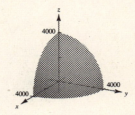

65. $z = \dfrac{y^2}{b^2} - \dfrac{x^2}{a^2},\ z = bx + ay$

$$bx + ay = \dfrac{y^2}{b^2} - \dfrac{x^2}{a^2}$$

$$\dfrac{1}{a^2}\left(x^2 + a^2 bx + \dfrac{a^4 b^2}{4}\right) = \dfrac{1}{b^2}\left(y^2 - ab^2 y + \dfrac{a^2 b^4}{4}\right)$$

$$\dfrac{\left(x + \dfrac{a^2 b}{2}\right)^2}{a^2} = \dfrac{\left(y - \dfrac{ab^2}{2}\right)^2}{b^2}$$

$$y = \pm\dfrac{b}{a}\left(x + \dfrac{a^2 b}{2}\right) + \dfrac{ab^2}{2}$$

Letting $x = at$, you obtain the two intersecting lines
$x = at,\ y = -bt,\ z = 0$ and $x = at,\ y = bt + ab^2$
$z = 2abt + a^2 b^2$.

67. The Klein bottle *does not* have both an "inside" and an "outside." It is formed by inserting the small open end through the side of the bottle and making it contiguous with the top of the bottle.

Section 10.7   Cylindrical and Spherical Coordinates

1. $(5, 0, 2)$, cylindrical

$x = 5 \cos 0 = 5$

$y = 5 \sin 0 = 0$

$z = 2$

$(5, 0, 2)$, rectangular

3. $\left(2, \dfrac{\pi}{3}, 2\right)$, cylindrical

$x = 2 \cos \dfrac{\pi}{3} = 1$

$y = 2 \sin \dfrac{\pi}{3} = \sqrt{3}$

$z = 2$

$\left(1, \sqrt{3}, 2\right)$, rectangular

5. $\left(4, \dfrac{7\pi}{6}, 3\right)$, cylindrical

$x = 4 \cos \dfrac{7\pi}{6} = -2\sqrt{3}$

$y = 4 \sin \dfrac{7\pi}{6} = -2$

$z = 3$

$\left(-2\sqrt{3}, -2, 3\right)$, rectangular

7. $(0, 5, 1)$, rectangular

$r = \sqrt{(0)^2 + (5)^2} = 5$

$\theta = \arctan \dfrac{5}{0} = \dfrac{\pi}{2}$

$z = 1$

$\left(5, \dfrac{\pi}{2}, 1\right)$, cylindrical

9. $\left(1, \sqrt{3}, 4\right)$, rectangular

$r = \sqrt{1^2 + \left(\sqrt{3}\right)^2} = 2$

$\theta = \arctan \sqrt{3} = \dfrac{\pi}{3}$

$z = 4$

$\left(2, \dfrac{\pi}{3}, 4\right)$, cylindrical

11. $(2, -2, -4)$, rectangular

$r = \sqrt{2^2 + (-2)^2} = 2\sqrt{2}$

$\theta = \arctan(-1) = -\dfrac{\pi}{4}$

$z = -4$

$\left(2\sqrt{2}, \dfrac{-\pi}{4}, -4\right)$, cylindrical

13. $x^2 + y^2 + z^2 = 10$ rectangular equation

$\qquad r^2 + z^2 = 10$ cylindrical equation

15. $y = x^2$ $\qquad\qquad\qquad$ rectangular equation

$\qquad r \sin \theta = (r \cos \theta)^2$

$\qquad \sin \theta = r \cos^2 \theta$

$\qquad\qquad r = \sec \theta \cdot \tan \theta$ cylindrical equation

17. $r = 2$

$\sqrt{x^2 + y^2} = 2$

$x^2 + y^2 = 4$

19. $\theta = \dfrac{\pi}{6}$

$\tan \dfrac{\pi}{6} = \dfrac{y}{x}$

$\dfrac{1}{\sqrt{3}} = \dfrac{y}{x}$

$x = \sqrt{3}y$

$x - \sqrt{3}y = 0$

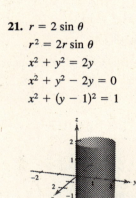

21. $r = 2 \sin \theta$

$r^2 = 2r \sin \theta$

$x^2 + y^2 = 2y$

$x^2 + y^2 - 2y = 0$

$x^2 + (y - 1)^2 = 1$

23. $r^2 + z^2 = 4$

$x^2 + y^2 + z^2 = 4$

25. $(4, 0, 0)$, rectangular

$\rho = \sqrt{4^2 + 0^2 + 0^2} = 4$

$\theta = \arctan 0 = 0$

$\phi = \arccos 0 = \dfrac{\pi}{2}$

$\left(4, 0, \dfrac{\pi}{2}\right)$, spherical

27. $\left(-2, 2\sqrt{3}, 4\right)$, rectangular

$\rho = \sqrt{(-2)^2 + \left(2\sqrt{3}\right)^2 + 4^2} = 4\sqrt{2}$

$\theta = \arctan\left(-\sqrt{3}\right) = \dfrac{2\pi}{3}$

$\phi = \arccos \dfrac{1}{\sqrt{2}} = \dfrac{\pi}{4}$

$\left(4\sqrt{2}, \dfrac{2\pi}{3}, \dfrac{\pi}{4}\right)$, spherical

29. $\left(\sqrt{3}, 1, 2\sqrt{3}\right)$, rectangular

$\rho = \sqrt{3 + 1 + 12} = 4$

$\theta = \arctan \dfrac{1}{\sqrt{3}} = \dfrac{\pi}{6}$

$\phi = \arccos \dfrac{\sqrt{3}}{2} = \dfrac{\pi}{6}$

$\left(4, \dfrac{\pi}{6}, \dfrac{\pi}{6}\right)$, spherical

31. $\left(4, \dfrac{\pi}{6}, \dfrac{\pi}{4}\right)$, spherical

$x = 4 \sin \dfrac{\pi}{4} \cos \dfrac{\pi}{6} = \sqrt{6}$

$y = 4 \sin \dfrac{\pi}{4} \sin \dfrac{\pi}{6} = \sqrt{2}$

$z = 4 \cos \dfrac{\pi}{4} = 2\sqrt{2}$

$\left(\sqrt{6}, \sqrt{2}, 2\sqrt{2}\right)$, rectangular

33. $\left(12, \dfrac{-\pi}{4}, 0\right)$, spherical

$x = 12 \sin 0 \cos\left(\dfrac{-\pi}{4}\right) = 0$

$y = 12 \sin 0 \sin\left(\dfrac{-\pi}{4}\right) = 0$

$z = 12 \cos 0 = 12$

$(0, 0, 12)$, rectangular

35. $\left(5, \dfrac{\pi}{4}, \dfrac{3\pi}{4}\right)$, spherical

$x = 5 \sin \dfrac{3\pi}{4} \cos \dfrac{\pi}{4} = \dfrac{5}{2}$

$y = 5 \sin \dfrac{3\pi}{4} \sin \dfrac{\pi}{4} = \dfrac{5}{2}$

$z = 5 \cos \dfrac{3\pi}{4} = -\dfrac{5\sqrt{2}}{2}$

$\left(\dfrac{5}{2}, \dfrac{5}{2}, -\dfrac{5\sqrt{2}}{2}\right)$, rectangular

37. (a) Programs will vary.

     (b) $(x, y, z) = (3, -4, 2)$

        $(\rho, \theta, \phi) \approx (5.385, -0.927, 1.190)$

39. $x^2 + y^2 + z^2 = 36$ rectangular equation

    $\rho^2 = 36$ spherical equation

41. $x^2 + y^2 = 9$ rectangular equation

$\rho^2 \sin^2 \phi \cos^2 \theta + \rho^2 \sin^2 \phi \sin^2 \theta = 9$

$\rho^2 \sin^2 \phi = 9$

$\rho \sin \phi = 3$

$\rho = 3 \csc \phi$ spherical equation

43. $\rho = 2$

    $x^2 + y^2 + z^2 = 4$

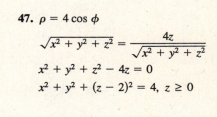

45. $\phi = \dfrac{\pi}{6}$

$\cos \phi = \dfrac{z}{\sqrt{x^2 + y^2 + z^2}}$

$\dfrac{\sqrt{3}}{2} = \dfrac{z}{\sqrt{x^2 + y^2 + z^2}}$

$\dfrac{3}{4} = \dfrac{z^2}{x^2 + y^2 + z^2}$

$3x^2 + 3y^2 - z^2 = 0, \ z \geq 0$

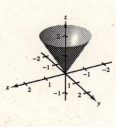

47. $\rho = 4 \cos \phi$

$\sqrt{x^2 + y^2 + z^2} = \dfrac{4z}{\sqrt{x^2 + y^2 + z^2}}$

$x^2 + y^2 + z^2 - 4z = 0$

$x^2 + y^2 + (z - 2)^2 = 4, \ z \geq 0$

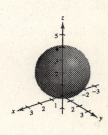

49. $\rho = \csc \phi$

$\rho \sin \phi = 1$

$\sqrt{x^2 + y^2} = 1$

$x^2 + y^2 = 1$

51. $\left(4, \dfrac{\pi}{4}, 0\right)$, cylindrical

$\rho = \sqrt{4^2 + 0^2} = 4$

$\theta = \dfrac{\pi}{4}$

$\phi = \arccos 0 = \dfrac{\pi}{2}$

$\left(4, \dfrac{\pi}{4}, \dfrac{\pi}{2}\right)$, spherical

53. $\left(4, \dfrac{\pi}{2}, 4\right)$, cylindrical

$\rho = \sqrt{4^2 + 4^2} = 4\sqrt{2}$

$\theta = \dfrac{\pi}{2}$

$\phi = \arccos\left(\dfrac{4}{4\sqrt{2}}\right) = \dfrac{\pi}{4}$

$\left(4\sqrt{2}, \dfrac{\pi}{2}, \dfrac{\pi}{4}\right)$, spherical

55. $\left(4, \dfrac{-\pi}{6}, 6\right)$, cylindrical

$\rho = \sqrt{4^2 + 6^2} = 2\sqrt{13}$

$\theta = \dfrac{-\pi}{6}$

$\phi = \arccos \dfrac{3}{\sqrt{13}}$

$\left(2\sqrt{13}, \dfrac{-\pi}{6}, \arccos \dfrac{3}{\sqrt{13}}\right)$,

spherical

57. $(12, \pi, 5)$, cylindrical

$\rho = \sqrt{12^2 + 5^2} = 13$

$\theta = \pi$

$\phi = \arccos \dfrac{5}{13}$

$\left(13, \pi, \arccos \dfrac{5}{13}\right)$, spherical

59. $\left(10, \dfrac{\pi}{6}, \dfrac{\pi}{2}\right)$, spherical

$r = 10 \sin \dfrac{\pi}{2} = 10$

$\theta = \dfrac{\pi}{6}$

$z = 10 \cos \dfrac{\pi}{2} = 0$

$\left(10, \dfrac{\pi}{6}, 0\right)$, cylindrical

61. $\left(36, \pi, \dfrac{\pi}{2}\right)$, spherical

$r = \rho \sin \phi = 36 \sin \dfrac{\pi}{2} = 36$

$\theta = \pi$

$z = \rho \cos \phi = 36 \cos \dfrac{\pi}{2} = 0$

$(36, \pi, 0)$, cylindrical

63. $\left(6, -\dfrac{\pi}{6}, \dfrac{\pi}{3}\right)$, spherical

$r = 6 \sin \dfrac{\pi}{3} = 3\sqrt{3}$

$\theta = -\dfrac{\pi}{6}$

$z = 6 \cos \dfrac{\pi}{3} = 3$

$\left(3\sqrt{3}, -\dfrac{\pi}{6}, 3\right)$, cylindrical

65. $\left(8, \dfrac{7\pi}{6}, \dfrac{\pi}{6}\right)$, spherical

$r = 8 \sin \dfrac{\pi}{6} = 4$

$\theta = \dfrac{7\pi}{6}$

$z = 8 \cos \dfrac{\pi}{6} = \dfrac{8\sqrt{3}}{2}$

$\left(4, \dfrac{7\pi}{6}, 4\sqrt{3}\right)$, cylindrical

	Rectangular	*Cylindrical*	*Spherical*
67.	$(4, 6, 3)$	$(7.211, 0.983, 3)$	$(7.810, 0.983, 1.177)$
69.	$(4.698, 1.710, 8)$	$\left(5, \dfrac{\pi}{9}, 8\right)$	$(9.434, 0.349, 0.559)$
71.	$(-7.071, 12.247, 14.142)$	$(14.142, 2.094, 14.142)$	$\left(20, \dfrac{2\pi}{3}, \dfrac{\pi}{4}\right)$
73.	$(3, -2, 2)$	$(3.606, -0.588, 2)$	$(4.123, -0.588, 1.064)$
75.	$\left(\dfrac{5}{2}, \dfrac{4}{3}, \dfrac{-3}{2}\right)$	$(2.833, 0.490, -1.5)$	$(3.206, 0.490, 2.058)$
77.	$(-3.536, 3.536, -5)$	$\left(5, \dfrac{3\pi}{4}, -5\right)$	$(7.071, 2.356, 2.356)$
79.	$(2.804, -2.095, 6)$	$(-3.5, 2.5, 6)$	$(6.946, 5.642, 0.528)$

[Note: Use the cylindrical coordinates $(3.5, 5.642, 6)$]

81. $r = 5$

Cylinder

Matches graph (d)

83. $\rho = 5$

Sphere

Matches graph (c)

85. $r^2 = z, x^2 + y^2 = z$

Paraboloid

Matches graph (f)

87. Rectangular to cylindrical: $r^2 = x^2 + y^2$

$\tan \theta = \dfrac{y}{x}$

$z = z$

Cylindrical to rectangular: $x = r \cos \theta$

$y = r \sin \theta$

$z = z$

89. Rectangular to spherical: $\rho^2 = x^2 + y^2 + z^2$

$\tan \theta = \dfrac{y}{x}$

$\phi = \arccos \left(\dfrac{z}{\sqrt{x^2 + y^2 + z^2}}\right)$

Spherical to rectangular: $x = \rho \sin \phi \cos \theta$

$y = \rho \sin \phi \sin \theta$

$z = \rho \cos \phi$

91. $x^2 + y^2 + z^2 = 16$

    (a) $r^2 + z^2 = 16$

    (b) $\rho^2 = 16, \rho = 4$

93. $x^2 + y^2 + z^2 - 2z = 0$

    (a) $r^2 + z^2 - 2z = 0, r^2 + (z - 1)^2 = 1$

    (b) $\rho^2 - 2\rho \cos \phi = 0, \rho(\rho - 2 \cos \phi) = 0,$

        $\rho = 2 \cos \phi$

95. $x^2 + y^2 = 4y$

    (a) $r^2 = 4r \sin \theta, \ r = 4 \sin \theta$

    (b) $\rho^2 \sin^2 \phi = 4\rho \sin \phi \sin \theta,$

        $\rho \sin \phi (\rho \sin \phi - 4 \sin \theta) = 0,$

        $\rho = \dfrac{4 \sin \theta}{\sin \phi}, \ \rho = 4 \sin \theta \csc \phi$

97. $x^2 - y^2 = 9$

    (a) $r^2 \cos^2 \theta - r^2 \sin^2 \theta = 9,$

        $r^2 = \dfrac{9}{\cos^2 \theta - \sin^2 \theta}$

    (b) $\rho^2 \sin^2 \phi \cos^2 \theta - \rho^2 \sin^2 \phi \sin^2 \theta = 9,$

        $\rho^2 \sin^2 \phi = \dfrac{9}{\cos^2 \theta - \sin^2 \theta},$

        $\rho^2 = \dfrac{9 \csc^2 \phi}{\cos^2 \theta - \sin^2 \theta}$

99. $0 \le \theta \le \dfrac{\pi}{2}$

    $0 \le r \le 2$

    $0 \le z \le 4$

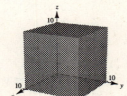

101. $0 \le \theta \le 2\pi$

    $0 \le r \le a$

    $r \le z \le a$

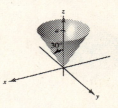

103. $0 \le \theta \le 2\pi$

    $0 \le \phi \le \dfrac{\pi}{6}$

    $0 \le \rho \le a \sec \phi$

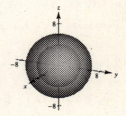

105. Rectangular

    $0 \le x \le 10$

    $0 \le y \le 10$

    $0 \le z \le 10$

107. Spherical

    $4 \le \rho \le 6$

109. $z = \sin \theta, r = 1$

    $z = \dfrac{y}{r} = \dfrac{y}{1} = y$

The curve of intersection is the ellipse formed by the intersection of the plane $z = y$ and the cylinder $r = 1$.

Review Exercises for Chapter 10

1. $P = (1, 2), \ Q = (4, 1), \ R = (5, 4)$

    (a) $\mathbf{u} = \overrightarrow{PQ} = \langle 3, -1 \rangle = 3\mathbf{i} - \mathbf{j},$

        $\mathbf{v} = \overrightarrow{PR} = \langle 4, 2 \rangle = 4\mathbf{i} + 2\mathbf{j}$

    (b) $\|\mathbf{v}\| = \sqrt{4^2 + 2^2} = 2\sqrt{5}$

    (c) $2\mathbf{u} + \mathbf{v} = \langle 6, -2 \rangle + \langle 4, 2 \rangle = \langle 10, 0 \rangle = 10\mathbf{i}$

3. $\mathbf{v} = \|\mathbf{v}\| \cos \theta \, \mathbf{i} + \|\mathbf{v}\| \sin \theta \, \mathbf{j} = 8 \cos 120° \, \mathbf{i} + 8 \sin 120° \, \mathbf{j}$

        $= -4\mathbf{i} + 4\sqrt{3}\mathbf{j}$

5. $120 \cos \theta = 100$

$$\theta = \arccos\left(\frac{5}{6}\right)$$

$$\tan \theta = \frac{2}{y} \implies y = \frac{2}{\tan \theta}$$

$$y = \frac{2}{\tan[\arccos(5/6)]} = \frac{2}{\sqrt{11}/5} = \frac{10}{\sqrt{11}} \approx 3.015 \text{ ft}$$

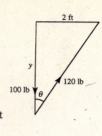

7. $z = 0,\ y = 4,\ x = -5$: $(-5, 4, 0)$

9. Looking down from the positive x-axis towards the yz-plane, the point is either in the first quadrant $(y > 0, z > 0)$ or in the third quadrant $(y < 0, z < 0)$. The x-coordinate can be any number.

11. $(x - 3)^2 + (y + 2)^2 + (z - 6)^2 = \left(\frac{15}{2}\right)^2$

13. $(x^2 - 4x + 4) + (y^2 - 6y + 9) + z^2 = -4 + 4 + 9$

$(x - 2)^2 + (y - 3)^2 + z^2 = 9$

Center: $(2, 3, 0)$

Radius: 3

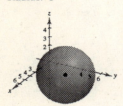

15. $\mathbf{v} = \langle 4 - 2, 4 + 1, -7 - 3 \rangle = \langle 2, 5, -10 \rangle$

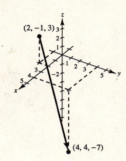

17. $\mathbf{v} = \langle -1 - 3, 6 - 4, 9 + 1 \rangle = \langle -4, 2, 10 \rangle$

$\mathbf{w} = \langle 5 - 3, 3 - 4, -6 + 1 \rangle = \langle 2, -1, -5 \rangle$

Since $-2\mathbf{w} = \mathbf{v}$, the points lie in a straight line.

19. Unit vector: $\dfrac{\mathbf{u}}{\|\mathbf{u}\|} = \dfrac{\langle 2, 3, 5 \rangle}{\sqrt{38}} = \left\langle \dfrac{2}{\sqrt{38}}, \dfrac{3}{\sqrt{38}}, \dfrac{5}{\sqrt{38}} \right\rangle$

21. $P = (5, 0, 0),\ Q = (4, 4, 0),\ R = (2, 0, 6)$

   (a) $\mathbf{u} = \overrightarrow{PQ} = \langle -1, 4, 0 \rangle = -\mathbf{i} + 4\mathbf{j}$,

      $\mathbf{v} = \overrightarrow{PR} = \langle -3, 0, 6 \rangle = -3\mathbf{i} + 6\mathbf{k}$

   (b) $\mathbf{u} \cdot \mathbf{v} = (-1)(-3) + 4(0) + 0(6) = 3$

   (c) $\mathbf{v} \cdot \mathbf{v} = 9 + 36 = 45$

23. $\mathbf{u} = \langle 7, -2, 3 \rangle,\ \mathbf{v} = \langle -1, 4, 5 \rangle$

Since $\mathbf{u} \cdot \mathbf{v} = 0$, the vectors are orthogonal.

25. $\mathbf{u} = 5\left(\cos\dfrac{3\pi}{4}\mathbf{i} + \sin\dfrac{3\pi}{4}\mathbf{j}\right) = \dfrac{5\sqrt{2}}{2}[-\mathbf{i} + \mathbf{j}]$

$\mathbf{v} = 2\left(\cos\dfrac{2\pi}{3}\mathbf{i} + \sin\dfrac{2\pi}{3}\mathbf{j}\right) = -\mathbf{i} + \sqrt{3}\mathbf{j}$

$\mathbf{u} \cdot \mathbf{v} = \dfrac{5\sqrt{2}}{2}(1 + \sqrt{3})$

$\|\mathbf{u}\| = 5$

$\|\mathbf{v}\| = 2$

$\cos \theta = \dfrac{|\mathbf{u} \cdot \mathbf{v}|}{\|\mathbf{u}\|\,\|\mathbf{v}\|} = \dfrac{\left(5\sqrt{2}/2\right)\left(1 + \sqrt{3}\right)}{5(2)} = \dfrac{\sqrt{2} + \sqrt{6}}{4}$

$\theta = \arccos\dfrac{\sqrt{2} + \sqrt{6}}{4} = 15° \text{ or } \dfrac{3\pi}{4} - \dfrac{2\pi}{3} = \dfrac{\pi}{12} \text{ or } 15°$

27. $\mathbf{u} = \langle 10, -5, 15 \rangle,\ \mathbf{v} = \langle -2, 1, -3 \rangle$

$\mathbf{u} = -5\mathbf{v} \implies \mathbf{u}$ is parallel to $\mathbf{v}$ and in the opposite direction.

$\theta = \pi$

29. There are many correct answers. For example: $\mathbf{v} = \pm\langle 6, -5, 0\rangle$.

In Exercises 31–39, $\mathbf{u} = \langle 3, -2, 1\rangle$, $\mathbf{v} = \langle 2, -4, -3\rangle$, $\mathbf{w} = \langle -1, 2, 2\rangle$.

31. $\mathbf{u} \cdot \mathbf{u} = 3(3) + (-2)(-2) + (1)(1)$

$\qquad = 14 = \left(\sqrt{14}\right)^2 = \|\mathbf{u}\|^2$

33. $\text{proj}_\mathbf{u}\mathbf{w} = \left(\dfrac{\mathbf{u} \cdot \mathbf{w}}{\|\mathbf{u}\|^2}\right)\mathbf{u}$

$\qquad = -\dfrac{5}{14}\langle 3, -2, 1\rangle$

$\qquad = \left\langle -\dfrac{15}{14}, \dfrac{10}{14}, -\dfrac{5}{14}\right\rangle$

$\qquad = \left\langle -\dfrac{15}{14}, \dfrac{5}{7}, -\dfrac{5}{14}\right\rangle$

35. $\mathbf{n} = \mathbf{v} \times \mathbf{w} = \begin{vmatrix} \mathbf{i} & \mathbf{j} & \mathbf{k} \\ 2 & -4 & -3 \\ -1 & 2 & 2 \end{vmatrix} = -2\mathbf{i} - \mathbf{j}$

$\|\mathbf{n}\| = \sqrt{5}$

$\dfrac{\mathbf{n}}{\|\mathbf{n}\|} = \dfrac{1}{\sqrt{5}}(-2\mathbf{i} - \mathbf{j})$, unit vector OR $\dfrac{1}{\sqrt{5}}(2\mathbf{i} + \mathbf{j})$

37. $V = |\mathbf{u} \cdot (\mathbf{v} \times \mathbf{w})|$

$\qquad = |\langle 3, -2, 1\rangle \cdot \langle -2, -1, 0\rangle| = |-4| = 4$

39. Area parallelogram $= \|\mathbf{u} \times \mathbf{v}\| = \|\langle 10, 11, -8\rangle\| = \sqrt{10^2 + 11^2 + (-8)^2}$ (See Exercises 36, 38)

$\qquad\qquad\qquad\qquad = \sqrt{285}$

41. $\mathbf{F} = c(\cos 20°\mathbf{j} + \sin 20°\mathbf{k})$

$\overrightarrow{PQ} = 2\mathbf{k}$

$\overrightarrow{PQ} \times \mathbf{F} = \begin{vmatrix} \mathbf{i} & \mathbf{j} & \mathbf{k} \\ 0 & 0 & 2 \\ 0 & c\cos 20° & c\sin 20° \end{vmatrix} = -2c\cos 20°\mathbf{i}$

$200 = \|\overrightarrow{PQ} \times \mathbf{F}\| = 2c\cos 20°$

$c = \dfrac{100}{\cos 20°}$

$\mathbf{F} = \dfrac{100}{\cos 20°}(\cos 20°\mathbf{j} + \sin 20°\mathbf{k}) = 100(\mathbf{j} + \tan 20°\mathbf{k})$

$\|\mathbf{F}\| = 100\sqrt{1 + \tan^2 20°} = 100\sec 20° \approx 106.4 \text{ lb}$

43. $\mathbf{v} = \mathbf{j}$

(a) $x = 1$, $y = 2 + t$, $z = 3$

(b) None

45. $3x - 3y - 7z = -4$, $x - y + 2z = 3$

Solving simultaneously, we have $z = 1$. Substituting $z = 1$ into the second equation we have $y = x - 1$. Substituting for x in this equation we obtain two points on the line of intersection, $(0, -1, 1)$, $(1, 0, 1)$. The direction vector of the line of intersection is $\mathbf{v} = \mathbf{i} + \mathbf{j}$.

(a) $x = t$, $y = -1 + t$, $z = 1$

(b) $x = y + 1$, $z = 1$

47. The two lines are parallel as they have the same direction numbers, $-2, 1, 1$. Therefore, a vector parallel to the plane is $\mathbf{v} = -2\mathbf{i} + \mathbf{j} + \mathbf{k}$. A point on the first line is $(1, 0, -1)$ and a point on the second line is $(-1, 1, 2)$. The vector $\mathbf{u} = 2\mathbf{i} - \mathbf{j} - 3\mathbf{k}$ connecting these two points is also parallel to the plane. Therefore, a normal to the plane is

$$\mathbf{v} \times \mathbf{u} = \begin{vmatrix} \mathbf{i} & \mathbf{j} & \mathbf{k} \\ -2 & 1 & 1 \\ 2 & -1 & -3 \end{vmatrix}$$

$$= -2\mathbf{i} - 4\mathbf{j} = -2(\mathbf{i} + 2\mathbf{j}).$$

Equation of the plane: $(x - 1) + 2y = 0$

$$x + 2y = 1$$

49. $Q(1, 0, 2)$ point

$2x - 3y + 6z = 6$

A point P on the plane is $(3, 0, 0)$.

$$\overrightarrow{PQ} = \langle -2, 0, 2 \rangle$$

$\mathbf{n} = \langle 2, -3, 6 \rangle$ normal to plane

$$D = \frac{|\overrightarrow{PQ} \cdot \mathbf{n}|}{\|\mathbf{n}\|} = \frac{8}{7}$$

51. $Q(3, -2, 4)$ point

$P(5, 0, 0)$ point on plane

$\mathbf{n} = \langle 2, -5, 1 \rangle$ normal to plane

$\overrightarrow{PQ} = \langle -2, -2, 4 \rangle$

$$D = \frac{|\overrightarrow{PQ} \cdot \mathbf{n}|}{\|\mathbf{n}\|} = \frac{10}{\sqrt{30}} = \frac{\sqrt{30}}{3}$$

53. $x + 2y + 3z = 6$

Plane

Intercepts: $(6, 0, 0)$, $(0, 3, 0)$, $(0, 0, 2)$

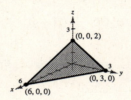

55. $y = \frac{1}{2}z$

Plane with rulings parallel to the x-axis

57. $\dfrac{x^2}{16} + \dfrac{y^2}{9} + z^2 = 1$

Ellipsoid

xy-trace: $\dfrac{x^2}{16} + \dfrac{y^2}{9} = 1$

xz-trace: $\dfrac{x^2}{16} + z^2 = 1$

yz-trace: $\dfrac{y^2}{9} + z^2 = 1$

59. $\dfrac{x^2}{16} - \dfrac{y^2}{9} + z^2 = -1$

$$\frac{y^2}{9} - \frac{x^2}{16} - z^2 = 1$$

Hyperboloid of two sheets

xy-trace: $\dfrac{y^2}{9} - \dfrac{x^2}{16} = 1$

xz-trace: None

yz-trace: $\dfrac{y^2}{9} - z^2 = 1$

61. (a)
$$x^2 + y^2 = [r(z)]^2$$
$$= \left[\sqrt{2(z-1)}\right]^2$$
$$x^2 + y^2 - 2z + 2 = 0$$

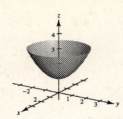

(b) $V = 2\pi \displaystyle\int_0^2 x\left[3 - \left(\frac{1}{2}x^2 + 1\right)\right] dx$

$\qquad = 2\pi \displaystyle\int_0^2 \left(2x - \frac{1}{2}x^3\right) dx$

$\qquad = 2\pi \left[x^2 - \dfrac{x^4}{8}\right]_0^2$

$\qquad = 4\pi \approx 12.6 \text{ cm}^3$

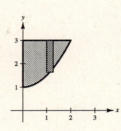

(c) $V = 2\pi \displaystyle\int_{1/2}^2 x\left[3 - \left(\frac{1}{2}x^2 + 1\right)\right] dx$

$\qquad = 2\pi \displaystyle\int_{1/2}^2 \left(2x - \frac{1}{2}x^3\right) dx$

$\qquad = 2\pi \left[x^2 - \dfrac{x^4}{8}\right]_{1/2}^2$

$\qquad = 4\pi - \dfrac{31\pi}{64} = \dfrac{225\pi}{64} \approx 11.04 \text{ cm}^3$

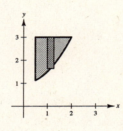

63. $\left(-2\sqrt{2}, 2\sqrt{2}, 2\right)$, rectangular

$\quad$ **(a)** $r = \sqrt{\left(-2\sqrt{2}\right)^2 + \left(2\sqrt{2}\right)^2} = 4$, $\theta = \arctan(-1) = \dfrac{3\pi}{4}$, $z = 2$, $\left(4, \dfrac{3\pi}{4}, 2\right)$, cylindrical

$\quad$ **(b)** $\rho = \sqrt{\left(-2\sqrt{2}\right)^2 + \left(2\sqrt{2}\right)^2 + (2)^2} = 2\sqrt{5}$, $\theta = \dfrac{3\pi}{4}$, $\phi = \arccos\dfrac{2}{2\sqrt{5}} = \arccos\dfrac{1}{\sqrt{5}}$, $\left(2\sqrt{5}, \dfrac{3\pi}{4}, \arccos\dfrac{\sqrt{5}}{5}\right)$, spherical

65. $\left(100, -\dfrac{\pi}{6}, 50\right)$, cylindrical

$\rho = \sqrt{100^2 + 50^2} = 50\sqrt{5}$

$\theta = -\dfrac{\pi}{6}$

$\phi = \arccos\left(\dfrac{50}{50\sqrt{5}}\right) = \arccos\left(\dfrac{1}{\sqrt{5}}\right) \approx 63.4° \text{ OR } 1.107$

$\left(50\sqrt{5}, -\dfrac{\pi}{6}, 63.4°\right)$, spherical

67. $\left(25, -\dfrac{\pi}{4}, \dfrac{3\pi}{4}\right)$, spherical

$r^2 = \left(25\sin\left(\dfrac{3\pi}{4}\right)\right)^2 \Rightarrow r = 25\dfrac{\sqrt{2}}{2}$

$\theta = -\dfrac{\pi}{4}$

$z = \rho\cos\phi = 25\cos\dfrac{3\pi}{4} = -25\dfrac{\sqrt{2}}{2}$

$\left(25\dfrac{\sqrt{2}}{2}, -\dfrac{\pi}{4}, -\dfrac{25\sqrt{2}}{2}\right)$, cylindrical

69. $x^2 - y^2 = 2z$

$\quad$ **(a)** Cylindrical: $r^2\cos^2\theta - r^2\sin^2\theta = 2z$, $r^2\cos 2\theta = 2z$

$\quad$ **(b)** Spherical: $\rho^2\sin^2\phi\cos^2\theta - \rho^2\sin^2\phi\sin^2\theta = 2\rho\cos\phi$, $\rho\sin^2\phi\cos 2\theta - 2\cos\phi = 0$, $\rho = 2\sec 2\theta\cos\phi\csc^2\phi$

Problem Solving for Chapter 10

1.
$$\mathbf{a} + \mathbf{b} + \mathbf{c} = 0$$
$$\mathbf{b} \times (\mathbf{a} + \mathbf{b} + \mathbf{c}) = 0$$
$$(\mathbf{b} \times \mathbf{a}) + (\mathbf{b} \times \mathbf{c}) = 0$$
$$\|\mathbf{a} \times \mathbf{b}\| = \|\mathbf{b} \times \mathbf{c}\|$$
$$\|\mathbf{b} \times \mathbf{c}\| = \|\mathbf{b}\|\,\|\mathbf{c}\| \sin A$$
$$\|\mathbf{a} \times \mathbf{b}\| = \|\mathbf{a}\|\,\|\mathbf{b}\| \sin C$$

Then,

$$\frac{\sin A}{\|\mathbf{a}\|} = \frac{\|\mathbf{b} \times \mathbf{c}\|}{\|\mathbf{a}\|\,\|\mathbf{b}\|\,\|\mathbf{c}\|}$$

$$= \frac{\|\mathbf{a} \times \mathbf{b}\|}{\|\mathbf{a}\|\,\|\mathbf{b}\|\,\|\mathbf{c}\|}$$

$$= \frac{\sin C}{\|\mathbf{c}\|}.$$

The other case, $\dfrac{\sin A}{\|\mathbf{a}\|} = \dfrac{\sin B}{\|\mathbf{b}\|}$ is similar.

3. Label the figure as indicated.

From the figure, you see that

$$\overrightarrow{SP} = \frac{1}{2}\mathbf{a} - \frac{1}{2}\mathbf{b} = \overrightarrow{RQ} \text{ and}$$

$$\overrightarrow{SR} = \frac{1}{2}\mathbf{a} + \frac{1}{2}\mathbf{b} = \overrightarrow{PQ}.$$

Since $\overrightarrow{SP} = \overrightarrow{RQ}$ and $\overrightarrow{SR} = \overrightarrow{PQ}$, $PSRQ$ is a parallelogram.

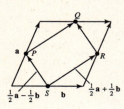

5. (a) $\mathbf{u} = \langle 0, 1, 1 \rangle$ direction vector of line determined by P_1 and P_2.

$$D = \frac{\|\overrightarrow{P_1Q} \times \mathbf{u}\|}{\|\mathbf{u}\|}$$

$$= \frac{\|\langle 2, 0, -1 \rangle \times \langle 0, 1, 1 \rangle\|}{\sqrt{2}}$$

$$= \frac{\|\langle 1, -2, 2 \rangle\|}{\sqrt{2}} = \frac{3}{\sqrt{2}} = \frac{3\sqrt{2}}{2}$$

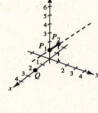

(b) The shortest distance to the line **segment** is $\|P_1Q\| = \|\langle 2, 0, -1 \rangle\| = \sqrt{5}$.

7. (a) $V = \pi \displaystyle\int_0^1 \left(\sqrt{z}\right)^2 dz = \left[\pi \dfrac{z^2}{2}\right]_0^1 = \dfrac{1}{2}\pi$

   Note: $\dfrac{1}{2}(\text{base})(\text{altitude}) = \dfrac{1}{2}\pi(1) = \dfrac{1}{2}\pi$

(b) $\dfrac{x^2}{a^2} + \dfrac{y^2}{b^2} = z$: (slice at $z = c$)

   $$\frac{x^2}{\left(\sqrt{ca}\right)^2} + \frac{y^2}{\left(\sqrt{cb}\right)^2} = 1$$

   At $z = c$, figure is ellipse of area

   $$\pi\left(\sqrt{ca}\right)\left(\sqrt{cb}\right) = \pi abc.$$

   $$V = \int_0^k \pi abc \cdot dc = \left[\frac{\pi abc^2}{2}\right]_0^k = \frac{\pi abk^2}{2}$$

(c) $V = \dfrac{1}{2}(\pi abk)k = \dfrac{1}{2}(\text{base})(\text{height})$

9. (a) $\rho = 2 \sin \phi$

   Torus

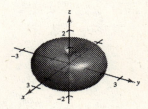

(b) $\rho = 2 \cos \phi$

   Sphere

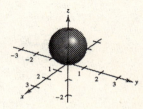

11. From Exercise 64, Section 10.4, $(\mathbf{u} \times \mathbf{v}) \times (\mathbf{w} \times \mathbf{z}) = [(\mathbf{u} \times \mathbf{v}) \cdot \mathbf{z}]\mathbf{w} - [(\mathbf{u} \times \mathbf{v}) \cdot \mathbf{w}]\mathbf{z}$.

13. (a) $\mathbf{u} = \|\mathbf{u}\|(\cos 0\,\mathbf{i} + \sin 0\,\mathbf{j}) = \|\mathbf{u}\|\mathbf{i}$

Downward force $\mathbf{w} = -\mathbf{j}$

$\mathbf{T} = \|\mathbf{T}\|(\cos(90° + \theta)\mathbf{i} + \sin(90° + \theta)\mathbf{j})$

$\quad = \|\mathbf{T}\|(-\sin\theta\,\mathbf{i} + \cos\theta\,\mathbf{j})$

$\mathbf{0} = \mathbf{u} + \mathbf{w} + \mathbf{T} = \|\mathbf{u}\|\mathbf{i} - \mathbf{j} + \|\mathbf{T}\|(-\sin\theta\,\mathbf{i} + \cos\theta\,\mathbf{j})$

$\quad \|\mathbf{u}\| = \sin\theta\,\|\mathbf{T}\|$

$\quad 1 = \cos\theta\,\|\mathbf{T}\|$

If $\theta = 30°$, $\|\mathbf{u}\| = (1/2)\|\mathbf{T}\|$ and $1 = \left(\sqrt{3}/2\right)\|\mathbf{T}\|$

$\Rightarrow \|\mathbf{T}\| = \dfrac{2}{\sqrt{3}} \approx 1.1547\ \text{lb}$

and

$\|\mathbf{u}\| = \dfrac{1}{2}\left(\dfrac{2}{\sqrt{3}}\right) \approx 0.5774\ \text{lb}$

(b) From part (a), $\|\mathbf{u}\| = \tan\theta$ and $\|\mathbf{T}\| = \sec\theta$.

Domain: $0 \le \theta \le 90°$

(c)

θ	0°	10°	20°	30°	40°	50°	60°
T	1	1.0154	1.0642	1.1547	1.3054	1.5557	2
$\|\mathbf{u}\|$	0	0.1763	0.3640	0.5774	0.8391	1.1918	1.7321

(d)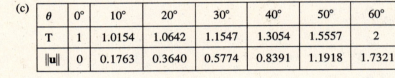

(e) Both are increasing functions.

(f) $\displaystyle\lim_{\theta \to \pi/2^-} T = \infty$ and $\displaystyle\lim_{\theta \to \pi/2^-} \|\mathbf{u}\| = \infty.$

15. Let $\theta = \alpha - \beta$, the angle between $\mathbf{u}$ and $\mathbf{v}$. Then

$\sin(\alpha - \beta) = \dfrac{\|\mathbf{u} \times \mathbf{v}\|}{\|\mathbf{u}\|\,\|\mathbf{v}\|} = \dfrac{\|\mathbf{v} \times \mathbf{u}\|}{\|\mathbf{u}\|\,\|\mathbf{v}\|}.$

For $\mathbf{u} = \langle\cos\alpha, \sin\alpha, 0\rangle$ and $\mathbf{v} = \langle\cos\beta, \sin\beta, 0\rangle$, $\|\mathbf{u}\| = \|\mathbf{v}\| = 1$ and

$\mathbf{v} \times \mathbf{u} = \begin{vmatrix} \mathbf{i} & \mathbf{j} & \mathbf{k} \\ \cos\beta & \sin\beta & 0 \\ \cos\alpha & \sin\alpha & 0 \end{vmatrix} = (\sin\alpha\cos\beta - \cos\alpha\sin\beta)\mathbf{k}.$

Thus, $\sin(\alpha - \beta) = \|\mathbf{v} \times \mathbf{u}\| = \sin\alpha\cos\beta - \cos\alpha\sin\beta.$

17. From Theorem 10.13 and Theorem 10.7 (6) we have

$D = \dfrac{|\overrightarrow{PQ} \cdot \mathbf{n}|}{\|\mathbf{n}\|}$

$\quad = \dfrac{|\mathbf{w} \cdot (\mathbf{u} \times \mathbf{v})|}{\|\mathbf{u} \times \mathbf{v}\|} = \dfrac{|(\mathbf{u} \times \mathbf{v}) \cdot \mathbf{w}|}{\|\mathbf{u} \times \mathbf{v}\|} = \dfrac{|\mathbf{u} \cdot (\mathbf{v} \times \mathbf{w})|}{\|\mathbf{u} \times \mathbf{v}\|}.$

19. $a_1, b_1, c_1,$ and a_2, b_2, c_2 are two sets of direction numbers for the same line. The line is parallel to both $\mathbf{u} = a_1\mathbf{i} + b_1\mathbf{j} + c_1\mathbf{k}$ and $\mathbf{v} = a_2\mathbf{i} + b_2\mathbf{j} + c_2\mathbf{k}$. Therefore, $\mathbf{u}$ and $\mathbf{v}$ are parallel, and there exists a scalar d such that $\mathbf{u} = d\mathbf{v}$, $a_1\mathbf{i} + b_1\mathbf{j} + c_1\mathbf{k} = d(a_2\mathbf{i} + b_2\mathbf{j} + c_2\mathbf{k})$, $a_1 = a_2 d, b_1 = b_2 d, c_1 = c_2 d.$

CHAPTER 11
Vector-Valued Functions

CHAPTER 11
Vector-Valued Functions

Section 11.1 Vector-Valued Functions

Solutions to Odd-Numbered Exercises

1. $\mathbf{r}(t) = 5t\mathbf{i} - 4t\mathbf{j} - \dfrac{1}{t}\mathbf{k}$

Component functions: $f(t) = 5t$

$\qquad\qquad\qquad g(t) = -4t$

$\qquad\qquad\qquad h(t) = -\dfrac{1}{t}$

Domain: $(-\infty, 0) \cup (0, \infty)$

3. $\mathbf{r}(t) = \ln t\mathbf{i} - e^t\mathbf{j} - t\mathbf{k}$

Component functions: $f(t) = \ln t$

$\qquad\qquad\qquad g(t) = -e^t$

$\qquad\qquad\qquad h(t) = -t$

Domain: $(0, \infty)$

5. $\mathbf{r}(t) = \mathbf{F}(t) + \mathbf{G}(t) = \left(\cos t\mathbf{i} - \sin t\mathbf{j} + \sqrt{t}\mathbf{k}\right) + (\cos t\mathbf{i} + \sin t\mathbf{j}) = 2\cos t\mathbf{i} + \sqrt{t}\mathbf{k}$

Domain: $[0, \infty)$

7. $\mathbf{r}(t) = \mathbf{F}(t) \times \mathbf{G}(t) = \begin{vmatrix} \mathbf{i} & \mathbf{j} & \mathbf{k} \\ \sin t & \cos t & 0 \\ 0 & \sin t & \cos t \end{vmatrix} = \cos^2 t\mathbf{i} - \sin t \cos t\mathbf{j} + \sin^2 t\mathbf{k}$

Domain: $(-\infty, \infty)$

9. $\mathbf{r}(t) = \frac{1}{2}t^2\mathbf{i} - (t - 1)\mathbf{j}$

(a) $\mathbf{r}(1) = \frac{1}{2}\mathbf{i}$

(b) $\mathbf{r}(0) = \mathbf{j}$

(c) $\mathbf{r}(s + 1) = \frac{1}{2}(s + 1)^2\mathbf{i} - (s + 1 - 1)\mathbf{j} = \frac{1}{2}(s + 1)^2\mathbf{i} - s\mathbf{j}$

(d) $\mathbf{r}(2 + \Delta t) - \mathbf{r}(2) = \frac{1}{2}(2 + \Delta t)^2\mathbf{i} - (2 + \Delta t - 1)\mathbf{j} - (2\mathbf{i} - \mathbf{j})$

$\qquad\qquad\qquad\qquad = \left(2 + 2\Delta t + \tfrac{1}{2}(\Delta t)^2\right)\mathbf{i} - (1 + \Delta t)\mathbf{j} - 2\mathbf{i} + \mathbf{j}$

$\qquad\qquad\qquad\qquad = \left(2\Delta t + \tfrac{1}{2}(\Delta t)^2\right)\mathbf{i} - (\Delta t)\mathbf{j}$

11. $\mathbf{r}(t) = \ln t\mathbf{i} + \dfrac{1}{t}\mathbf{j} + 3t\mathbf{k}$

(a) $\mathbf{r}(2) = \ln 2\mathbf{i} + \dfrac{1}{2}\mathbf{j} + 6\mathbf{k}$

(b) $\mathbf{r}(-3)$ is not defined.   ($\ln(-3)$ does not exist.)

(c) $\mathbf{r}(t - 4) = \ln(t - 4)\mathbf{i} + \dfrac{1}{t - 4}\mathbf{j} + 3(t - 4)\mathbf{k}$

(d) $\mathbf{r}(1 + \Delta t) - \mathbf{r}(1) = \ln(1 + \Delta t)\mathbf{i} + \dfrac{1}{1 + \Delta t}\mathbf{j} + 3(1 + \Delta t)\mathbf{k} - (0\mathbf{i} + \mathbf{j} + 3\mathbf{k})$

$\qquad\qquad\qquad\qquad\qquad = \ln(1 + \Delta t)\mathbf{i} + \left(\dfrac{1}{1 + \Delta t} - 1\right)\mathbf{j} + (3\Delta t)\mathbf{k}$

13. $\mathbf{r}(t) = \sin 3t\mathbf{i} + \cos 3t\mathbf{j} + t\mathbf{k}$

$\|\mathbf{r}(t)\| = \sqrt{(\sin 3t)^2 + (\cos 3t)^2 + t^2} = \sqrt{1 + t^2}$

15. $\mathbf{r}(t) \cdot \mathbf{u}(t) = (3t - 1)(t^2) + \left(\frac{1}{4}t^3\right)(-8) + 4(t^3)$

$= 3t^3 - t^2 - 2t^3 + 4t^3 = 5t^3 - t^2$, a scalar.

The dot product is a scalar-valued function.

17. $\mathbf{r}(t) = t\mathbf{i} + 2t\mathbf{j} + t^2\mathbf{k}, -2 \le t \le 2$

$x = t, y = 2t, z = t^2$

Thus, $z = x^2$. Matches (b)

19. $\mathbf{r}(t) = t\mathbf{i} + t^2\mathbf{j} + e^{0.75t}\mathbf{k}, -2 \le t \le 2$

$x = t, y = t^2, z = e^{0.75t}$

Thus, $y = x^2$. Matches (d)

21. (a) View from the negative x-axis: $(-20, 0, 0)$

(c) View from the z-axis: $(0, 0, 20)$

(b) View from above the first octant: $(10, 20, 10)$

(d) View from the positive x-axis: $(20, 0, 0)$

23. $x = 3t$

$y = t - 1$

$y = \dfrac{x}{3} - 1$

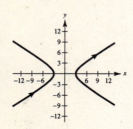

25. $x = t^3, y = t^2$

$y = x^{2/3}$

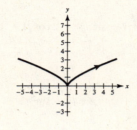

27. $x = \cos \theta, y = 3 \sin \theta$

$x^2 + \dfrac{y^2}{9} = 1$ Ellipse

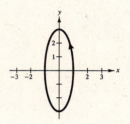

29. $x = 3 \sec \theta, y = 2 \tan \theta$

$\dfrac{x^2}{9} = \dfrac{y^2}{4} + 1$ Hyperbola

31. $x = -t + 1$

$y = 4t + 2$

$z = 2t + 3$

Line passing through the points:

$(0, 6, 5), \ (1, 2, 3)$

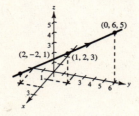

33. $x = 2 \cos t, \ y = 2 \sin t, \ z = t$

$\dfrac{x^2}{4} + \dfrac{y^2}{4} = 1$

$z = t$

Circular helix

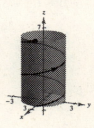

35. $x = 2 \sin t, \ y = 2 \cos t, \ z = e^{-t}$

$x^2 + y^2 = 4$

$z = e^{-t}$

37. $x = t, \ y = t^2, \ z = \frac{2}{3}t^3$

$y = x^2, \ z = \frac{2}{3}x^3$

t	-2	-1	0	1	2
x	-2	-1	0	1	2
y	4	1	0	1	4
z	$-\frac{16}{3}$	$-\frac{2}{3}$	0	$\frac{2}{3}$	$\frac{16}{3}$

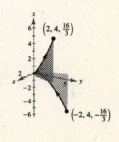

39. $\mathbf{r}(t) = -\dfrac{1}{2}t^2\mathbf{i} + t\mathbf{j} - \dfrac{\sqrt{3}}{2}t^2\mathbf{k}$

Parabola

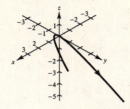
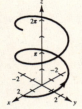
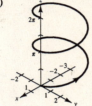

41. $\mathbf{r}(t) = \sin t\,\mathbf{i} + \left(\dfrac{\sqrt{3}}{2}\cos t - \dfrac{1}{2}t\right)\mathbf{j} + \left(\dfrac{1}{2}\cos t + \dfrac{\sqrt{3}}{2}\right)\mathbf{k}$

Helix

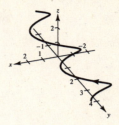

43.

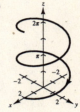
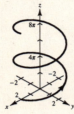

(a)

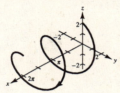

The helix is translated 2 units back on the x-axis.

(b)

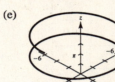

The height of the helix increases at a faster rate.

(c)

The orientation of the helix is reversed.

(d)

The axis of the helix is the x-axis.

(e)

The radius of the helix is increased from 2 to 6.

45. $y = 4 - x$

Let $x = t$, then $y = 4 - t$.

$\mathbf{r}(t) = t\mathbf{i} + (4 - t)\mathbf{j}$

47. $y = (x - 2)^2$

Let $x = t$, then $y = (t - 2)^2$.

$\mathbf{r}(t) = t\mathbf{i} + (t - 2)^2\,\mathbf{j}$

49. $x^2 + y^2 = 25$

Let $x = 5\cos t$, then $y = 5\sin t$.

$\mathbf{r}(t) = 5\cos t\,\mathbf{i} + 5\sin t\,\mathbf{j}$

51. $\dfrac{x^2}{16} - \dfrac{y^2}{4} = 1$

Let $x = 4\sec t$, $y = 2\tan t$.

$\mathbf{r}(t) = 4\sec t\,\mathbf{i} + 2\tan t\,\mathbf{j}$

53. The parametric equations for the line are

$x = 2 - 2t$, $y = 3 + 5t$, $z = 8t$.

One possible answer is

$\mathbf{r}(t) = (2 - 2t)\mathbf{i} + (3 + 5t)\mathbf{j} + 8t\mathbf{k}$.

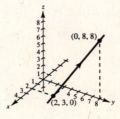

55. $\mathbf{r}_1(t) = t\mathbf{i}$, $\qquad\qquad 0 \le t \le 4 \quad (\mathbf{r}_1(0) = \mathbf{0}, \mathbf{r}_1(4) = 4\mathbf{i})$

$\mathbf{r}_2(t) = (4 - 4t)\mathbf{i} + 6t\mathbf{j}, \quad 0 \le t \le 1 \quad (\mathbf{r}_2(0) = 4\mathbf{i}, \mathbf{r}_2(1) = 6\mathbf{j})$

$\mathbf{r}_3(t) = (6 - t)\mathbf{j}, \qquad\quad 0 \le t \le 6 \quad (\mathbf{r}_3(0) = 6\mathbf{j}, \mathbf{r}_3(6) = \mathbf{0})$

(Other answers possible)

57. $\mathbf{r}_1(t) = t\mathbf{i} + t^2\mathbf{j}, \quad 0 \le t \le 2 \;\; (y = x^2)$

$\mathbf{r}_2(t) = (2 - t)\mathbf{i}, \quad 0 \le t \le 2$

$\mathbf{r}_3(t) = (4 - t)\mathbf{j}, \quad 0 \le t \le 4$

(Other answers possible)

59. $z = x^2 + y^2, \; x + y = 0$

Let $x = t$, then $y = -x = -t$ and $z = x^2 + y^2 = 2t^2$. Therefore,

$$x = t, \; y = -t, \; z = 2t^2.$$

$$\mathbf{r}(t) = t\mathbf{i} - t\mathbf{j} + 2t^2\mathbf{k}$$

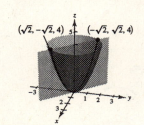

61. $x^2 + y^2 = 4, \; z = x^2$

$x = 2\sin t, \; y = 2\cos t$

$z = x^2 = 4\sin^2 t$

t	0	$\dfrac{\pi}{6}$	$\dfrac{\pi}{4}$	$\dfrac{\pi}{2}$	$\dfrac{3\pi}{4}$	π
x	0	1	$\sqrt{2}$	2	$\sqrt{2}$	0
y	2	$\sqrt{3}$	$\sqrt{2}$	0	$-\sqrt{2}$	-2
z	0	1	2	4	2	0

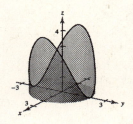

$\mathbf{r}(t) = 2\sin t\,\mathbf{i} + 2\cos t\,\mathbf{j} + 4\sin^2 t\,\mathbf{k}$

63. $x^2 + y^2 + z^2 = 4, \; x + z = 2$

Let $x = 1 + \sin t$, then $z = 2 - x = 1 - \sin t$ and $x^2 + y^2 + z^2 = 4$.

$(1 + \sin t)^2 + y^2 + (1 - \sin t)^2 = 2 + 2\sin^2 t + y^2 = 4$

$y^2 = 2\cos^2 t, \quad y = \pm\sqrt{2}\cos t$

$x = 1 + \sin t, \; y = \pm\sqrt{2}\cos t$

$z = 1 - \sin t$

t	$-\dfrac{\pi}{2}$	$-\dfrac{\pi}{6}$	0	$\dfrac{\pi}{6}$	$\dfrac{\pi}{2}$
x	0	$\dfrac{1}{2}$	1	$\dfrac{3}{2}$	2
y	0	$\pm\dfrac{\sqrt{6}}{2}$	$\pm\sqrt{2}$	$\pm\dfrac{\sqrt{6}}{2}$	0
z	2	$\dfrac{3}{2}$	1	$\dfrac{1}{2}$	0

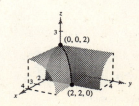

$\mathbf{r}(t) = (1 + \sin t)\mathbf{i} + \sqrt{2}\cos t\,\mathbf{j} - (1 - \sin t)\mathbf{k}$ and

$\mathbf{r}(t) = (1 + \sin t)\mathbf{i} - \sqrt{2}\cos t\,\mathbf{j} + (1 - \sin t)\mathbf{k}$

65. $x^2 + z^2 = 4, \; y^2 + z^2 = 4$

Subtracting, we have $x^2 - y^2 = 0$ or $y = \pm x$.

Therefore, in the first octant, if we let $x = t$, then $x = t, \; y = t, \; z = \sqrt{4 - t^2}$.

$\mathbf{r}(t) = t\mathbf{i} + t\mathbf{j} + \sqrt{4 - t^2}\,\mathbf{k}$

67. $y^2 + z^2 = (2t \cos t)^2 + (2t \sin t)^2 = 4t^2 = 4x^2$

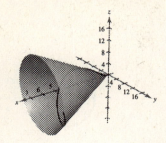

69. $\lim\limits_{t \to 2} \left[t\mathbf{i} + \dfrac{t^2 - 4}{t^2 - 2t}\mathbf{j} + \dfrac{1}{t}\mathbf{k} \right] = 2\mathbf{i} + 2\mathbf{j} + \dfrac{1}{2}\mathbf{k}$

since

$$\lim\limits_{t \to 2} \dfrac{t^2 - 4}{t^2 - 2t} = \lim\limits_{t \to 2} \dfrac{2t}{2t - 2} = 2. \quad \text{(L'Hôpital's Rule)}$$

71. $\lim\limits_{t \to 0} \left[t^2\mathbf{i} + 3t\mathbf{j} + \dfrac{1 - \cos t}{t}\mathbf{k} \right] = \mathbf{0}$

since

$$\lim\limits_{t \to 0} \dfrac{1 - \cos t}{t} = \lim\limits_{t \to 0} \dfrac{\sin t}{1} = 0. \quad \text{(L'Hôpital's Rule)}$$

73. $\lim\limits_{t \to 0} \left[\dfrac{1}{t}\mathbf{i} + \cos t\mathbf{j} + \sin t\mathbf{k} \right]$

does not exist since $\lim\limits_{t \to 0} \dfrac{1}{t}$ does not exist.

75. $\mathbf{r}(t) = t\mathbf{i} + \dfrac{1}{t}\mathbf{j}$

Continuous on $(-\infty, 0), \ (0, \infty)$

77. $\mathbf{r}(t) = t\mathbf{i} + \arcsin t\mathbf{j} + (t - 1)\mathbf{k}$

Continuous on $[-1, 1]$

79. $\mathbf{r}(t) = \langle e^{-t}, t^2, \tan t \rangle$

Discontinuous at $t = \dfrac{\pi}{2} + n\pi$

Continuous on $\left(-\dfrac{\pi}{2} + n\pi, \dfrac{\pi}{2} + n\pi \right)$

81. See the definition on page 786.

83. $\mathbf{r}(t) = t^2\mathbf{i} + (t - 3)\mathbf{j} + t\mathbf{k}$

    (a) $\mathbf{s}(t) = \mathbf{r}(t) + 2\mathbf{k} = t^2\mathbf{i} + (t - 3)\mathbf{j} + (t + 3)\mathbf{k}$

    (b) $\mathbf{s}(t) = \mathbf{r}(t) - 2\mathbf{i} = (t^2 - 2)\mathbf{i} + (t - 3)\mathbf{j} + t\mathbf{k}$

    (c) $\mathbf{s}(t) = \mathbf{r}(t) + 5\mathbf{j} = t^2\mathbf{i} + (t + 2)\mathbf{j} + t\mathbf{k}$

85. Let $\mathbf{r}(t) = x_1(t) + y_1(t)\mathbf{j} + z_1(t)\mathbf{k}$ and $\mathbf{u}(t) = x_2(t)\mathbf{i} + y_2(t)\mathbf{j} + z_2(t)\mathbf{k}$. Then:

$$\lim\limits_{t \to c} [\mathbf{r}(t) \times \mathbf{u}(t)] = \lim\limits_{t \to c} \{ [y_1(t)z_2(t) - y_2(t)z_1(t)]\mathbf{i} - [x_1(t)z_2(t) - x_2(t)z_1(t)]\mathbf{j} + [x_1(t)y_2(t) - x_2(t)y_1(t)]\mathbf{k} \}$$

$$= \left[\lim\limits_{t \to c} y_1(t) \lim\limits_{t \to c} z_2(t) - \lim\limits_{t \to c} y_2(t) \lim\limits_{t \to c} z_1(t) \right]\mathbf{i} - \left[\lim\limits_{t \to c} x_1(t) \lim\limits_{t \to c} z_2(t) - \lim\limits_{t \to c} x_2(t) \lim\limits_{t \to c} z_1(t) \right]\mathbf{j}$$

$$+ \left[\lim\limits_{t \to c} x_1(t) \lim\limits_{t \to c} y_2(t) - \lim\limits_{t \to c} x_2(t) \lim\limits_{t \to c} y_1(t) \right]\mathbf{k}$$

$$= \left[\lim\limits_{t \to c} x_1(t)\mathbf{i} + \lim\limits_{t \to c} y_1(t)\mathbf{j} + \lim\limits_{t \to c} z_1(t)\mathbf{k} \right] \times \left[\lim\limits_{t \to c} x_2(t)\mathbf{i} + \lim\limits_{t \to c} y_2(t)\mathbf{j} + \lim\limits_{t \to c} z_2(t)\mathbf{k} \right]$$

$$= \lim\limits_{t \to c} \mathbf{r}(t) \times \lim\limits_{t \to c} \mathbf{u}(t)$$

87. Let $\mathbf{r}(t) = x(t)\mathbf{i} + y(t)\mathbf{j} + z(t)\mathbf{k}$. Since $\mathbf{r}$ is continuous at $t = c$, then $\lim\limits_{t \to c} \mathbf{r}(t) = \mathbf{r}(c)$.

    $\mathbf{r}(c) = x(c)\mathbf{i} + y(c)\mathbf{j} + z(c)\mathbf{k} \Longrightarrow x(c), \ y(c), \ z(c)$

are defined at c.

$$\|\mathbf{r}\| = \sqrt{(x(t))^2 + (y(t))^2 + (z(t))^2}$$

$$\lim\limits_{t \to c} \|\mathbf{r}\| = \sqrt{(x(c))^2 + (y(c))^2 + (z(c))^2} = \|\mathbf{r}(c)\|$$

Therefore, $\|\mathbf{r}\|$ is continuous at c.

89. True

Section 11.2 Differentiation and Integration of Vector-Valued Functions

1. $\mathbf{r}(t) = t^2\mathbf{i} + t\mathbf{j}, \ t_0 = 2$

$x(t) = t^2, \ y(t) = t$

$x = y^2$

$\qquad \mathbf{r}(2) = 4\mathbf{i} + 2\mathbf{j}$

$\qquad \mathbf{r}'(t) = 2t\mathbf{i} + \mathbf{j}$

$\qquad \mathbf{r}'(2) = 4\mathbf{i} + \mathbf{j}$

$\mathbf{r}'(t_0)$ is tangent to the curve.

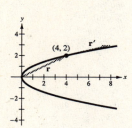

3. $\mathbf{r}(t) = \cos t\mathbf{i} + \sin t\mathbf{j}, \ t_0 = \dfrac{\pi}{2}$

$x(t) = \cos t, \ y(t) = \sin t$

$x^2 + y^2 = 1$

$\qquad \mathbf{r}\left(\dfrac{\pi}{2}\right) = \mathbf{j}$

$\qquad \mathbf{r}'(t) = -\sin t\mathbf{i} + \cos t\mathbf{j}$

$\qquad \mathbf{r}'\left(\dfrac{\pi}{2}\right) = -\mathbf{i}$

$\mathbf{r}'(t_0)$ is tangent to the curve.

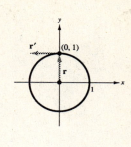

5. $\mathbf{r}(t) = t\mathbf{i} + t^2\mathbf{j}$

(a)

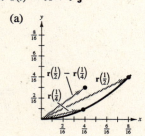

(b) $\qquad \mathbf{r}\left(\dfrac{1}{4}\right) = \dfrac{1}{4}\mathbf{i} + \dfrac{1}{16}\mathbf{j}$

$\qquad \mathbf{r}\left(\dfrac{1}{2}\right) = \dfrac{1}{2}\mathbf{i} + \dfrac{1}{4}\mathbf{j}$

$\quad \mathbf{r}\left(\dfrac{1}{2}\right) - \mathbf{r}\left(\dfrac{1}{4}\right) = \dfrac{1}{4}\mathbf{i} + \dfrac{3}{16}\mathbf{j}$

(c) $\qquad \mathbf{r}'(t) = \mathbf{i} + 2t\mathbf{j}$

$\qquad \mathbf{r}'\left(\dfrac{1}{4}\right) = \mathbf{i} + \dfrac{1}{2}\mathbf{j}$

$\dfrac{\mathbf{r}(1/2) - \mathbf{r}(1/4)}{(1/2) - (1/4)} = \dfrac{(1/4)\mathbf{i} + (3/16)\mathbf{j}}{1/4} = \mathbf{i} + \dfrac{3}{4}\mathbf{j}$

This vector approximates $\mathbf{r}'\left(\tfrac{1}{4}\right)$.

7. $\mathbf{r}(t) = 2\cos t\mathbf{i} + 2\sin t\mathbf{j} + t\mathbf{k}, \ t_0 = \dfrac{3\pi}{2}$

$x^2 + y^2 = 4, \ z = t$

$\qquad \mathbf{r}'(t) = -2\sin t\mathbf{i} + 2\cos t\mathbf{j} + \mathbf{k}$

$\qquad \mathbf{r}\left(\dfrac{3\pi}{2}\right) = -2\mathbf{j} + \dfrac{3\pi}{2}\mathbf{k}$

$\qquad \mathbf{r}'\left(\dfrac{3\pi}{2}\right) = 2\mathbf{i} + \mathbf{k}$

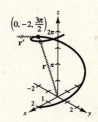

9. $\mathbf{r}(t) = 6t\mathbf{i} - 7t^2\mathbf{j} + t^3\mathbf{k}$

$\quad \mathbf{r}'(t) = 6\mathbf{i} - 14t\mathbf{j} + 3t^2\mathbf{k}$

11. $\mathbf{r}(t) = a\cos^3 t\mathbf{i} + a\sin^3 t\mathbf{j} + \mathbf{k}$

$\quad \mathbf{r}'(t) = -3a\cos^2 t\sin t\mathbf{i} + 3a\sin^2 t\cos t\mathbf{j}$

13. $\mathbf{r}(t) = e^{-t}\mathbf{i} + 4\mathbf{j}$

$\quad \mathbf{r}'(t) = -e^{-t}\mathbf{i}$

15. $\mathbf{r}(t) = \langle t\sin t, t\cos t, t\rangle$

$\quad \mathbf{r}'(t) = \langle \sin t + t\cos t, \cos t - t\sin t, 1\rangle$

17. $\mathbf{r}(t) = t^3\mathbf{i} + \dfrac{1}{2}t^2\mathbf{j}$

(a) $\mathbf{r}'(t) = 3t^2\mathbf{i} + t\mathbf{j}$

$\quad \mathbf{r}''(t) = 6t\mathbf{i} + \mathbf{j}$

(b) $\mathbf{r}'(t) \cdot \mathbf{r}''(t) = 3t^2(6t) + t = 18t^3 + t$

19. $\mathbf{r}(t) = 4\cos t\mathbf{i} + 4\sin t\mathbf{j}$

(a) $\mathbf{r}'(t) = -4\sin t\mathbf{i} + 4\cos t\mathbf{j}$

$\mathbf{r}''(t) = -4\cos t\mathbf{i} - 4\sin t\mathbf{j}$

(b) $\mathbf{r}'(t) \cdot \mathbf{r}''(t) = (-4\sin t)(-4\cos t) + 4\cos t(-4\sin t)$

$= 0$

21. $\mathbf{r}(t) = \frac{1}{2}t^2\mathbf{i} - t\mathbf{j} + \frac{1}{6}t^3\mathbf{k}$

(a) $\mathbf{r}'(t) = t\mathbf{i} - \mathbf{j} + \frac{1}{2}t^2\mathbf{k}$

$\mathbf{r}''(t) = \mathbf{i} + t\mathbf{k}$

(b) $\mathbf{r}'(t) \cdot \mathbf{r}''(t) = t(1) - 1(0) + \frac{1}{2}t^2(t) = t + \frac{t^3}{2}$

23. $\mathbf{r}(t) = \langle \cos t + t\sin t, \sin t - t\cos t, t \rangle$

(a) $\mathbf{r}'(t) = \langle -\sin t + \sin t + t\cos t, \cos t - \cos t + t\sin t, 1 \rangle$

$= \langle t\cos t, t\sin t, 1 \rangle$

$\mathbf{r}''(t) = \langle \cos t - t\sin t, \sin t + t\cos t, 0 \rangle$

(b) $\mathbf{r}'(t) \cdot \mathbf{r}''(t) = (t\cos t)(\cos t - t\sin t) + (t\sin t)(\sin t + t\cos t) = t$

25.

$$\mathbf{r}(t) = \cos(\pi t)\mathbf{i} + \sin(\pi t)\mathbf{j} + t^2\mathbf{k}, \ t_0 = -\frac{1}{4}$$

$$\mathbf{r}'(t) = -\pi\sin(\pi t)\mathbf{i} + \pi\cos(\pi t)\mathbf{j} + 2t\mathbf{k}$$

$$\mathbf{r}'\left(-\frac{1}{4}\right) = \frac{\sqrt{2}\pi}{2}\mathbf{i} + \frac{\sqrt{2}\pi}{2}\mathbf{j} - \frac{1}{2}\mathbf{k}$$

$$\left\|\mathbf{r}'\left(\frac{1}{4}\right)\right\| = \sqrt{\left(\frac{\sqrt{2}\pi}{2}\right)^2 + \left(\frac{\sqrt{2}\pi}{2}\right)^2 + \left(-\frac{1}{2}\right)^2} = \sqrt{\pi^2 + \frac{1}{4}} = \frac{\sqrt{4\pi^2 + 1}}{2}$$

$$\frac{\mathbf{r}'(-1/4)}{\|\mathbf{r}'(-1/4)\|} = \frac{1}{\sqrt{4\pi^2 + 1}}(\sqrt{2}\pi\mathbf{i} + \sqrt{2}\pi\mathbf{j} - \mathbf{k})$$

$$\mathbf{r}''(t) = -\pi^2\cos(\pi t)\mathbf{i} - \pi^2\sin(\pi t)\mathbf{j} + 2\mathbf{k}$$

$$\mathbf{r}''\left(-\frac{1}{4}\right) = -\frac{\sqrt{2}\pi^2}{2}\mathbf{i} + \frac{\sqrt{2}\pi^2}{2}\mathbf{j} + 2\mathbf{k}$$

$$\left\|\mathbf{r}''\left(-\frac{1}{4}\right)\right\| = \sqrt{\left(-\frac{\sqrt{2}\pi^2}{2}\right)^2 + \left(\frac{\sqrt{2}\pi^2}{2}\right)^2 + (2)^2} = \sqrt{\pi^4 + 4}$$

$$\frac{\mathbf{r}''(-1/4)}{\|\mathbf{r}''(-1/4)\|} = \frac{1}{2\sqrt{\pi^4 + 4}}(-\sqrt{2}\pi^2\mathbf{i} + \sqrt{2}\pi^2\mathbf{j} + 4\mathbf{k})$$

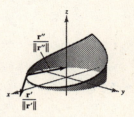

27. $\mathbf{r}(t) = t^2\mathbf{i} + t^3\mathbf{j}$

$\mathbf{r}'(t) = 2t\mathbf{i} + 3t^2\mathbf{j}$

$\mathbf{r}'(0) = \mathbf{0}$

Smooth on $(-\infty, 0), \ (0, \infty)$

29. $\mathbf{r}(\theta) = 2\cos^3\theta\mathbf{i} + 3\sin^3\theta\mathbf{j}$

$\mathbf{r}'(\theta) = -6\cos^2\theta\sin\theta\mathbf{i} + 9\sin^2\theta\cos\theta\mathbf{j}$

$\mathbf{r}'\left(\frac{n\pi}{2}\right) = \mathbf{0}$

Smooth on $\left(\frac{n\pi}{2}, \frac{(n+1)\pi}{2}\right)$, n any integer.

31. $\mathbf{r}(\theta) = (\theta - 2\sin\theta)\mathbf{i} + (1 - 2\cos\theta)\mathbf{j}$

$\mathbf{r}'(\theta) = (1 - 2\cos\theta)\mathbf{i} + (1 + 2\sin\theta)\mathbf{j}$

$\mathbf{r}'(\theta) \ne \mathbf{0}$ for any value of θ

Smooth on $(-\infty, \infty)$

33. $\mathbf{r}(t) = (t - 1)\mathbf{i} + \frac{1}{t}\mathbf{j} - t^2\mathbf{k}$

$\mathbf{r}'(t) = \mathbf{i} - \frac{1}{t^2}\mathbf{j} - 2t\mathbf{k} \ne \mathbf{0}$

$\mathbf{r}$ is smooth for all $t \ne 0$: $(-\infty, 0), (0, \infty)$

35. $\mathbf{r}(t) = t\mathbf{i} - 3t\mathbf{j} + \tan t\mathbf{k}$

$\mathbf{r}'(t) = \mathbf{i} - 3\mathbf{j} + \sec^2 t\mathbf{k} \neq \mathbf{0}$

$\mathbf{r}$ is smooth for all $t \neq \dfrac{\pi}{2} + n\pi = \dfrac{2n+1}{2}\pi$.

Smooth on intervals of form $\left(-\dfrac{\pi}{2} + n\pi, \dfrac{\pi}{2} + n\pi \right)$

37. $\mathbf{r}(t) = t\mathbf{i} + 3t\mathbf{j} + t^2\mathbf{k}, \ \mathbf{u}(t) = 4t\mathbf{i} + t^2\mathbf{j} + t^3\mathbf{k}$

(a) $\mathbf{r}'(t) = \mathbf{i} + 3\mathbf{j} + 2t\mathbf{k}$ (b) $\mathbf{r}''(t) = 2\mathbf{k}$

(c) $\mathbf{r}(t) \cdot \mathbf{u}(t) = 4t^2 + 3t^3 + t^5$ (d) $3\mathbf{r}(t) - \mathbf{u}(t) = -t\mathbf{i} + (9t - t^2)\mathbf{j} + (3t^2 - t^3)\mathbf{k}$

$\quad D_t[\mathbf{r}(t) \cdot \mathbf{u}(t)] = 8t + 9t^2 + 5t^4$ $\quad D_t[3\mathbf{r}(t) - \mathbf{u}(t)] = -\mathbf{i} + (9 - 2t)\mathbf{j} + (6t - 3t^2)\mathbf{k}$

(e) $\mathbf{r}(t) \times \mathbf{u}(t) = 2t^4\mathbf{i} - (t^4 - 4t^3)\mathbf{j} + (t^3 - 12t^2)\mathbf{k}$ (f) $\|\mathbf{r}(t)\| = \sqrt{10t^2 + t^4} = t\sqrt{10 + t^2}$

$\quad D_t[\mathbf{r}(t) \times \mathbf{u}(t)] = 8t^3\mathbf{i} + (12t^2 - 4t^3)\mathbf{j} + (3t^2 - 24t)\mathbf{k}$ $\quad D_t[\|\mathbf{r}(t)\|] = \dfrac{10 + 2t^2}{\sqrt{10 + t^2}}$

39. $\mathbf{r}(t) = 3\sin t\mathbf{i} + 4\cos t\mathbf{j}$

       $\mathbf{r}'(t) = 3\cos t\mathbf{i} - 4\sin t\mathbf{j}$

$\mathbf{r}(t) \cdot \mathbf{r}'(t) = 9\sin t\cos t - 16\cos t\sin t = -7\sin t\cos t$

$\qquad \cos\theta = \dfrac{\mathbf{r}(t) \cdot \mathbf{r}'(t)}{\|\mathbf{r}(t)\|\,\|\mathbf{r}'(t)\|} = \dfrac{-7\sin t\cos t}{\sqrt{9\sin^2 t + 16\cos^2 t}\sqrt{9\cos^2 t + 16\sin^2 t}}$

$\qquad\qquad \theta = \arccos\left[\dfrac{-7\sin t\cos t}{\sqrt{(9\sin^2 t + 16\cos^2 t)(9\cos^2 t + 16\sin^2 t)}} \right]$

$\theta = 1.855$ maximum at $t = 3.927\left(\dfrac{5\pi}{4}\right)$ and $t = 0.785\left(\dfrac{\pi}{4}\right)$.

$\theta = 1.287$ minimum at $t = 2.356\left(\dfrac{3\pi}{4}\right)$ and $t = 5.498\left(\dfrac{7\pi}{4}\right)$.

$\theta = \dfrac{\pi}{2}(1.571)$ for $t = \dfrac{n\pi}{2}, n = 0, 1, 2, 3, \ldots$

41. $\mathbf{r}'(t) = \lim\limits_{\Delta t \to 0} \dfrac{\mathbf{r}(t + \Delta t) - \mathbf{r}(t)}{\Delta t}$

$\qquad = \lim\limits_{\Delta t \to 0} \dfrac{[3(t + \Delta t) + 2]\mathbf{i} + [1 - (t + \Delta t)^2]\mathbf{j} - (3t + 2)\mathbf{i} - (1 - t^2)\mathbf{j}}{\Delta t}$

$\qquad = \lim\limits_{\Delta t \to 0} \dfrac{(3\Delta t)\mathbf{i} - (2t(\Delta t) + (\Delta t)^2)\mathbf{j}}{\Delta t}$

$\qquad = \lim\limits_{\Delta t \to 0} 3\mathbf{i} - (2t + \Delta t)\mathbf{j} = 3\mathbf{i} - 2t\mathbf{j}$

43. $\displaystyle\int (2t\mathbf{i} + \mathbf{j} + \mathbf{k})\,dt = t^2\mathbf{i} + t\mathbf{j} + t\mathbf{k} + \mathbf{C}$ **45.** $\displaystyle\int \left(\dfrac{1}{t}\mathbf{i} + \mathbf{j} - t^{3/2}\mathbf{k}\right)dt = \ln t\mathbf{i} + t\mathbf{j} - \dfrac{2}{5}t^{5/2}\mathbf{k} + \mathbf{C}$

47. $\displaystyle\int \left[(2t - 1)\mathbf{i} + 4t^3\mathbf{j} + 3\sqrt{t}\mathbf{k}\right]dt = (t^2 - t)\mathbf{i} + t^4\mathbf{j} + 2t^{3/2}\mathbf{k} + \mathbf{C}$

49. $\displaystyle\int \left[\sec^2 t\mathbf{i} + \dfrac{1}{1 + t^2}\mathbf{j}\right]dt = \tan t\mathbf{i} + \arctan t\mathbf{j} + \mathbf{C}$

51. $\int_0^1 (8t\mathbf{i} + t\mathbf{j} - \mathbf{k})\, dt = \left[4t^2\mathbf{i} \right]_0^1 + \left[\frac{t^2}{2}\mathbf{j} \right]_0^1 - \left[t\mathbf{k} \right]_0^1 = 4\mathbf{i} + \frac{1}{2}\mathbf{j} - \mathbf{k}$

53. $\int_0^{\pi/2} [(a\cos t)\mathbf{i} + (a\sin t)\mathbf{j} + \mathbf{k}]\, dt = \left[a\sin t\,\mathbf{i} \right]_0^{\pi/2} - \left[a\cos t\,\mathbf{j} \right]_0^{\pi/2} + \left[t\mathbf{k} \right]_0^{\pi/2} = a\mathbf{i} + a\mathbf{j} + \frac{\pi}{2}\mathbf{k}$

55. $\mathbf{r}(t) = \int (4e^{2t}\mathbf{i} + 3e^t\mathbf{j})\, dt = 2e^{2t}\mathbf{i} + 3e^t\mathbf{j} + \mathbf{C}$

$\mathbf{r}(0) = 2\mathbf{i} + 3\mathbf{j} + \mathbf{C} = 2\mathbf{i} \implies \mathbf{C} = -3\mathbf{j}$

$\mathbf{r}(t) = 2e^{2t}\mathbf{i} + 3(e^t - 1)\mathbf{j}$

57. $\mathbf{r}'(t) = \int -32\mathbf{j}\, dt = -32t\mathbf{j} + \mathbf{C}_1$

$\mathbf{r}'(0) = \mathbf{C}_1 = 600\sqrt{3}\mathbf{i} + 600\mathbf{j}$

$\mathbf{r}'(t) = 600\sqrt{3}\mathbf{i} + (600 - 32t)\mathbf{j}$

$\mathbf{r}(t) = \int \left[600\sqrt{3}\mathbf{i} + (600 - 32t)\mathbf{j} \right] dt$

$\quad = 600\sqrt{3}\,t\mathbf{i} + (600t - 16t^2)\mathbf{j} + \mathbf{C}$

$\mathbf{r}(0) = \mathbf{C} = \mathbf{0}$

$\mathbf{r}(t) = 600\sqrt{3}\,t\mathbf{i} + (600t - 16t^2)\mathbf{j}$

59. $\mathbf{r}(t) = \int (te^{-t^2}\mathbf{i} - e^{-t}\mathbf{j} + \mathbf{k})\, dt = -\frac{1}{2}e^{-t^2}\mathbf{i} + e^{-t}\mathbf{j} + t\mathbf{k} + \mathbf{C}$

$\mathbf{r}(0) = -\frac{1}{2}\mathbf{i} + \mathbf{j} + \mathbf{C} = \frac{1}{2}\mathbf{i} - \mathbf{j} + \mathbf{k} \implies \mathbf{C} = \mathbf{i} - 2\mathbf{j} + \mathbf{k}$

$\mathbf{r}(t) = \left(1 - \frac{1}{2}e^{-t^2} \right)\mathbf{i} + (e^{-t} - 2)\mathbf{j} + (t + 1)\mathbf{k} = \left(\frac{2 - e^{-t^2}}{2} \right)\mathbf{i} + (e^{-t} - 2)\mathbf{j} + (t + 1)\mathbf{k}$

61. See "Definition of the Derivative of a Vector-Valued Function" and Figure 11.8 on page 794.

63. At $t = t_0$, the graph of $\mathbf{u}(t)$ is increasing in the x, y, and z directions simultaneously.

65. Let $\mathbf{r}(t) = x(t)\mathbf{i} + y(t)\mathbf{j} + z(t)\mathbf{k}$. Then $c\mathbf{r}(t) = cx(t)\mathbf{i} + cy(t)\mathbf{j} + cz(t)\mathbf{k}$ and

$D_t[c\mathbf{r}(t)] = cx'(t)\mathbf{i} + cy'(t)\mathbf{j} + cz'(t)\mathbf{k}$

$\quad = c[x'(t)\mathbf{i} + y'(t)\mathbf{j} + z'(t)\mathbf{k}] = c\mathbf{r}'(t).$

67. Let $\mathbf{r}(t) = x(t)\mathbf{i} + y(t)\mathbf{j} + z(t)\mathbf{k}$, then $f(t)\mathbf{r}(t) = f(t)x(t)\mathbf{i} + f(t)y(t)\mathbf{j} + f(t)z(t)\mathbf{k}$.

$D_t[f(t)\mathbf{r}(t)] = [f(t)x'(t) + f'(t)x(t)]\mathbf{i} + [f(t)y'(t) + f'(t)y(t)]\mathbf{j} + [f(t)z'(t) + f'(t)z(t)]\mathbf{k}$

$\quad = f(t)[x'(t)\mathbf{i} + y'(t)\mathbf{j} + z'(t)\mathbf{k}] + f'(t)[x(t)\mathbf{i} + y(t)\mathbf{j} + z(t)\mathbf{k}]$

$\quad = f(t)\mathbf{r}'(t) + f'(t)\mathbf{r}(t)$

69. Let $\mathbf{r}(t) = x(t)\mathbf{i} + y(t)\mathbf{j} + z(t)\mathbf{k}$. Then $\mathbf{r}(f(t)) = x(f(t))\mathbf{i} + y(f(t))\mathbf{j} + z(f(t))\mathbf{k}$ and

$D_t[\mathbf{r}(f(t))] = x'(f(t))f'(t)\mathbf{i} + y'(f(t))f'(t)\mathbf{j} + z'(f(t))f'(t)\mathbf{k}$ (Chain Rule)

$\quad = f'(t)[x'(f(t))\mathbf{i} + y'(f(t))\mathbf{j} + z'(f(t))\mathbf{k}] = f'(t)\mathbf{r}'(f(t)).$

71. Let $\mathbf{r}(t) = x_1(t)\mathbf{i} + y_1(t)\mathbf{j} + z_1(t)\mathbf{k}$, $\mathbf{u}(t) = x_2(t)\mathbf{i} + y_2(t)\mathbf{j} + z_2(t)\mathbf{k}$, and $\mathbf{v}(t) = x_3(t)\mathbf{i} + y_3(t)\mathbf{j} + z_3(t)\mathbf{k}$. Then:

$$\mathbf{r}(t) \cdot [\mathbf{u}(t) \times \mathbf{v}(t)] = x_1(t)[y_2(t)z_3(t) - z_2(t)y_3(t)] - y_1(t)[x_2(t)z_3(t) - z_2(t)x_3(t)] + z_1(t)[x_2(t)y_3(t) - y_2(t)x_3(t)]$$

$$D_t[\mathbf{r}(t) \cdot (\mathbf{u}(t) \times \mathbf{v}(t))] = x_1(t)y_2(t)z_3{}'(t) + x_1(t)y_2{}'(t)z_3(t) + x_1{}'(t)y_2(t)z_3(t) - x_1(t)y_3(t)z_2{}'(t) -$$

$$x_1(t)y_3{}'(t)z_2(t) - x_1{}'(t)y_3(t)z_2(t) - y_1(t)x_2(t)z_3{}'(t) - y_1(t)x_2{}'(t)z_3(t) - y_1{}'(t)x_2(t)z_3(t) +$$

$$y_1(t)z_2(t)x_3{}'(t) + y_1(t)z_2{}'(t)x_3(t) + y_1{}'(t)z_2(t)x_3(t) + z_1(t)x_2(t)y_3{}'(t) + z_1(t)x_2{}'(t)y_3(t) +$$

$$z_1{}'(t)x_2(t)y_3(t) - z_1(t)y_2(t)x_3{}'(t) - z_1(t)y_2{}'(t)x_3(t) - z_1{}'(t)y_2(t)x_3(t)$$

$$= \{x_1{}'(t)[y_2(t)z_3(t) - y_3(t)z_2(t)] + y_1{}'(t)[-x_2(t)z_3(t) + z_2(t)x_3(t)] + z_1{}'(t)[x_2(t)y_3(t) - y_2(t)x_3(t)]\} +$$

$$\{x_1(t)[y_2{}'(t)z_3(t) - y_3(t)z_2{}'(t)] + y_1(t)[-x_2{}'(t)z_3(t) + z_2{}'(t)x_3(t)] + z_1(t)[x_2{}'(t)y_3(t) - y_2{}'(t)x_3(t)]\} +$$

$$\{x_1(t)[y_2(t)z_3{}'(t) - y_3{}'(t)z_2(t)] + y_1(t)[-x_2(t)z_3{}'(t) + z_2(t)x_3{}'(t)] + z_1(t)[x_2(t)y_3{}'(t) - y_2(t)x_3{}'(t)]\}$$

$$= \mathbf{r}'(t) \cdot [\mathbf{u}(t) \times \mathbf{v}(t)] + \mathbf{r}(t) \cdot [\mathbf{u}'(t) \times \mathbf{v}(t)] + \mathbf{r}(t) \cdot [\mathbf{u}(t) \times \mathbf{v}'(t)]$$

73. False. Let $\mathbf{r}(t) = \cos t\mathbf{i} + \sin t\mathbf{j} + \mathbf{k}$.

$$\|\mathbf{r}(t)\| = \sqrt{2}$$

$$\frac{d}{dt}[\|\mathbf{r}(t)\|] = 0$$

$$\mathbf{r}'(t) = -\sin t\mathbf{i} + \cos t\mathbf{j}$$

$$\|\mathbf{r}'(t)\| = 1$$

Section 11.3   Velocity and Acceleration

1. $\mathbf{r}(t) = 3t\mathbf{i} + (t - 1)\mathbf{j}$

$\mathbf{v}(t) = \mathbf{r}'(t) = 3\mathbf{i} + \mathbf{j}$

$\mathbf{a}(t) = \mathbf{r}''(t) = \mathbf{0}$

$x = 3t, \ y = t - 1, \ y = \dfrac{x}{3} - 1$

At $(3, 0)$, $t = 1$.

$\quad \mathbf{v}(1) = 3\mathbf{i} + \mathbf{j}, \mathbf{a}(1) = \mathbf{0}$

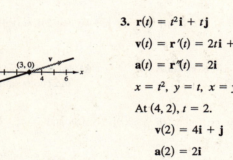

3. $\mathbf{r}(t) = t^2\mathbf{i} + t\mathbf{j}$

$\mathbf{v}(t) = \mathbf{r}'(t) = 2t\mathbf{i} + \mathbf{j}$

$\mathbf{a}(t) = \mathbf{r}''(t) = 2\mathbf{i}$

$x = t^2, \ y = t, \ x = y^2$

At $(4, 2)$, $t = 2$.

$\quad \mathbf{v}(2) = 4\mathbf{i} + \mathbf{j}$

$\quad \mathbf{a}(2) = 2\mathbf{i}$

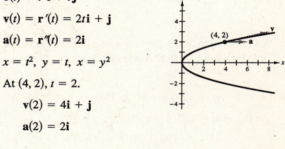

5. $\mathbf{r}(t) = 2\cos t\mathbf{i} + 2\sin t\mathbf{j}$

$\mathbf{v}(t) = \mathbf{r}'(t) = -2\sin t\mathbf{i} + 2\cos t\mathbf{j}$

$\mathbf{a}(t) = \mathbf{r}''(t) = -2\cos t\mathbf{i} - 2\sin t\mathbf{j}$

$x = 2\cos t, \ y = 2\sin t, \ x^2 + y^2 = 4$

At $\left(\sqrt{2}, \sqrt{2}\right)$, $t = \dfrac{\pi}{4}$.

$\quad \mathbf{v}\left(\dfrac{\pi}{4}\right) = -\sqrt{2}\mathbf{i} + \sqrt{2}\mathbf{j}$

$\quad \mathbf{a}\left(\dfrac{\pi}{4}\right) = -\sqrt{2}\mathbf{i} - \sqrt{2}\mathbf{j}$

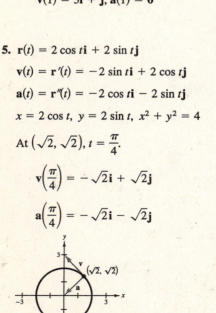

7. $\mathbf{r}(t) = \langle t - \sin t, 1 - \cos t \rangle$

$\mathbf{v}(t) = \mathbf{r}'(t) = \langle 1 - \cos t, \sin t \rangle$

$\mathbf{a}(t) = \mathbf{r}''(t) = \langle \sin t, \cos t \rangle$

$x = t - \sin t, \ y = 1 - \cos t \quad \text{(cycloid)}$

At $(\pi, 2)$, $t = \pi$.

$\mathbf{v}(\pi) = \langle 2, 0 \rangle = 2\mathbf{i}$

$\mathbf{a}(\pi) = \langle 0, -1 \rangle = -\mathbf{j}$

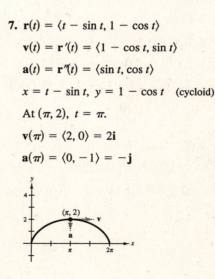

9. $\mathbf{r}(t) = t\mathbf{i} + (2t - 5)\mathbf{j} + 3t\mathbf{k}$

    $\mathbf{v}(t) = \mathbf{i} + 2\mathbf{j} + 3\mathbf{k}$

    $s(t) = \|\mathbf{v}(t)\| = \sqrt{1 + 4 + 9} = \sqrt{14}$

    $\mathbf{a}(t) = \mathbf{0}$

11. $\mathbf{r}(t) = t\mathbf{i} + t^2\mathbf{j} + \dfrac{t^2}{2}\mathbf{k}$

    $\mathbf{v}(t) = \mathbf{i} + 2t\mathbf{j} + t\mathbf{k}$

    $s(t) = \sqrt{1 + 4t^2 + t^2} = \sqrt{1 + 5t^2}$

    $\mathbf{a}(t) = 2\mathbf{j} + \mathbf{k}$

13. $\mathbf{r}(t) = t\mathbf{i} + t\mathbf{j} + \sqrt{9 - t^2}\,\mathbf{k}$

    $\mathbf{v}(t) = \mathbf{i} + \mathbf{j} - \dfrac{t}{\sqrt{9 - t^2}}\mathbf{k}$

    $s(t) = \sqrt{1 + 1 + \dfrac{t^2}{9 - t^2}} = \sqrt{\dfrac{18 - t^2}{9 - t^2}}$

    $\mathbf{a}(t) = -\dfrac{9}{(9 - t^2)^{3/2}}\mathbf{k}$

15. $\mathbf{r}(t) = \langle 4t, 3\cos t, 3\sin t \rangle$

    $\mathbf{v}(t) = \langle 4, -3\sin t, 3\cos t \rangle = 4\mathbf{i} - 3\sin t\mathbf{j} + 3\cos t\mathbf{k}$

    $s(t) = \sqrt{16 + 9\sin^2 t + 9\cos^2 t} = 5$

    $\mathbf{a}(t) = \langle 0, -3\cos t, -3\sin t \rangle = -3\cos t\mathbf{j} - 3\sin t\mathbf{k}$

17. (a) $\mathbf{r}(t) = \left\langle t, -t^2, \dfrac{t^3}{4} \right\rangle, \ t_0 = 1$

    $\mathbf{r}'(t) = \left\langle 1, -2t, \dfrac{3t^2}{4} \right\rangle$

    $\mathbf{r}'(1) = \left\langle 1, -2, \dfrac{3}{4} \right\rangle$

    $x = 1 + t, \ y = -1 - 2t, \ z = \dfrac{1}{4} + \dfrac{3}{4}t$

(b) $\mathbf{r}(1 + 0.1) \approx \left\langle 1 + 0.1, -1 - 2(0.1), \dfrac{1}{4} + \dfrac{3}{4}(0.1) \right\rangle$

         $= \langle 1.100, -1.200, 0.325 \rangle$

19. $\mathbf{a}(t) = \mathbf{i} + \mathbf{j} + \mathbf{k}, \ \mathbf{v}(0) = \mathbf{0}, \ \mathbf{r}(0) = \mathbf{0}$

    $\mathbf{v}(t) = \displaystyle\int (\mathbf{i} + \mathbf{j} + \mathbf{k})\,dt = t\mathbf{i} + t\mathbf{j} + t\mathbf{k} + \mathbf{C}$

    $\mathbf{v}(0) = \mathbf{C} = \mathbf{0}, \ \mathbf{v}(t) = t\mathbf{i} + t\mathbf{j} + t\mathbf{k}, \ \mathbf{v}(t) = t(\mathbf{i} + \mathbf{j} + \mathbf{k})$

    $\mathbf{r}(t) = \displaystyle\int (t\mathbf{i} + t\mathbf{j} + t\mathbf{k})\,dt = \dfrac{t^2}{2}(\mathbf{i} + \mathbf{j} + \mathbf{k}) + \mathbf{C}$

    $\mathbf{r}(0) = \mathbf{C} = \mathbf{0}, \ \mathbf{r}(t) = \dfrac{t^2}{2}(\mathbf{i} + \mathbf{j} + \mathbf{k}),$

    $\mathbf{r}(2) = 2(\mathbf{i} + \mathbf{j} + \mathbf{k}) = 2\mathbf{i} + 2\mathbf{j} + 2\mathbf{k}$

21. $\mathbf{a}(t) = t\mathbf{j} + t\mathbf{k}, \ \mathbf{v}(1) = 5\mathbf{j}, \ \mathbf{r}(1) = \mathbf{0}$

    $\mathbf{v}(t) = \displaystyle\int (t\mathbf{j} + t\mathbf{k})\,dt = \dfrac{t^2}{2}\mathbf{j} + \dfrac{t^2}{2}\mathbf{k} + \mathbf{C}$

    $\mathbf{v}(1) = \dfrac{1}{2}\mathbf{j} + \dfrac{1}{2}\mathbf{k} + \mathbf{C} = 5\mathbf{j} \implies \mathbf{C} = \dfrac{9}{2}\mathbf{j} - \dfrac{1}{2}\mathbf{k}$

    $\mathbf{v}(t) = \left(\dfrac{t^2}{2} + \dfrac{9}{2}\right)\mathbf{j} + \left(\dfrac{t^2}{2} - \dfrac{1}{2}\right)\mathbf{k}$

    $\mathbf{r}(t) = \displaystyle\int \left[\left(\dfrac{t^2}{2} + \dfrac{9}{2}\right)\mathbf{j} + \left(\dfrac{t^2}{2} - \dfrac{1}{2}\right)\mathbf{k} \right]\,dt$

      $= \left(\dfrac{t^3}{6} + \dfrac{9}{2}t\right)\mathbf{j} + \left(\dfrac{t^3}{6} - \dfrac{1}{2}t\right)\mathbf{k} + \mathbf{C}$

    $\mathbf{r}(1) = \dfrac{14}{3}\mathbf{j} - \dfrac{1}{3}\mathbf{k} + \mathbf{C} = \mathbf{0} \implies \mathbf{C} = -\dfrac{14}{3}\mathbf{j} + \dfrac{1}{3}\mathbf{k}$

    $\mathbf{r}(t) = \left(\dfrac{t^3}{6} + \dfrac{9}{2}t - \dfrac{14}{3}\right)\mathbf{j} + \left(\dfrac{t^3}{6} - \dfrac{1}{2}t + \dfrac{1}{3}\right)\mathbf{k}$

    $\mathbf{r}(2) = \dfrac{17}{3}\mathbf{j} + \dfrac{2}{3}\mathbf{k}$

23. The velocity of an object involves both magnitude and direction of motion, whereas speed involves only magnitude.

25. $\mathbf{r}(t) = (88\cos 30°)t\mathbf{i} + [10 + (88\sin 30°)t - 16t^2]\mathbf{j}$

    $= 44\sqrt{3}\,t\mathbf{i} + (10 + 44t - 16t^2)\mathbf{j}$

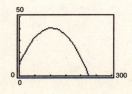

27. $\mathbf{r}(t) = (v_0 \cos \theta)t\mathbf{i} + \left[h + (v_0 \sin \theta)t - \frac{1}{2}gt^2 \right]\mathbf{j} = \frac{v_0}{\sqrt{2}}t\mathbf{i} + \left(3 + \frac{v_0}{\sqrt{2}}t - 16t^2 \right)\mathbf{j}$

$\frac{v_0}{\sqrt{2}}t = 300$ when $3 + \frac{v_0}{\sqrt{2}}t - 16t^2 = 3$.

$$t = \frac{300\sqrt{2}}{v_0}, \quad \frac{v_0}{\sqrt{2}}\left(\frac{300\sqrt{2}}{v_0} \right) - 16\left(\frac{300\sqrt{2}}{v_0} \right)^2 = 0, \quad 300 - \frac{300^2(32)}{v_0^2} = 0$$

$$v_0^2 = 300(32), \quad v_0 = \sqrt{9600} = 40\sqrt{6}, \quad v_0 = 40\sqrt{6} \approx 97.98 \text{ ft/sec}$$

The maximum height is reached when the derivative of the vertical component is zero.

$$y(t) = 3 + \frac{tv_0}{\sqrt{2}} - 16t^2 = 3 + \frac{40\sqrt{6}}{\sqrt{2}}t - 16t^2 = 3 + 40\sqrt{3}t - 16t^2$$

$$y'(t) = 40\sqrt{3} - 32t = 0$$

$$t = \frac{40\sqrt{3}}{32} = \frac{5\sqrt{3}}{4}$$

Maximum height: $y\left(\frac{5\sqrt{3}}{4} \right) = 3 + 40\sqrt{3}\left(\frac{5\sqrt{3}}{4} \right) - 16\left(\frac{5\sqrt{3}}{4} \right)^2 = 78$ feet

29. $x(t) = t(v_0 \cos \theta)$ or $t = \frac{x}{v_0 \cos \theta}$

$$y(t) = t(v_0 \sin \theta) - 16t^2 + h$$

$$y = \frac{x}{v_0 \cos \theta}(v_0 \sin \theta) - 16\left(\frac{x^2}{v_0^2 \cos^2 \theta} \right) + h = (\tan \theta)x - \left(\frac{16}{v_0^2} \sec^2 \theta \right)x^2 + h$$

31. $\mathbf{r}(t) = t\mathbf{i} + (-0.004t^2 + 0.3667t + 6)\mathbf{j}$

(a) $y = -0.004x^2 + 0.3667x + 6$

(b)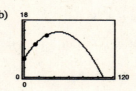

(c) $y' = -0.008x + 0.3667 = 0 \Rightarrow x = 45.8375$ and

$y(45.8375) \approx 14.4$ feet.

(d) From Exercise 29,

$\tan \theta = 0.3667 \Rightarrow \theta \approx 20.14°$

$\frac{16 \sec^2 \theta}{v_0^2} = 0.004 \Rightarrow v_0^2 = \frac{16 \sec^2 \theta}{0.004} = \frac{4000}{\cos^2 \theta}$

$\Rightarrow v_0 \approx 67.4$ ft/sec.

33. $100 \text{ mph} = \left(100 \frac{\text{miles}}{\text{hr}} \right)\left(5280 \frac{\text{feet}}{\text{mile}} \right)/(3600 \text{ sec/hour}) = \frac{440}{3}$ ft/sec

(a) $\mathbf{r}(t) = \left(\frac{440}{3} \cos \theta_0 \right)t\mathbf{i} + \left[3 + \left(\frac{440}{3} \sin \theta_0 \right)t - 16t^2 \right]\mathbf{j}$

(b) 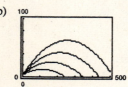   Graphing these curves together with $y = 10$ shows that $\theta_0 = 20°$.

—CONTINUED—

33. —CONTINUED—

(c) We want

$$x(t) = \left(\frac{440}{3} \cos \theta\right)t \geq 400 \quad \text{and} \quad y(t) = 3 + \left(\frac{440}{3} \sin \theta\right)t - 16t^2 \geq 10.$$

From $x(t)$, the minimum angle occurs when $t = 30/(11 \cos \theta)$. Substituting this for t in $y(t)$ yields:

$$3 + \left(\frac{440}{3} \sin \theta\right)\left(\frac{30}{11 \cos \theta}\right) - 16\left(\frac{30}{11 \cos \theta}\right)^2 = 10$$

$$400 \tan \theta - \frac{14,400}{121} \sec^2 \theta = 7$$

$$\frac{14,400}{121}(1 + \tan^2 \theta) - 400 \tan \theta + 7 = 0$$

$$14,400 \tan^2 \theta - 48,400 \tan \theta + 15,247 = 0$$

$$\tan \theta = \frac{48,400 \pm \sqrt{48,400^2 - 4(14,400)(15,247)}}{2(14,400)}$$

$$\theta = \tan^{-1}\left(\frac{48,400 - \sqrt{1,464,332,800}}{28,800}\right) \approx 19.38°$$

35. $\mathbf{r}(t) = (v \cos \theta)t\mathbf{i} + [(v \sin \theta)t - 16t^2]\mathbf{j}$

(a) We want to find the minimum initial speed v as a function of the angle θ. Since the bale must be thrown to the position $(16, 8)$, we have

$$16 = (v \cos \theta)t$$

$$8 = (v \sin \theta)t - 16t^2.$$

$t = 16/(v \cos \theta)$ from the first equation. Substituting into the second equation and solving for v, we obtain:

$$8 = (v \sin \theta)\left(\frac{16}{v \cos \theta}\right) - 16\left(\frac{16}{v \cos \theta}\right)^2$$

$$1 = 2 \frac{\sin \theta}{\cos \theta} - 512\left(\frac{1}{v^2 \cos^2 \theta}\right)$$

$$512 \frac{1}{v^2 \cos^2 \theta} = 2 \frac{\sin \theta}{\cos \theta} - 1$$

$$\frac{1}{v^2} = \left(2 \frac{\sin \theta}{\cos \theta} - 1\right)\frac{\cos^2 \theta}{512} = \frac{2 \sin \theta \cos \theta - \cos^2 \theta}{512}$$

$$v^2 = \frac{512}{2 \sin \theta \cos \theta - \cos^2 \theta}$$

We minimize $f(\theta) = \dfrac{512}{2 \sin \theta \cos \theta - \cos^2 \theta}$.

$$f'(\theta) = -512 \frac{2 \cos^2 \theta - 2 \sin^2 \theta + 2 \sin \theta \cos \theta}{(2 \sin \theta \cos \theta - \cos^2 \theta)^2}$$

$$f'(\theta) = 0 \implies 2 \cos(2\theta) + \sin(2\theta) = 0$$

$$\tan(2\theta) = -2$$

$$\theta \approx 1.01722 \approx 58.28°$$

Substituting into the equation for v, $v \approx 28.78$ feet per second.

(b) If $\theta = 45°$,

$$16 = (v \cos \theta)t = v\frac{\sqrt{2}}{2}t$$

$$8 = (v \sin \theta)t - 16t^2 = v\frac{\sqrt{2}}{2}t - 16t^2$$

From part (a), $v^2 = \dfrac{512}{2(\sqrt{2}/2)(\sqrt{2}/2) - (\sqrt{2}/2)^2} = \dfrac{512}{1/2} = 1024 \implies v = 32$ ft/sec.

37. $\mathbf{r}(t) = (v_0 \cos \theta)t\mathbf{i} + [(v_0 \sin \theta)t - 16t^2]\mathbf{j}$

$(v_0 \sin \theta)t - 16t^2 = 0$ when $t = 0$ and $t = \dfrac{v_0 \sin \theta}{16}$.

The range is

$x = (v_0 \cos \theta)t = (v_0 \cos \theta)\dfrac{v_0 \sin \theta}{16} = \dfrac{v_0^2}{32} \sin 2\theta.$

Hence,

$x = \dfrac{1200^2}{32} \sin(2\theta) = 3000 \implies \sin 2\theta = \dfrac{1}{15} \implies \theta \approx 1.91°.$

39. (a) $\theta = 10°$, $v_0 = 66$ ft/sec

$\mathbf{r}(t) = (66 \cos 10°)t\mathbf{i} + [0 + (66 \sin 10°)t - 16t^2]\mathbf{j}$

$\mathbf{r}(t) \approx (65t)\mathbf{i} + (11.46t - 16t^2)\mathbf{j}$

Maximum height: 2.052 feet

Range: 46.557 feet

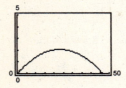

(c) $\theta = 45°$, $v_0 = 66$ ft/sec

$\mathbf{r}(t) = (66 \cos 45°)t\mathbf{i} + [0 + (66 \sin 45°)t - 16t^2]\mathbf{j}$

$\mathbf{r}(t) \approx (46.67t)\mathbf{i} + (46.67t - 16t^2)\mathbf{j}$

Maximum height: 34.031 feet

Range: 136.125 feet

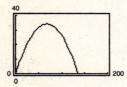

(e) $\theta = 60°$, $v_0 = 66$ ft/sec

$\mathbf{r}(t) = (66 \cos 60°)t\mathbf{i} + [0 + (66 \sin 60°)t - 16t^2]\mathbf{j}$

$\mathbf{r}(t) \approx (33t)\mathbf{i} + (57.16t - 16t^2)\mathbf{j}$

Maximum height: 51.074 feet

Range: 117.888 feet

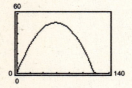

(b) $\theta = 10°$, $v_0 = 146$ ft/sec

$\mathbf{r}(t) = (146 \cos 10°)t\mathbf{i} + [0 + (146 \sin 10°)t - 16t^2]\mathbf{j}$

$\mathbf{r}(t) \approx (143.78t)\mathbf{i} + (25.35t - 16t^2)\mathbf{j}$

Maximum height: 10.043 feet

Range: 227.828 feet

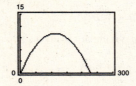

(d) $\theta = 45°$, $v_0 = 146$ ft/sec

$\mathbf{r}(t) = (146 \cos 45°)t\mathbf{i} + [0 + (146 \sin 45°)t - 16t^2]\mathbf{j}$

$\mathbf{r}(t) \approx (103.24t)\mathbf{i} + (103.24t - 16t^2)\mathbf{j}$

Maximum height: 166.531 feet

Range: 666.125 feet

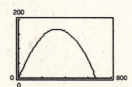

(f) $\theta = 60°$, $v_0 = 146$ ft/sec

$\mathbf{r}(t) = (146 \cos 60°)t\mathbf{i} + [0 + (146 \sin 60°)t - 16t^2]\mathbf{j}$

$\mathbf{r}(t) \approx (73t)\mathbf{i} + (126.44t - 16t^2)\mathbf{j}$

Maximum height: 249.797 feet

Range: 576.881 feet

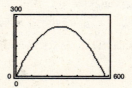

41. $\mathbf{r}(t) = (v_0 \cos \theta)t\mathbf{i} + [h + (v_0 \sin \theta)t - 4.9t^2]\mathbf{j}$

$\qquad = (100 \cos 30°)t\mathbf{i} + [1.5 + (100 \sin 30°)t - 4.9t^2]\mathbf{j}$

The projectile hits the ground when $-4.9t^2 + 100(\frac{1}{2})t + 1.5 = 0 \implies t \approx 10.234$ seconds.

The range is therefore $(100 \cos 30°)(10.234) \approx 886.3$ meters.

The maximum height occurs when $dy/dt = 0$.

$\qquad 100 \sin 30 = 9.8t \implies t \approx 5.102$ sec

The maximum height is

$\qquad y = 1.5 + (100 \sin 30°)(5.102) - 4.9(5.102)^2 \approx 129.1$ meters.

43. $\mathbf{r}(t) = b(\omega t - \sin \omega t)\mathbf{i} + b(1 - \cos \omega t)\mathbf{j}$

$\mathbf{v}(t) = b(\omega - \omega \cos \omega t)\mathbf{i} + b\omega \sin \omega t\, \mathbf{j} = b\omega(1 - \cos \omega t)\mathbf{i} + b\omega \sin \omega t\mathbf{j}$

$\mathbf{a}(t) = (b\omega^2 \sin \omega t)\mathbf{i} + (b\omega^2 \cos \omega t)\mathbf{j} = b\omega^2[\sin(\omega t)\mathbf{i} + \cos(\omega t)\mathbf{j}]$

$\|\mathbf{v}(t)\| = \sqrt{2}\, b\omega \sqrt{1 - \cos(\omega t)}$

$\|\mathbf{a}(t)\| = b\omega^2$

(a) $\|\mathbf{v}(t)\| = 0$ when $\omega t = 0, 2\pi, 4\pi, \dots$ (b) $\|\mathbf{v}(t)\|$ is maximum when $\omega t = \pi, 3\pi, \dots$,
then $\|\mathbf{v}(t)\| = 2b\omega$.

45. $\qquad \mathbf{v}(t) = -b\omega \sin(\omega t)\mathbf{i} + b\omega \cos(\omega t)\mathbf{j}$

$\mathbf{r}(t) \cdot \mathbf{v}(t) = -b^2\omega \sin(\omega t) \cos(\omega t) + b^2\omega \sin(\omega t) \cos(\omega t) = 0$

Therefore, $\mathbf{r}(t)$ and $\mathbf{v}(t)$ are orthogonal.

47. $\mathbf{a}(t) = -b\omega^2 \cos(\omega t)\mathbf{i} - b\omega^2 \sin(\omega t)\mathbf{j} = -b\omega^2[\cos(\omega t)\mathbf{i} + \sin(\omega t)\mathbf{j}] = -\omega^2 \mathbf{r}(t)$

$\mathbf{a}(t)$ is a negative multiple of a unit vector from $(0, 0)$ to $(\cos \omega t, \sin \omega t)$ and thus $\mathbf{a}(t)$ is directed toward the origin.

49. $\|\mathbf{a}(t)\| = \omega^2 b, \ b = 2$

$1 = m(32)$

$F = m(\omega^2 b) = \dfrac{1}{32}(2\omega^2) = 10$

$\omega = 4\sqrt{10}$ rad/sec

$\|\mathbf{v}(t)\| = b\omega = 8\sqrt{10}$ ft/sec

51. To find the range, set $y(t) = h + (v_0 \sin \theta)t - \frac{1}{2}gt^2 = 0$ then $0 = (\frac{1}{2}g)t^2 - (v_0 \sin \theta)t - h$.
By the Quadratic Formula, (discount the negative value)

$\qquad t = \dfrac{v_0 \sin \theta + \sqrt{(-v_0 \sin \theta)^2 - 4[(1/2)g](-h)}}{2[(1/2)g]} = \dfrac{v_0 \sin \theta + \sqrt{v_0^2 \sin^2 \theta + 2gh}}{g}$ seconds

At this time,

$\qquad x(t) = v_0 \cos \theta \left(\dfrac{v_0 \sin \theta + \sqrt{v_0^2 \sin^2 \theta + 2gh}}{g} \right) = \dfrac{v_0 \cos \theta}{g} \left(v_0 \sin \theta + \sqrt{v_0^2 \left(\sin^2 \theta + \dfrac{2gh}{v_0^2} \right)} \right)$

$\qquad\qquad\qquad = \dfrac{v_0^2 \cos \theta}{g} \left(\sin \theta + \sqrt{\sin^2 \theta + \dfrac{2gh}{v_0^2}} \right)$ feet

53. $\mathbf{r}(t) = x(t)\mathbf{i} + y(t)\mathbf{j} + z(t)\mathbf{k}$ Position vector

$\mathbf{v}(t) = x'(t)\mathbf{i} + y'(t)\mathbf{j} + z'(t)\mathbf{k}$ Velocity vector

$\mathbf{a}(t) = x''(t)\mathbf{i} + y''(t)\mathbf{j} + z''(t)\mathbf{k}$ Acceleration vector

$\text{Speed} = \|\mathbf{v}(t)\| = \sqrt{(x'(t)^2 + y'(t)^2 + z'(t)^2}$

$= C, \; C$ is a constant.

$$\frac{d}{dt}[x'(t)^2 + y'(t)^2 + z'(t)^2] = 0$$

$$2x'(t)x''(t) + 2y'(t)y''(t) + 2z'(t)z''(t) = 0$$

$$2[x'(t)x''(t) + y'(t)y''(t) + z'(t)z''(t)] = 0$$

$$\mathbf{v}(t) \cdot \mathbf{a}(t) = 0$$

Orthogonal

55. $\mathbf{r}(t) = 6\cos t\mathbf{i} + 3\sin t\mathbf{j}$

(a) $\mathbf{v}(t) = \mathbf{r}'(t) = -6\sin t\mathbf{i} + 3\cos t\mathbf{j}$

$\|\mathbf{v}(t)\| = \sqrt{36\sin^2 t + 9\cos^2 t}$

$= 3\sqrt{4\sin^2 t + \cos^2 t}$

$= 3\sqrt{3\sin^2 t + 1}$

$\mathbf{a}(t) = \mathbf{v}'(t) = -6\cos t\mathbf{i} - 3\sin t\mathbf{j}$

(b)

t	0	$\dfrac{\pi}{4}$	$\dfrac{\pi}{2}$	$\dfrac{2\pi}{3}$	π
Speed	3	$\dfrac{3}{2}\sqrt{10}$	6	$\dfrac{3}{2}\sqrt{13}$	3

(c)

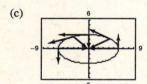

(d) The speed is increasing when the angle between $\mathbf{v}$ and $\mathbf{a}$ is in the interval

$$\left[0, \frac{\pi}{2}\right).$$

The speed is decreasing when the angle is in the interval

$$\left(\frac{\pi}{2}, \pi\right].$$

Section 11.4   Tangent Vectors and Normal Vectors

1. $\mathbf{r}(t) = t^2\mathbf{i} + 2t\mathbf{j}, \; t = 1$

$\mathbf{r}'(t) = 2t\mathbf{i} + 2\mathbf{j}, \|\mathbf{r}'(t)\| = \sqrt{4t^2 + 4} = 2\sqrt{t^2 + 1}$

$\mathbf{T}(t) = \dfrac{\mathbf{r}'(t)}{\|\mathbf{r}'(t)\|} = \dfrac{2t\mathbf{i} + 2\mathbf{j}}{2\sqrt{t^2 + 1}} = \dfrac{1}{\sqrt{t^2 + 1}}(t\mathbf{i} + \mathbf{j})$

$\mathbf{T}(1) = \dfrac{1}{\sqrt{2}}(\mathbf{i} + \mathbf{j}) = \dfrac{\sqrt{2}}{2}\mathbf{i} + \dfrac{\sqrt{2}}{2}\mathbf{j}$

3. $\mathbf{r}(t) = 4\cos t\mathbf{i} + 4\sin t\mathbf{j}, t = \dfrac{\pi}{4}$

$\mathbf{r}'(t) = -4\sin t\mathbf{i} + 4\cos t\mathbf{j}$

$\|\mathbf{r}'(t)\| = \sqrt{16\sin^2 t + 16\cos^2 t} = 4$

$\mathbf{T}(t) = \dfrac{\mathbf{r}'(t)}{\|\mathbf{r}'(t)\|} = -\sin t\mathbf{i} + \cos t\mathbf{j}$

$\mathbf{T}\left(\dfrac{\pi}{4}\right) = -\dfrac{\sqrt{2}}{2}\mathbf{i} + \dfrac{\sqrt{2}}{2}\mathbf{j}$

5. $\mathbf{r}(t) = t\mathbf{i} + t^2\mathbf{j} + t\mathbf{k}, \; p(0, 0, 0)$

$\mathbf{r}'(t) = \mathbf{i} + 2t\mathbf{j} + \mathbf{k}$

When $t = 0$, $\mathbf{r}'(0) = \mathbf{i} + \mathbf{k}$, $[t = 0$ at $(0, 0, 0)]$.

$\mathbf{T}(0) = \dfrac{\mathbf{r}'(0)}{\|\mathbf{r}'(0)\|} = \dfrac{\sqrt{2}}{2}(\mathbf{i} + \mathbf{k})$

Direction numbers: $a = 1, \; b = 0, \; c = 1$

Parametric equations: $x = t, \; y = 0, \; z = t$

7. $\mathbf{r}(t) = 2\cos t\mathbf{i} + 2\sin t\mathbf{j} + t\mathbf{k}, \; P(2, 0, 0)$

$\mathbf{r}'(t) = -2\sin t\mathbf{i} + 2\cos t\mathbf{j} + \mathbf{k}$

When $t = 0$, $\mathbf{r}'(0) = 2\mathbf{j} + \mathbf{k}$, $[t = 0$ at $(2, 0, 0)]$.

$\mathbf{T}(0) = \dfrac{\mathbf{r}'(0)}{\|\mathbf{r}'(0)\|} = \dfrac{\sqrt{5}}{5}(2\mathbf{j} + \mathbf{k})$

Direction numbers: $a = 0, \; b = 2, \; c = 1$

Parametric equations: $x = 2, \; y = 2t, \; z = t$

9. $\mathbf{r}(t) = \langle 2\cos t, 2\sin t, 4 \rangle$, $P\left(\sqrt{2}, \sqrt{2}, 4\right)$

$\mathbf{r}'(t) = \langle -2\sin t, 2\cos t, 0 \rangle$

When $t = \dfrac{\pi}{4}$, $\mathbf{r}\left(\dfrac{\pi}{4}\right) = \langle -\sqrt{2}, \sqrt{2}, 0 \rangle$, $\left[t = \dfrac{\pi}{4} \text{ at } \left(\sqrt{2}, \sqrt{2}, 4\right)\right]$.

$\mathbf{T}\left(\dfrac{\pi}{4}\right) = \dfrac{\mathbf{r}'(\pi/4)}{\|\mathbf{r}'(\pi/4)\|} = \dfrac{1}{2}\langle -\sqrt{2}, \sqrt{2}, 0 \rangle$

Direction numbers: $a = -\sqrt{2}$, $b = \sqrt{2}$, $c = 0$

Parametric equations: $x = -\sqrt{2}t + \sqrt{2}$, $y = \sqrt{2}t + \sqrt{2}$, $z = 4$

11. $\mathbf{r}(t) = \left\langle t, t^2, \dfrac{2}{3}t^3 \right\rangle$

$\mathbf{r}'(t) = \langle 1, 2t, 2t^2 \rangle$

When $t = 3$, $\mathbf{r}'(3) = \langle 1, 6, 18 \rangle$, $[t = 3 \text{ at } (3, 9, 18)]$.

$\mathbf{T}(3) = \dfrac{\mathbf{r}'(3)}{\|\mathbf{r}'(3)\|} = \dfrac{1}{19}\langle 1, 6, 18 \rangle$

Direction numbers: $a = 1$, $b = 6$, $c = 18$

Parametric equations: $x = t + 3$, $y = 6t + 9$, $z = 18t + 18$

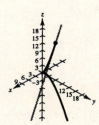

13. $\mathbf{r}(t) = t\mathbf{i} + \ln t\,\mathbf{j} + \sqrt{t}\,\mathbf{k}$, $t_0 = 1$

$\mathbf{r}'(t) = \mathbf{i} + \dfrac{1}{t}\mathbf{j} + \dfrac{1}{2\sqrt{t}}\mathbf{k}$; $\mathbf{r}'(1) = \mathbf{i} + \mathbf{j} + \dfrac{1}{2}\mathbf{k}$

$\mathbf{T}(1) = \dfrac{\mathbf{r}'(t)}{\|\mathbf{r}'(t)\|} = \dfrac{\mathbf{i} + \mathbf{j} + (1/2)\mathbf{k}}{\sqrt{1 + 1 + (1/4)}} = \dfrac{2}{3}\mathbf{i} + \dfrac{2}{3}\mathbf{j} + \dfrac{1}{3}\mathbf{k}$

Tangent line: $x = 1 + t$, $y = t$, $z = 1 + \dfrac{1}{2}t$

$\mathbf{r}(t_0 + 0.1) = \mathbf{r}(1.1) \approx 1.1\mathbf{i} + 0.1\mathbf{j} + 1.05\mathbf{k}$

$\qquad\qquad\qquad = \langle 1.1, 0.1, 1.05 \rangle$

15. $\mathbf{r}(4) = \langle 2, 16, 2 \rangle$

$\mathbf{u}(8) = \langle 2, 16, 2 \rangle$

Hence the curves intersect.

$\mathbf{r}'(t) = \left\langle 1, 2t, \dfrac{1}{2} \right\rangle$, $\mathbf{r}'(4) = \left\langle 1, 8, \dfrac{1}{2} \right\rangle$

$\mathbf{u}'(s) = \left\langle \dfrac{1}{4}, 2, \dfrac{1}{3}s^{-2/3} \right\rangle$, $\mathbf{u}'(8) = \left\langle \dfrac{1}{4}, 2, \dfrac{1}{12} \right\rangle$

$\cos\theta = \dfrac{\mathbf{r}'(4)\cdot\mathbf{u}'(8)}{\|\mathbf{r}'(4)\|\,\|\mathbf{u}'(8)\|} \approx \dfrac{16.29167}{16.29513} \Rightarrow \theta \approx 1.2°$

17. $\mathbf{r}(t) = t\mathbf{i} + \dfrac{1}{2}t^2\mathbf{j}$, $t = 2$

$\mathbf{r}'(t) = \mathbf{i} + t\mathbf{j}$

$\mathbf{T}(t) = \dfrac{\mathbf{r}'(t)}{\|\mathbf{r}'(t)\|} = \dfrac{\mathbf{i} + t\mathbf{j}}{\sqrt{1 + t^2}}$

$\mathbf{T}'(t) = \dfrac{-t}{(t^2 + 1)^{3/2}}\mathbf{i} + \dfrac{1}{(t^2 + 1)^{3/2}}\mathbf{j}$

$\mathbf{T}'(2) = \dfrac{-2}{5^{3/2}}\mathbf{i} + \dfrac{1}{5^{3/2}}\mathbf{j}$

$\mathbf{N}(2) = \dfrac{\mathbf{T}'(2)}{\|\mathbf{T}'(2)\|} = \dfrac{1}{\sqrt{5}}(-2\mathbf{i} + \mathbf{j}) = \dfrac{-2\sqrt{5}}{5}\mathbf{i} + \dfrac{\sqrt{5}}{5}\mathbf{j}$

19. $\mathbf{r}(t) = 6\cos t\,\mathbf{i} + 6\sin t\,\mathbf{j} + \mathbf{k}$, $t = \dfrac{3\pi}{4}$

$\mathbf{r}'(t) = -6\sin t\,\mathbf{i} + 6\cos t\,\mathbf{j}$

$\mathbf{T}(t) = \dfrac{\mathbf{r}'(t)}{\|\mathbf{r}'(t)\|} = -\sin t\,\mathbf{i} + \cos t\,\mathbf{j}$

$\mathbf{T}'(t) = -\cos t\,\mathbf{i} - \sin t\,\mathbf{j}$, $\|\mathbf{T}'(t)\| = 1$

$\mathbf{N}\left(\dfrac{3\pi}{4}\right) = \dfrac{\mathbf{T}'(3\pi/4)}{\|\mathbf{T}'(3\pi/4)\|} = \dfrac{\sqrt{2}}{2}\mathbf{i} - \dfrac{\sqrt{2}}{2}\mathbf{j}$

21. $\mathbf{r}(t) = 4t\mathbf{i}$

$\mathbf{v}(t) = 4\mathbf{i}$

$\mathbf{a}(t) = \mathbf{O}$

$\mathbf{T}(t) = \dfrac{\mathbf{v}(t)}{\|\mathbf{v}(t)\|} = \dfrac{4\mathbf{i}}{4} = \mathbf{i}$

$\mathbf{T}'(t) = \mathbf{O}$

$\mathbf{N}(t) = \dfrac{\mathbf{T}'(t)}{\|\mathbf{T}'(t)\|}$ is undefined.

The path is a line and the speed is constant.

23. $\mathbf{r}(t) = 4t^2\mathbf{i}$

$\mathbf{v}(t) = 8t\mathbf{i}$

$\mathbf{a}(t) = 8\mathbf{i}$

$\mathbf{T}(t) = \dfrac{\mathbf{v}(t)}{\|\mathbf{v}(t)\|} = \dfrac{8t\mathbf{i}}{8t} = \mathbf{i}$

$\mathbf{T}'(t) = \mathbf{O}$

$\mathbf{N}(t) = \dfrac{\mathbf{T}'(t)}{\|\mathbf{T}'(t)\|}$ is undefined.

The path is a line and the speed is variable.

25. $\mathbf{r}(t) = t\mathbf{i} + \dfrac{1}{t}\mathbf{j}, \ \mathbf{v}(t) = \mathbf{i} - \dfrac{1}{t^2}\mathbf{j}, \ \mathbf{v}(1) = \mathbf{i} - \mathbf{j},$

$\mathbf{a}(t) = \dfrac{2}{t^3}\mathbf{j}, \ \mathbf{a}(1) = 2\mathbf{j}$

$\mathbf{T}(t) = \dfrac{\mathbf{v}(t)}{\|\mathbf{v}(t)\|} = \dfrac{t^2}{\sqrt{t^4 + 1}}\left(\mathbf{i} - \dfrac{1}{t^2}\mathbf{j}\right) = \dfrac{1}{\sqrt{t^4 + 1}}(t^2\mathbf{i} - \mathbf{j})$

$\mathbf{T}(1) = \dfrac{1}{\sqrt{2}}(\mathbf{i} - \mathbf{j}) = \dfrac{\sqrt{2}}{2}(\mathbf{i} - \mathbf{j})$

$\mathbf{N}(t) = \dfrac{\mathbf{T}'(t)}{\|\mathbf{T}'(t)\|} = \dfrac{\dfrac{2t}{(t^4 + 1)^{3/2}}\mathbf{i} + \dfrac{2t^3}{(t^4 + 1)^{3/2}}\mathbf{j}}{\dfrac{2t}{(t^4 + 1)}}$

$\qquad = \dfrac{1}{\sqrt{t^4 + 1}}(\mathbf{i} + t^2\mathbf{j})$

$\mathbf{N}(1) = \dfrac{1}{\sqrt{2}}(\mathbf{i} + \mathbf{j}) = \dfrac{\sqrt{2}}{2}(\mathbf{i} + \mathbf{j})$

$a_\mathbf{T} = \mathbf{a} \cdot \mathbf{T} = -\sqrt{2}$

$a_\mathbf{N} = \mathbf{a} \cdot \mathbf{N} = \sqrt{2}$

27. $\mathbf{r}(t) = (e^t \cos t)\mathbf{i} + (e^t \sin t)\mathbf{j}$

$\mathbf{v}(t) = e^t(\cos t - \sin t)\mathbf{i} + e^t(\cos t + \sin t)\mathbf{j}$

$\mathbf{a}(t) = e^t(-2 \sin t)\mathbf{i} + e^t(2 \cos t)\mathbf{j}$

At $t = \dfrac{\pi}{2}, \ \mathbf{T} = \dfrac{\mathbf{v}}{\|\mathbf{v}\|} = \dfrac{1}{\sqrt{2}}(-\mathbf{i} + \mathbf{j}) = \dfrac{\sqrt{2}}{2}(-\mathbf{i} + \mathbf{j}).$

Motion along $\mathbf{r}$ is counterclockwise. Therefore,

$\mathbf{N} = \dfrac{1}{\sqrt{2}}(-\mathbf{i} - \mathbf{j}) = -\dfrac{\sqrt{2}}{2}(\mathbf{i} + \mathbf{j}).$

$a_\mathbf{T} = \mathbf{a} \cdot \mathbf{T} = \sqrt{2}e^{\pi/2}$

$a_\mathbf{N} = \mathbf{a} \cdot \mathbf{N} = \sqrt{2}e^{\pi/2}$

29. $\mathbf{r}(t_0) = (\cos \omega t_0 + \omega t_0 \sin \omega t_0)\mathbf{i} + (\sin \omega t_0 - \omega t_0 \cos \omega t_0)\mathbf{j}$

$\mathbf{v}(t_0) = (\omega^2 t_0 \cos \omega t_0)\mathbf{i} + (\omega^2 t_0 \sin \omega t_0)\mathbf{j}$

$\mathbf{a}(t_0) = \omega^2[(\cos \omega t_0 - \omega t_0 \sin \omega t_0)\mathbf{i} + (\omega t_0 \cos \omega t_0 + \sin \omega t_0)\mathbf{j}]$

$\mathbf{T}(t_0) = \dfrac{\mathbf{v}}{\|\mathbf{v}\|} = (\cos \omega t_0)\mathbf{i} + (\sin \omega t_0)\mathbf{j}$

Motion along $\mathbf{r}$ is counterclockwise. Therefore

$\mathbf{N}(t_0) = (-\sin \omega t_0)\mathbf{i} + (\cos \omega t_0)\mathbf{j}.$

$a_\mathbf{T} = \mathbf{a} \cdot \mathbf{T} = \omega^2$

$a_\mathbf{N} = \mathbf{a} \cdot \mathbf{N} = \omega^2(\omega t_0) = \omega^3 t_0$

31. $\mathbf{r}(t) = a \cos \omega t \, \mathbf{i} + a \sin \omega t \, \mathbf{j}$

$\mathbf{v}(t) = -a\omega \sin \omega t \, \mathbf{i} + a\omega \cos \omega t \, \mathbf{j}$

$\mathbf{a}(t) = -a\omega^2 \cos \omega t \, \mathbf{i} - a\omega^2 \sin \omega t \, \mathbf{j}$

$\mathbf{T}(t) = \dfrac{\mathbf{v}(t)}{\|\mathbf{v}(t)\|} = -\sin \omega t \, \mathbf{i} + \cos \omega t \, \mathbf{j}$

$\mathbf{N}(t) = \dfrac{\mathbf{T}'(t)}{\|\mathbf{T}'(t)\|} = -\cos \omega t \, \mathbf{i} - \sin \omega t \, \mathbf{j}$

$a_\mathbf{T} = \mathbf{a} \cdot \mathbf{T} = 0$

$a_\mathbf{N} = \mathbf{a} \cdot \mathbf{N} = a\omega^2$

33. Speed: $\|\mathbf{v}(t)\| = a\omega$

The speed is constant since $a_\mathbf{T} = 0$.

35. $\mathbf{r}(t) = t\mathbf{i} + \dfrac{1}{t}\mathbf{j}$, $t_0 = 2$

$x = t, \ y = \dfrac{1}{t} \implies xy = 1$

$\mathbf{r}'(t) = \mathbf{i} - \dfrac{1}{t^2}\mathbf{j}$

$\mathbf{T}(t) = \dfrac{t^2\mathbf{i} - \mathbf{j}}{\sqrt{t^4 + 1}}$

$\mathbf{N}(t) = \dfrac{\mathbf{i} + t^2\mathbf{j}}{\sqrt{t^4 + 1}}$

$\mathbf{r}(2) = 2\mathbf{i} + \dfrac{1}{2}\mathbf{j}$

$\mathbf{T}(2) = \dfrac{\sqrt{17}}{17}(4\mathbf{i} - \mathbf{j})$

$\mathbf{N}(2) = \dfrac{\sqrt{17}}{17}(\mathbf{i} + 4\mathbf{j})$

37. $\mathbf{r}(t) = t\mathbf{i} + 2t\mathbf{j} - 3t\mathbf{k}$, $t = 1$

$\mathbf{v}(t) = \mathbf{i} + 2\mathbf{j} - 3\mathbf{k}$

$\mathbf{a}(t) = \mathbf{0}$

$\mathbf{T}(t) = \dfrac{\mathbf{v}}{\|\mathbf{v}\|} = \dfrac{1}{\sqrt{14}}(\mathbf{i} + 2\mathbf{j} - 3\mathbf{k}) = \dfrac{\sqrt{14}}{14}(\mathbf{i} + 2\mathbf{j} - 3\mathbf{k})$

$\mathbf{N}(t) = \dfrac{\mathbf{T}'}{\|\mathbf{T}'\|}$ is undefined.

$a_\mathbf{T}, a_\mathbf{N}$ are not defined.

39. $\mathbf{r}(t) = t\mathbf{i} + t^2\mathbf{j} + \dfrac{t^2}{2}\mathbf{k}$, $t = 1$

$\mathbf{v}(t) = \mathbf{i} + 2t\mathbf{j} + t\mathbf{k}$

$\mathbf{v}(1) = \mathbf{i} + 2\mathbf{j} + \mathbf{k}$

$\mathbf{a}(t) = 2\mathbf{j} + \mathbf{k}$

$\mathbf{T}(t) = \dfrac{\mathbf{v}}{\|\mathbf{v}\|} = \dfrac{1}{\sqrt{1 + 5t^2}}(\mathbf{i} + 2t\mathbf{j} + t\mathbf{k})$

$\mathbf{T}(1) = \dfrac{\sqrt{6}}{6}(\mathbf{i} + 2\mathbf{j} + \mathbf{k})$

$\mathbf{N}(t) = \dfrac{\mathbf{T}'}{\|\mathbf{T}'\|} = \dfrac{\dfrac{-5t\mathbf{i} + 2\mathbf{j} + \mathbf{k}}{(1 + 5t^2)^{3/2}}}{\dfrac{\sqrt{5}}{1 + 5t^2}} = \dfrac{-5t\mathbf{i} + 2\mathbf{j} + \mathbf{k}}{\sqrt{5}\sqrt{1 + 5t^2}}$

$\mathbf{N}(1) = \dfrac{\sqrt{30}}{30}(-5\mathbf{i} + 2\mathbf{j} + \mathbf{k})$

$a_\mathbf{T} = \mathbf{a} \cdot \mathbf{T} = \dfrac{5\sqrt{6}}{6}$

$a_\mathbf{N} = \mathbf{a} \cdot \mathbf{N} = \dfrac{\sqrt{30}}{6}$

41. $\mathbf{r}(t) = 4t\mathbf{i} + 3 \cos t\mathbf{j} + 3 \sin t\mathbf{k}$, $t = \dfrac{\pi}{2}$

$\mathbf{v}(t) = 4\mathbf{i} - 3 \sin t\mathbf{j} + 3 \cos t\mathbf{k}$

$\mathbf{v}\left(\dfrac{\pi}{2}\right) = 4\mathbf{i} - 3\mathbf{j}$

$\mathbf{a}(t) = -3 \cos t\mathbf{j} - 3 \sin t\mathbf{k}$

$\mathbf{a}\left(\dfrac{\pi}{2}\right) = -3\mathbf{k}$

$\mathbf{T}(t) = \dfrac{\mathbf{v}}{\|\mathbf{v}\|} = \dfrac{1}{5}(4\mathbf{i} - 3 \sin t\mathbf{j} + 3 \cos t\mathbf{k})$

$\mathbf{T}\left(\dfrac{\pi}{2}\right) = \dfrac{1}{5}(4\mathbf{i} - 3\mathbf{j})$

$\mathbf{N}(t) = \dfrac{\mathbf{T}'}{\|\mathbf{T}'\|} = -\cos t\mathbf{j} - \sin t\mathbf{k}$

$\mathbf{N}\left(\dfrac{\pi}{2}\right) = -\mathbf{k}$

$a_\mathbf{T} = \mathbf{a} \cdot \mathbf{T} = 0$

$a_\mathbf{N} = \mathbf{a} \cdot \mathbf{N} = 3$

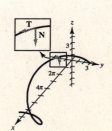

43. $\mathbf{T}(t) = \dfrac{\mathbf{r}'(t)}{\|\mathbf{r}'(t)\|}$

$\mathbf{N}(t) = \dfrac{\mathbf{T}'(t)}{\|\mathbf{T}'(t)\|}$

If $a(t) = a_{\mathbf{T}}\mathbf{T}(t) + a_{\mathbf{N}}\mathbf{N}(t)$, then $a_{\mathbf{T}}$ is the tangential component of acceleration and $a_{\mathbf{N}}$ is the normal component of acceleration.

45. If $a_{\mathbf{N}} = 0$, then the motion is in a straight line.

47. $\mathbf{r}(t) = \langle \pi t - \sin \pi t, 1 - \cos \pi t \rangle$

The graph is a cycloid.

(a) $\mathbf{r}(t) = \langle \pi t - \sin \pi t, 1 - \cos \pi t \rangle$

$\mathbf{v}(t) = \langle \pi - \pi \cos \pi t, \pi \sin \pi t \rangle$

$\mathbf{a}(t) = \langle \pi^2 \sin \pi t, \pi^2 \cos \pi t \rangle$

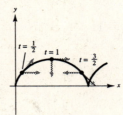

$\mathbf{T}(t) = \dfrac{\mathbf{v}(t)}{\|\mathbf{v}(t)\|} = \dfrac{1}{\sqrt{2(1 - \cos \pi t)}}\langle 1 - \cos \pi t, \sin \pi t \rangle$

$\mathbf{N}(t) = \dfrac{\mathbf{T}'(t)}{\|\mathbf{T}'(t)\|} = \dfrac{1}{\sqrt{2(1 - \cos \pi t)}}\langle \sin \pi t, -1 + \cos \pi t \rangle$

$a_{\mathbf{T}} = \mathbf{a} \cdot \mathbf{T} = \dfrac{1}{\sqrt{2(1 - \cos \pi t)}}[\pi^2 \sin \pi t(1 - \cos \pi t) + \pi^2 \cos \pi t \sin \pi t] = \dfrac{\pi^2 \sin \pi t}{\sqrt{2(1 - \cos \pi t)}}$

$a_{\mathbf{N}} = \mathbf{a} \cdot \mathbf{N} = \dfrac{1}{\sqrt{2(1 - \cos \pi t)}}[\pi^2 \sin^2 \pi t + \pi^2 \cos \pi t(-1 + \cos \pi t)] = \dfrac{\pi^2(1 - \cos \pi t)}{\sqrt{2(1 - \cos \pi t)}} = \dfrac{\pi^2\sqrt{2(1 - \cos \pi t)}}{2}$

When $t = \dfrac{1}{2}$: $a_{\mathbf{T}} = \dfrac{\pi^2}{\sqrt{2}} = \dfrac{\sqrt{2}\pi^2}{2}$, $a_{\mathbf{N}} = \dfrac{\sqrt{2}\pi^2}{2}$

When $t = 1$: $a_{\mathbf{T}} = 0$, $a_{\mathbf{N}} = \pi^2$

When $t = \dfrac{3}{2}$: $a_{\mathbf{T}} = -\dfrac{\sqrt{2}\pi^2}{2}$, $a_{\mathbf{N}} = \dfrac{\sqrt{2}\pi^2}{2}$

(b) Speed: $s = \|\mathbf{v}(t)\| = \pi\sqrt{2(1 - \cos \pi t)}$

$\dfrac{ds}{dt} = \dfrac{\pi^2 \sin \pi t}{\sqrt{2(1 - \cos \pi t)}} = a_{\mathbf{T}}$

When $t = \dfrac{1}{2}$: $a_{\mathbf{T}} = \dfrac{\sqrt{2}\pi^2}{2} > 0 \implies$ the speed in increasing.

When $t = 1$: $a_{\mathbf{T}} = 0 \implies$ the height is maximum.

When $t = \dfrac{3}{2}$: $a_{\mathbf{T}} = -\dfrac{\sqrt{2}\pi^2}{2} < 0 \implies$ the speed is decreasing.

49. $\mathbf{r}(t) = 2\cos t\mathbf{i} + 2\sin t\mathbf{j} + \dfrac{t}{2}\mathbf{k}, \ t_0 = \dfrac{\pi}{2}$

$\mathbf{r}'(t) = -2\sin t\mathbf{i} + 2\cos t\mathbf{j} + \dfrac{1}{2}\mathbf{k}$

$\mathbf{T}(t) = \dfrac{2\sqrt{17}}{17}\left(-2\sin t\mathbf{i} + 2\cos t\mathbf{j} + \dfrac{1}{2}\mathbf{k}\right)$

$\mathbf{N}(t) = -\cos t\mathbf{i} - \sin t\mathbf{j}$

$\mathbf{r}\!\left(\dfrac{\pi}{2}\right) = 2\mathbf{j} + \dfrac{\pi}{4}\mathbf{k}$

$\mathbf{T}\!\left(\dfrac{\pi}{2}\right) = \dfrac{2\sqrt{17}}{17}\left(-2\mathbf{i} + \dfrac{1}{2}\mathbf{k}\right) = \dfrac{\sqrt{17}}{17}(-4\mathbf{i} + \mathbf{k})$

$\mathbf{N}\!\left(\dfrac{\pi}{2}\right) = -\mathbf{j}$

$\mathbf{B}\!\left(\dfrac{\pi}{2}\right) = \mathbf{T}\!\left(\dfrac{\pi}{2}\right) \times \mathbf{N}\!\left(\dfrac{\pi}{2}\right) = \begin{vmatrix} \mathbf{i} & \mathbf{j} & \mathbf{k} \\ -\dfrac{4\sqrt{17}}{17} & 0 & \dfrac{\sqrt{17}}{17} \\ 0 & -1 & 0 \end{vmatrix} = \dfrac{\sqrt{17}}{17}\mathbf{i} + \dfrac{4\sqrt{17}}{17}\mathbf{k} = \dfrac{\sqrt{17}}{17}(\mathbf{i} + 4\mathbf{k})$

51. From Theorem 11.3 we have:

$\mathbf{r}(t) = (v_0 t\cos\theta)\mathbf{i} + (h + v_0 t\sin\theta - 16t^2)\mathbf{j}$

$\mathbf{v}(t) = v_0\cos\theta\mathbf{i} + (v_0\sin\theta - 32t)\mathbf{j}$

$\mathbf{a}(t) = -32\mathbf{j}$

$\mathbf{T}(t) = \dfrac{(v_0\cos\theta)\mathbf{i} + (v_0\sin\theta - 32t)\mathbf{j}}{\sqrt{v_0{}^2\cos^2\theta + (v_0\sin\theta - 32t)^2}}$

$\mathbf{N}(t) = \dfrac{(v_0\sin\theta - 32t)\mathbf{i} - v_0\cos\theta\mathbf{j}}{\sqrt{v_0{}^2\cos^2\theta + (v_0\sin\theta - 32t)^2}}$ (Motion is clockwise.)

$a_\mathbf{T} = \mathbf{a}\cdot\mathbf{T} = \dfrac{-32(v_0\sin\theta - 32t)}{\sqrt{v_0{}^2\cos^2\theta + (v_0\sin\theta - 32t)^2}}$

$a_\mathbf{N} = \mathbf{a}\cdot\mathbf{N} = \dfrac{32v_0\cos\theta}{\sqrt{v_0{}^2\cos^2\theta + (v_0\sin\theta - 32t)^2}}$

Maximum height when $v_0\sin\theta - 32t = 0$; (vertical component of velocity)

At maximum height, $a_\mathbf{T} = 0$ and $a_\mathbf{N} = 32$.

53. $\mathbf{r}(t) = \langle 10\cos 10\pi t, \ 10\sin 10\pi t, \ 4 + 4t\rangle, \ 0 \le t \le \tfrac{1}{20}$

(a) $\mathbf{r}'(t) = \langle -100\pi\sin(10\pi t), \ 100\pi\cos(10\pi t), \ 4\rangle$

$\|\mathbf{r}'(t)\| = \sqrt{(100\pi)^2\sin^2(10\pi t) + (100\pi)^2\cos^2(10\pi t) + 16}$

$\qquad = \sqrt{(100\pi)^2 + 16} = 4\sqrt{625\pi^2 + 1} \approx 314 \text{ mi/hr}$

(b) $a_\mathbf{T} = 0$ and $a_\mathbf{N} = 1000\pi^2$

$a_\mathbf{T} = 0$ because the speed is constant.

55. $\mathbf{r}(t) = (a\cos\omega t)\mathbf{i} + (a\sin\omega t)\mathbf{j}$

From Exercise 31, we know $\mathbf{a}\cdot\mathbf{T} = 0$ and $\mathbf{a}\cdot\mathbf{N} = a\omega^2$.

(a) Let $\omega_0 = 2\omega$. Then

$\mathbf{a}\cdot\mathbf{N} = a\omega_0{}^2 = a(2\omega)^2 = 4a\omega^2$

or the centripetal acceleration is increased by a factor of 4 when the velocity is doubled.

(b) Let $a_0 = a/2$. Then

$\mathbf{a}\cdot\mathbf{N} = a_0\omega^2 = \left(\dfrac{a}{2}\right)\omega^2 = \left(\dfrac{1}{2}\right)a\omega^2$

or the centripetal acceleration is halved when the radius is halved.

57. $v = \sqrt{\dfrac{9.56 \times 10^4}{4100}} \approx 4.83 \text{ mi/sec}$ **59.** $v = \sqrt{\dfrac{9.56 \times 10^4}{4385}} \approx 4.67 \text{ mi/sec}$

61. Let $\mathbf{T}(t) = \cos \phi \mathbf{i} + \sin \phi \mathbf{j}$ be the unit tangent vector. Then

$$\mathbf{T}'(t) = \frac{d\mathbf{T}}{dt} = \frac{d\mathbf{T}}{d\phi}\frac{d\phi}{dt} = -(\sin \phi \mathbf{i} - \cos \phi \mathbf{j})\frac{d\phi}{dt} = \mathbf{M}\frac{d\phi}{dt}.$$

$\mathbf{M} = -\sin \phi \mathbf{i} + \cos \phi \mathbf{j} = \cos[\phi + (\pi/2)]\mathbf{i} + \sin[\phi + (\pi/2)]\mathbf{j}$ and is rotated counterclockwise through an angle of $\pi/2$ from $\mathbf{T}$.

If $d\phi/dt > 0$, then the curve bends to the left and $\mathbf{M}$ has the same direction as $\mathbf{T}'$. Thus, $\mathbf{M}$ has the same direction as

$$\mathbf{N} = \frac{\mathbf{T}'}{\|\mathbf{T}'\|},$$

which is toward the concave side of the curve.

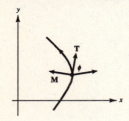

If $d\phi/dt < 0$, then the curve bends to the right and $\mathbf{M}$ has the opposite direction as $\mathbf{T}'$. Thus,

$$\mathbf{N} = \frac{\mathbf{T}'}{\|\mathbf{T}'\|}$$

again points to the concave side of the curve.

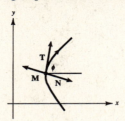

63. Using $\mathbf{a} = a_\mathbf{T}\mathbf{T} + a_\mathbf{N}\mathbf{N}$, $\mathbf{T} \times \mathbf{T} = \mathbf{O}$, and $\|\mathbf{T} \times \mathbf{N}\| = 1$, we have:

$$\mathbf{v} \times \mathbf{a} = \|\mathbf{v}\|\mathbf{T} \times (a_\mathbf{T}\mathbf{T} + a_\mathbf{N}\mathbf{N})$$
$$= \|\mathbf{v}\|a_\mathbf{T}(\mathbf{T} \times \mathbf{T}) + \|\mathbf{v}\|a_\mathbf{N}(\mathbf{T} \times \mathbf{N})$$
$$= \|\mathbf{v}\|a_\mathbf{N}(\mathbf{T} \times \mathbf{N})$$
$$\|\mathbf{v} \times \mathbf{a}\| = \|\mathbf{v}\|a_\mathbf{N}\|\mathbf{T} \times \mathbf{N}\|$$
$$= \|\mathbf{v}\|a_\mathbf{N}$$

Thus, $a_\mathbf{N} = \dfrac{\|\mathbf{v} \times \mathbf{a}\|}{\|\mathbf{v}\|}$.

Section 11.5   Arc Length and Curvature

1. $\mathbf{r}(t) = t\mathbf{i} + 3t\mathbf{j}$

$\dfrac{dx}{dt} = 1$, $\dfrac{dy}{dt} = 3$, $\dfrac{dz}{dt} = 0$

$s = \displaystyle\int_0^4 \sqrt{1 + 9}\, dt$

$= \sqrt{10}\displaystyle\int_0^4 dt$

$= \left[\sqrt{10}\,t\right]_0^4 = 4\sqrt{10}$

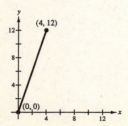

3. $\mathbf{r}(t) = a\cos^3 t\mathbf{i} + a\sin^3 t\mathbf{j}$

$\dfrac{dx}{dt} = -3a\cos^2 t \sin t$, $\dfrac{dy}{dt} = 3a\sin^2 t \cos t$

$s = 4\displaystyle\int_0^{\pi/2} \sqrt{[-3a\cos^2 t \sin t]^2 + [3a\sin^2 t \cos t]^2}\, dt$

$= 12a\displaystyle\int_0^{\pi/2} \sin t \cos t\, dt$

$= 3a\displaystyle\int_0^{\pi/2} 2\sin 2t\, dt = \left[-3a\cos 2t\right]_0^{\pi/2} = 6a$

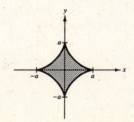

5. (a) $\mathbf{r}(t) = (v_0 \cos \theta)t\mathbf{i} + \left[h + (v_0 \sin \theta)t - \frac{1}{2}gt^2 \right]\mathbf{j}$

$$= (100 \cos 45°)t\mathbf{i} + \left[3 + (100 \sin 45°)t - \frac{1}{2}(32)t^2 \right]\mathbf{j}$$

$$= 50\sqrt{2}t\mathbf{i} + \left[3 + 50\sqrt{2}t - 16t^2 \right]\mathbf{j}$$

(b) $\mathbf{v}(t) = 50\sqrt{2}\mathbf{i} + \left(50\sqrt{2} - 32t\right)\mathbf{j}$

$$50\sqrt{2} - 32t = 0 \implies t = \frac{25\sqrt{2}}{16}$$

Maximum height: $3 + 50\sqrt{2}\left(\dfrac{25\sqrt{2}}{16}\right) - 16\left(\dfrac{25\sqrt{2}}{16}\right)^2 = 81.125$ ft

(c) $3 + 50\sqrt{2}t - 16t^2 = 0 \implies t \approx 4.4614$

Range: $50\sqrt{2}(4.4614) \approx 315.5$ feet

(d) $s = \displaystyle\int_0^{4.4614} \sqrt{\left(50\sqrt{2}\right)^2 + \left(50\sqrt{2} - 32t\right)^2}\,dt \approx 362.9$ feet

7. $\mathbf{r}(t) = 2t\mathbf{i} - 3t\mathbf{j} + t\mathbf{k}$

$\dfrac{dx}{dt} = 2\ \dfrac{dy}{dt} = -3, \ \dfrac{dz}{dt} = 1$

$s = \displaystyle\int_0^2 \sqrt{2^2 + (-3)^2 + 1^2}\,dt$

$= \displaystyle\int_0^2 \sqrt{14}\,dt = \left[\sqrt{14}\,t \right]_0^2 = 2\sqrt{14}$

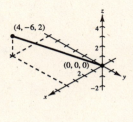

9. $\mathbf{r}(t) = a \cos t\mathbf{i} + a \sin t\mathbf{j} + bt\mathbf{k}$

$\dfrac{dx}{dt} = -a \sin t, \ \dfrac{dy}{dt} = a \cos t, \ \dfrac{dz}{dt} = b$

$s = \displaystyle\int_0^{2\pi} \sqrt{a^2 \sin^2 t + a^2 \cos^2 t + b^2}\,dt$

$= \displaystyle\int_0^{2\pi} \sqrt{a^2 + b^2}\,dt = \left[\sqrt{a^2 + b^2}\,t \right]_0^{2\pi} = 2\pi\sqrt{a^2 + b^2}$

11. $\mathbf{r}(t) = t^2\mathbf{i} + t\mathbf{j} + \ln t\mathbf{k}$

$\dfrac{dx}{dt} = 2t, \ \dfrac{dy}{dt} = 1, \ \dfrac{dz}{dt} = \dfrac{1}{t}$

$s = \displaystyle\int_1^3 \sqrt{(2t)^2 + (1)^2 + \left(\dfrac{1}{t}\right)^2}\,dt$

$= \displaystyle\int_1^3 \sqrt{\dfrac{4t^4 + t^2 + 1}{t^2}}\,dt$

$= \displaystyle\int_1^3 \dfrac{\sqrt{4t^4 + t^2 + 1}}{t}\,dt \approx 8.37$

13. $\mathbf{r}(t) = t\mathbf{i} + (4 - t^2)\mathbf{j} + t^3\mathbf{k}, \quad 0 \le t \le 2$

(a) $\mathbf{r}(0) = \langle 0, 4, 0 \rangle, \ \mathbf{r}(2) = \langle 2, 0, 8 \rangle$

distance $= \sqrt{2^2 + 4^2 + 8^2} = \sqrt{84} = 2\sqrt{21} \approx 9.165$

—CONTINUED—

13. **—CONTINUED—**

(b) $\mathbf{r}(0) = \langle 0, 4, 0 \rangle$

$\mathbf{r}(0.5) = \langle 0.5, 3.75, 0.125 \rangle$

$\mathbf{r}(1) = \langle 1, 3, 1 \rangle$

$\mathbf{r}(1.5) = \langle 1.5, 1.75, 3.375 \rangle$

$\mathbf{r}(2) = \langle 2, 0, 8 \rangle$

distance $\approx \sqrt{(0.5)^2 + (0.25)^2 + (0.125)^2} + \sqrt{(0.5)^2 + (0.75)^2 + (0.875)^2} + \sqrt{(0.5)^2 + (1.25)^2 + (2.375)^2} +$

$\sqrt{(0.5)^2 + (1.75)^2 + (4.625)^2}$

$\approx 0.5728 + 1.2562 + 2.7300 + 4.9702 \approx 9.529$

(c) Increase the number of line segments.

(d) Using a graphing utility, you obtain 9.57057.

15. $\mathbf{r}(t) = \langle 2 \cos t, 2 \sin t, t \rangle$

(a) $s = \displaystyle\int_0^t \sqrt{[x'(u)]^2 + [y'(u)]^2 + [z'(u)]^2}\, du$

$= \displaystyle\int_0^t \sqrt{(-2 \sin u)^2 + (2 \cos u)^2 + (1)^2}\, du$

$= \displaystyle\int_0^t \sqrt{5}\, du = \left[\sqrt{5}\, u\right]_0^t = \sqrt{5}\, t$

(b) $\dfrac{s}{\sqrt{5}} = t$

$x = 2 \cos\left(\dfrac{s}{\sqrt{5}}\right),\ y = 2 \sin\left(\dfrac{s}{\sqrt{5}}\right),\ z = \dfrac{s}{\sqrt{5}}$

$\mathbf{r}(s) = 2 \cos\left(\dfrac{s}{\sqrt{5}}\right)\mathbf{i} + 2 \sin\left(\dfrac{s}{\sqrt{5}}\right)\mathbf{j} + \dfrac{s}{\sqrt{5}}\mathbf{k}$

(c) When $s = \sqrt{5}$: $x = 2 \cos 1 \approx 1.081$

$y = 2 \sin 1 \approx 1.683$

$z = 1$

$(1.081, 1.683, 1.000)$

When $s = 4$: $x = 2 \cos \dfrac{4}{\sqrt{5}} \approx -0.433$

$y = 2 \sin \dfrac{4}{\sqrt{5}} \approx 1.953$

$z = \dfrac{4}{\sqrt{5}} \approx 1.789$

$(-0.433, 1.953, 1.789)$

(d) $\|\mathbf{r}'(s)\| = \sqrt{\left(-\dfrac{2}{\sqrt{5}} \sin\left(\dfrac{s}{\sqrt{5}}\right)\right)^2 + \left(\dfrac{2}{\sqrt{5}} \cos\left(\dfrac{s}{\sqrt{5}}\right)\right)^2 + \left(\dfrac{1}{\sqrt{5}}\right)^2} = \sqrt{\dfrac{4}{5} + \dfrac{1}{5}} = 1$

17. $\mathbf{r}(s) = \left(1 + \dfrac{\sqrt{2}}{2}s\right)\mathbf{i} + \left(1 - \dfrac{\sqrt{2}}{2}s\right)\mathbf{j}$

$\mathbf{r}'(s) = \dfrac{\sqrt{2}}{2}\mathbf{i} - \dfrac{\sqrt{2}}{2}\mathbf{j}$ and $\|\mathbf{r}'(s)\| = \sqrt{\dfrac{1}{2} + \dfrac{1}{2}} = 1$

$\mathbf{T}(s) = \dfrac{\mathbf{r}'(s)}{\|\mathbf{r}'(s)\|} = \mathbf{r}'(s)$

$\mathbf{T}'(s) = \mathbf{0} \implies K = \|\mathbf{T}'(s)\| = 0$ (The curve is a line.)

19. $\mathbf{r}(s) = 2 \cos\left(\dfrac{s}{\sqrt{5}}\right)\mathbf{i} + 2 \sin\left(\dfrac{s}{\sqrt{5}}\right)\mathbf{j} + \dfrac{s}{\sqrt{5}}\mathbf{k}$

$\mathbf{T}(s) = \mathbf{r}'(s) = -\dfrac{2}{\sqrt{5}} \sin\left(\dfrac{s}{\sqrt{5}}\right)\mathbf{i} + \dfrac{2}{\sqrt{5}} \cos\left(\dfrac{s}{\sqrt{5}}\right)\mathbf{j} + \dfrac{1}{\sqrt{5}}\mathbf{k}$

$\mathbf{T}'(s) = -\dfrac{2}{5} \cos\left(\dfrac{s}{\sqrt{5}}\right)\mathbf{i} - \dfrac{2}{5} \sin\left(\dfrac{s}{\sqrt{5}}\right)\mathbf{j}$

$K = \|\mathbf{T}'(s)\| = \dfrac{2}{5}$

21. $\mathbf{r}(t) = 4t\mathbf{i} - 2t\mathbf{j}$

$\mathbf{v}(t) = 4\mathbf{i} - 2\mathbf{j}$

$\mathbf{T}(t) = \dfrac{1}{\sqrt{5}}(2\mathbf{i} - \mathbf{j})$

$\mathbf{T}'(t) = \mathbf{0}$

$K = \dfrac{\|\mathbf{T}'(t)\|}{\|\mathbf{r}'(t)\|} = 0$ (The curve is a line.)

23. $\mathbf{r}(t) = t\mathbf{i} + \dfrac{1}{t}\mathbf{j}$

$\mathbf{v}(t) = \mathbf{i} - \dfrac{1}{t^2}\mathbf{j}$

$\mathbf{v}(1) = \mathbf{i} - \mathbf{j}$

$\mathbf{a}(t) = \dfrac{2}{t^3}\mathbf{j}$

$\mathbf{a}(1) = 2\mathbf{j}$

$\mathbf{T}(t) = \dfrac{t^2\mathbf{i} - \mathbf{j}}{\sqrt{t^4 + 1}}$

$\mathbf{N}(t) = \dfrac{1}{(t^4 + 1)^{1/2}}(\mathbf{i} + t^2\mathbf{j})$

$\mathbf{N}(1) = \dfrac{1}{\sqrt{2}}(\mathbf{i} + \mathbf{j})$

$K = \dfrac{\mathbf{a} \cdot \mathbf{N}}{\|\mathbf{v}\|^2} = \dfrac{\sqrt{2}}{2}$

25. $\mathbf{r}(t) = 4\cos 2\pi t\, \mathbf{i} + 4\sin 2\pi t\, \mathbf{j}$

$\mathbf{r}'(t) = -8\pi \sin 2\pi t\, \mathbf{i} + 8\pi \cos 2\pi t\, \mathbf{j}$

$\mathbf{T}(t) = -\sin 2\pi t\, \mathbf{i} + \cos 2\pi t\, \mathbf{j}$

$\mathbf{T}'(t) = -2\pi \cos 2\pi t\, \mathbf{i} - 2\pi \sin 2\pi t\, \mathbf{j}$

$K = \dfrac{\|\mathbf{T}'(t)\|}{\|\mathbf{r}'(t)\|} = \dfrac{2\pi}{8\pi} = \dfrac{1}{4}$

27. $\mathbf{r}(t) = a\cos \omega t\, \mathbf{i} + a\sin \omega t\, \mathbf{j}$

$\mathbf{r}'(t) = -a\omega \sin \omega t\, \mathbf{i} + a\omega \cos \omega t\, \mathbf{j}$

$\mathbf{T}(t) = -\sin \omega t\, \mathbf{i} + \cos \omega t\, \mathbf{j}$

$\mathbf{T}'(t) = -\omega \cos \omega t\, \mathbf{i} - \omega \sin \omega t\, \mathbf{j}$

$K = \dfrac{\|\mathbf{T}'(t)\|}{\|\mathbf{r}'(t)\|} = \dfrac{\omega}{a\omega} = \dfrac{1}{a}$

29. $\mathbf{r}(t) = e^t\cos t\mathbf{i} + e^t\sin t\mathbf{j}$

$\mathbf{r}'(t) = (-e^t \sin t + e^t \cos t)\mathbf{i} + (e^t \cos t + e^t \sin t)\mathbf{j}$

$\mathbf{T}(t) = \dfrac{1}{\sqrt{2}}[(-\sin t + \cos t)\mathbf{i} + (\cos t + \sin t)\mathbf{j}]$

$\mathbf{T}'(t) = \dfrac{1}{\sqrt{2}}[(-\cos t - \sin t)\mathbf{i} + (-\sin t + \cos t)\mathbf{j}]$

$K = \dfrac{\|\mathbf{T}'(t)\|}{\|\mathbf{r}'(t)\|} = \dfrac{1}{\sqrt{2}\,e^t} = \dfrac{\sqrt{2}}{2}e^{-t}$

31. $\mathbf{r}(t) = \langle \cos \omega t + \omega t \sin \omega t,\ \sin \omega t - \omega t \cos \omega t \rangle$

From Exercise 29, Section 11.4, we have:

$\mathbf{a} \cdot \mathbf{N} = \omega^3 t$

$K = \dfrac{\mathbf{a}(t) \cdot \mathbf{N}(t)}{\|\mathbf{v}\|^2} = \dfrac{\omega^3 t}{\omega^4 t^2} = \dfrac{1}{\omega t}$

33. $\mathbf{r}(t) = t\mathbf{i} + t^2\mathbf{j} + \dfrac{t^2}{2}\mathbf{k}$

$\mathbf{r}'(t) = \mathbf{i} + 2t\mathbf{j} + t\mathbf{k}$

$\mathbf{T}(t) = \dfrac{\mathbf{i} + 2t\mathbf{j} + t\mathbf{k}}{\sqrt{1 + 5t^2}}$

$\mathbf{T}'(t) = \dfrac{-5t\mathbf{i} + 2\mathbf{j} + \mathbf{k}}{(1 + 5t^2)^{3/2}}$

$K = \dfrac{\|\mathbf{T}'(t)\|}{\|\mathbf{r}'(t)\|}$

$= \dfrac{\dfrac{\sqrt{5}}{(1 + 5t^2)}}{\sqrt{1 + 5t^2}} = \dfrac{\sqrt{5}}{(1 + 5t^2)^{3/2}}$

35. $\mathbf{r}(t) = 4t\mathbf{i} + 3\cos t\mathbf{j} + 3\sin t\mathbf{k}$

$\mathbf{r}'(t) = 4\mathbf{i} - 3\sin t\mathbf{j} + 3\cos t\mathbf{k}$

$\mathbf{T}(t) = \dfrac{1}{5}[4\mathbf{i} - 3\sin t\mathbf{j} + 3\cos t\mathbf{k}]$

$\mathbf{T}'(t) = \dfrac{1}{5}[-3\cos t\mathbf{j} - 3\sin t\mathbf{k}]$

$K = \dfrac{\|\mathbf{T}'(t)\|}{\|\mathbf{r}'(t)\|} = \dfrac{3/5}{5} = \dfrac{3}{25}$

37. $y = 3x - 2$

Since $y'' = 0$, $K = 0$, and the radius of curvature is undefined.

39. $y = 2x^2 + 3$, $x = -1$

$y' = 4x$

$y'' = 4$

$K = \dfrac{4}{[1 + (-4)^2]^{3/2}} = \dfrac{4}{17^{3/2}} \approx 0.057$

$\dfrac{1}{K} = \dfrac{17^{3/2}}{4} \approx 17.523$ (radius of curvature)

41. $y = \sqrt{a^2 - x^2}$, $x = 0$

$y' = \dfrac{-x}{\sqrt{a^2 - x^2}}$

$y'' = \dfrac{a^2}{(a^2 - x^2)^{3/2}}$

At $x = 0$: $y' = 0$

$y'' = \dfrac{1}{a}$

$K = \dfrac{1/a}{(1 + 0^2)^{3/2}} = \dfrac{1}{a}$

$\dfrac{1}{K} = a$ (radius of curvature)

43. (a) Point on circle: $\left(\dfrac{\pi}{2}, 1\right)$

Center: $\left(\dfrac{\pi}{2}, 0\right)$

Equation: $\left(x - \dfrac{\pi}{2}\right)^2 + y^2 = 1$

(b) The circles have different radii since the curvature is different and

$r = \dfrac{1}{K}$.

45. $y = x + \dfrac{1}{x}$, $y' = 1 - \dfrac{1}{x^2}$, $y'' = \dfrac{2}{x^3}$

$K = \dfrac{2}{(1 + 0^2)^{3/2}} = 2$ at $(1, 2)$

Radius of curvature $= 1/2$. Since the tangent line is horizontal at $(1, 2)$, the normal line is vertical. The center of the circle is $1/2$ unit above the point $(1, 2)$ at $(1, 5/2)$.

Circle: $(x - 1)^2 + \left(y - \dfrac{5}{2}\right)^2 = \dfrac{1}{4}$

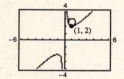

47. $y = e^x$, $x = 0$

$y' = e^x$, $y'' = e^x$

$y'(0) = 1$, $y''(0) = 1$

$K = \dfrac{1}{(1 + 1^2)^{3/2}} = \dfrac{1}{2^{3/2}} = \dfrac{1}{2\sqrt{2}}$, $r = \dfrac{1}{K} = 2\sqrt{2}$

The slope of the tangent line at $(0, 1)$ is $y'(0) = 1$.

The slope of the normal line is -1.

Equation of normal line: $y - 1 = -x$ or $y = -x + 1$

The center of the circle is on the normal line $2\sqrt{2}$ units away from the point $(0, 1)$.

$\sqrt{(0 - x)^2 + (1 - y)^2} = 2\sqrt{2}$

$x^2 + x^2 = 8$

$x^2 = 4$

$x = \pm 2$

Since the circle is above the curve, $x = -2$ and $y = 3$.

Center of circle: $(-2, 3)$

Equation of circle: $(x + 2)^2 + (y - 3)^2 = 8$

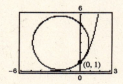

49.

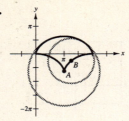

51. $y = (x - 1)^2 + 3$, $y' = 2(x - 1)$, $y'' = 2$

$$K = \frac{2}{(1 + [2(x - 1)]^2)^{3/2}} = \frac{2}{[1 + 4(x - 1)^2]^{3/2}}$$

(a) K is maximum when $x = 1$ or at the vertex $(1, 3)$.

(b) $\lim_{x \to \infty} K = 0$

53. $y = x^{2/3}$, $y' = \frac{2}{3}x^{-1/3}$, $y'' = -\frac{2}{9}x^{-4/3}$

$$K = \left| \frac{(-2/9)x^{-4/3}}{[1 + (4/9)x^{-2/3}]^{3/2}} \right| = \left| \frac{6}{x^{1/3}(9x^{2/3} + 4)^{3/2}} \right|$$

(a) $K \Rightarrow \infty$ as $x \Rightarrow 0$. No maximum

(b) $\lim_{x \to \infty} K = 0$

55. $y = (x - 1)^3 + 3$

$y' = 3(x - 1)^2$

$y'' = 6(x - 1)$

$$K = \frac{|y''|}{[1 + (y')^2]^{3/2}} = \frac{|6(x - 1)|}{[1 + 9(x - 1)^4]^{3/2}} = 0 \text{ at } x = 1.$$

Curvature is 0 at $(1, 3)$.

57. $K = \frac{|y''|}{[1 + (y')^2]^{3/2}}$

The curvature is zero when $y'' = 0$.

59. $s = \int_a^b \|\mathbf{r}'(t)\| \, dt$

61. The curve is a line.

63. Endpoints of the major axis: $(\pm 2, 0)$

Endpoints of the minor axis: $(0, \pm 1)$

$x^2 + 4y^2 = 4$

$2x + 8yy' = 0$

$$y' = -\frac{x}{4y}$$

$$y'' = \frac{(4y)(-1) - (-x)(4y')}{16y^2} = \frac{-4y - (x^2/y)}{16y^2} = \frac{-(4y^2 + x^2)}{16y^3} = \frac{-1}{4y^3}$$

$$K = \frac{|-1/4y^3|}{[1 + (-x/4y)^2]^{3/2}} = \frac{|-16|}{(16y^2 + x^2)^{3/2}} = \frac{16}{(12y^2 + 4)^{3/2}} = \frac{16}{(16 - 3x^2)^{3/2}}$$

Therefore, since $-2 \le x \le 2$, K is largest when $x = \pm 2$ and smallest when $x = 0$.

65. $f(x) = x^4 - x^2$

(a) $K = \frac{2|6x^2 - 1|}{[16x^6 - 16x^4 + 4x^2 + 1]^{3/2}}$

(b) For $x = 0$, $K = 2$. $f(0) = 0$. At $(0, 0)$, the circle of curvature has radius $\frac{1}{2}$. Using the symmetry of the graph of f, you obtain

$$x^2 + \left(y + \frac{1}{2}\right)^2 = \frac{1}{4}.$$

For $x = 1$, $K = (2\sqrt{5})/5$. $f(1) = 0$. At $(1, 0)$, the circle of curvature has radius

$$\frac{\sqrt{5}}{2} = \frac{1}{K}.$$

Using the graph of f, you see that the center of curvature is $(0, \frac{1}{2})$. Thus,

$$x^2 + \left(y - \frac{1}{2}\right)^2 = \frac{5}{4}.$$

To graph these circles, use

$$y = -\frac{1}{2} \pm \sqrt{\frac{1}{4} - x^2} \quad \text{and} \quad y = \frac{1}{2} \pm \sqrt{\frac{5}{4} - x^2}.$$

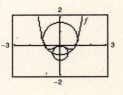

—CONTINUED—

65. —CONTINUED—

(c) The curvature tends to be greatest near the extrema of f, and K decreases as $x \to \pm\infty$. However, f and K do not have the same critical numbers.

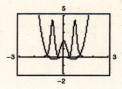

Critical numbers of f: $x = 0, \pm\dfrac{\sqrt{2}}{2} \approx \pm 0.7071$

Critical numbers of K: $x = 0, \pm 0.7647, \pm 0.4082$

67. (a) Imagine dropping the circle $x^2 + (y - k)^2 = 16$ into the parabola $y = x^2$. The circle will drop to the point where the tangents to the circle and parabola are equal.

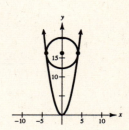

$$y = x^2 \quad \text{and} \quad x^2 + (y - k)^2 = 16 \implies x^2 + (x^2 - k)^2 = 16$$

Taking derivatives, $2x + 2(y - k)y' = 0$ and $y' = 2x$. Hence,

$$(y - k)y' = -x \implies y' = \frac{-x}{y - k}.$$

Thus,

$$\frac{-x}{y - k} = 2x \implies -x = 2x(y - k) \implies -1 = 2(x^2 - k) \implies x^2 - k = -\frac{1}{2}.$$

Thus,

$$x^2 + (x^2 - k)^2 = x^2 + \left(-\frac{1}{2}\right)^2 = 16 \implies x^2 = 15.75.$$

Finally, $k = x^2 + \frac{1}{2} = 16.25$, and the center of the circle is 16.25 units from the vertex of the parabola. Since the radius of the circle is 4, the circle is 12.25 units from the vertex.

(b) In 2-space, the parabola $z = y^2$ (or $z = x^2$) has a curvature of $K = 2$ at $(0, 0)$. The radius of the largest sphere that will touch the vertex has radius $= 1/K = \frac{1}{2}$.

69. Given $y = f(x)$: $K = \dfrac{|y''|}{(1 + [y']^2)^{3/2}}$

$$R = \frac{1}{K}$$

The center of the circle is on the normal line at a distance of R from (x, y).

Equation of normal line: $y - y_0 = -\dfrac{1}{y'}(x - x_0)$

$$\sqrt{(x - x_0)^2 + \left[-\frac{1}{y'}(x - x_0)\right]^2} = \frac{(1 + [y']^2)^{3/2}}{|y''|}$$

$$(x - x_0)^2\left[1 + \frac{1}{(y')^2}\right] = \frac{(1 + [y']^2)^3}{(y'')^2}$$

$$(x - x_0)^2 = \frac{(y')^2(1 + [y']^2)^2}{(y'')^2}$$

$$x - x_0 = \frac{y'(1 + [y']^2)}{y''} = y'z$$

$$x_0 = x - y'z$$

$$y - y_0 = -\frac{1}{y'}(x - (x - y'z)) = -z$$

$$y_0 = y + z$$

Thus, $(x_0, y_0) = (x - y'z, y + z)$.

For $y = e^x$, $y' = e^x$, $y'' = e^x$, $z = \dfrac{1 + e^{2x}}{e^x} = e^{-x} + e^x$.

When $x = 0$: $x_0 = x - y'z = 0 - (1)(2) = -2$

$\qquad\qquad y_0 = y + z = 1 + 2 = 3$

Center of curvature: $(-2, 3)$

(See Exercise 47)

71. $r = 1 + \sin \theta$

$r' = \cos \theta$

$r'' = -\sin \theta$

$K = \dfrac{|2(r')^2 - rr'' + r^2|}{[(r')^2 + r^2]^{3/2}}$

$\quad = \dfrac{|2\cos^2 \theta - (1 + \sin \theta)(-\sin \theta) + (1 + \sin \theta)^2|}{\sqrt{[\cos^2 \theta + (1 + \sin \theta)^2]^3}}$

$\quad = \dfrac{3(1 + \sin \theta)}{\sqrt{8(1 + \sin \theta)^3}} = \dfrac{3}{2\sqrt{2(1 + \sin \theta)}}$

73. $r = a \sin \theta$

$r' = a \cos \theta$

$r'' = -a \sin \theta$

$K = \dfrac{|2(r')^2 - rr'' + r^2|}{[(r')^2 + r^2]^{3/2}}$

$\quad = \dfrac{|2a^2 \cos^2 \theta + a^2 \sin^2 \theta + a^2 \sin^2 \theta|}{\sqrt{[a^2 \cos^2 \theta + a^2 \sin^2 \theta]^3}}$

$\quad = \dfrac{2a^2}{a^3} = \dfrac{2}{a}, a > 0$

75. $r = e^{a\theta}, a > 0$

$r' = ae^{a\theta}$

$r'' = a^2 e^{a\theta}$

$K = \dfrac{|2(r')^2 - rr'' + r^2|}{[(r')^2 + r^2]^{3/2}} = \dfrac{|2a^2 e^{2a\theta} - a^2 e^{2a\theta} + e^{2a\theta}|}{[a^2 e^{2a\theta} + e^{2a\theta}]^{3/2}}$

$\quad = \dfrac{1}{e^{a\theta}\sqrt{a^2 + 1}}$

(a) As $\theta \Longrightarrow \infty$, $K \Longrightarrow 0$.

(b) As $a \Longrightarrow \infty$, $K \Longrightarrow 0$.

77. $r = 4 \sin 2\theta$

$r' = 8 \cos 2\theta$

At the pole: $K = \dfrac{2}{|r'(0)|} = \dfrac{2}{8} = \dfrac{1}{4}$

79. $x = f(t)$

$\quad y = g(t)$

$y' = \dfrac{dy}{dx} = \dfrac{\frac{dy}{dt}}{\frac{dx}{dt}} = \dfrac{g'(t)}{f'(t)}$

$y'' = \dfrac{\dfrac{d}{dt}\left[\dfrac{g'(t)}{f'(t)}\right]}{\dfrac{dx}{dt}} = \dfrac{\dfrac{f'(t)g''(t) - g'(t)f''(t)}{[f'(t)]^2}}{f'(t)}$

$\quad = \dfrac{f'(t)g''(t) - g'(t)f''(t)}{[f'(t)]^3}$

$K = \dfrac{|y''|}{[1 + (y')^2]^{3/2}} = \dfrac{\left|\dfrac{f'(t)g''(t) - g'(t)f''(t)}{[f'(t)]^3}\right|}{\left[1 + \left(\dfrac{g'(t)}{f'(t)}\right)^2\right]^{3/2}}$

$\quad = \dfrac{\left|\dfrac{f'(t)g''(t) - g'(t)f''(t)}{[f'(t)]^3}\right|}{\sqrt{\left\{\dfrac{[f'(t)]^2 + [g'(t)]^2}{[f'(t)]^2}\right\}^3}}$

$\quad = \dfrac{|f'(t)g''(t) - g'(t)f''(t)|}{([f'(t)]^2 + [g'(t)]^2)^{3/2}}$

81. $x(\theta) = a(\theta - \sin \theta) \qquad y(\theta) = a(1 - \cos \theta)$

$\quad x'(\theta) = a(1 - \cos \theta) \qquad y'(\theta) = a \sin \theta$

$\quad x''(\theta) = a \sin \theta \qquad y''(\theta) = a \cos \theta$

$K = \dfrac{|x'(\theta)y''(\theta) - y'(\theta)x''(\theta)|}{[x'(\theta)^2 + y'(\theta)^2]^{3/2}}$

$\quad = \dfrac{|a^2(1 - \cos \theta)\cos \theta - a^2 \sin^2 \theta|}{[a^2(1 - \cos \theta)^2 + a^2 \sin^2 \theta]^{3/2}}$

$\quad = \dfrac{1}{a}\dfrac{|\cos \theta - 1|}{[2 - 2\cos \theta]^{3/2}}$

$\quad = \dfrac{1}{a}\dfrac{1 - \cos \theta}{2\sqrt{2}[1 - \cos \theta]^{3/2}} \quad (1 - \cos \geq 0)$

$\quad = \dfrac{1}{2a\sqrt{2 - 2\cos \theta}} = \dfrac{1}{4a}\csc\left(\dfrac{\theta}{2}\right)$

Minimum: $\dfrac{1}{4a}$ $\qquad (\theta = \pi)$

Maximum: none $\qquad (K \to \infty$ as $\theta \to 0)$

83. $a_N = mK\left(\dfrac{ds}{dt}\right)^2 = \left(\dfrac{5500 \text{ lb}}{32 \text{ ft/sec}^2}\right)\left(\dfrac{1}{100 \text{ ft}}\right)\left(\dfrac{30(5280) \text{ ft}}{3600 \text{ sec}}\right)^2 = 3327.5 \text{ lb}$

85. Let $\mathbf{r} = x(t)\mathbf{i} + y(t)\mathbf{j} + z(t)\mathbf{k}$. Then $r = \|\mathbf{r}\| = \sqrt{[x(t)]^2 + [y(t)]^2 + [z(t)]^2}$ and $\mathbf{r}' = x'(t)\mathbf{i} + y'(t)\mathbf{j} + z'(t)\mathbf{k}$. Then,

$$r\left(\frac{dr}{dt}\right) = \sqrt{[x(t)]^2 + [y(t)]^2 + [z(t)]^2}\left[\frac{1}{2}\{[x(t)]^2 + [y(t)]^2 + [z(t)]^2\}^{-1/2} \cdot (2x(t)x'(t) + 2y(t)y'(t) + 2z(t)z'(t))\right]$$

$$= x(t)x'(t) + y(t)y'(t) + z(t)z'(t) = \mathbf{r} \cdot \mathbf{r}'.$$

87. Let $\mathbf{r} = x\mathbf{i} + y\mathbf{j} + z\mathbf{k}$ where x, y, and z are functions of t, and $r = \|\mathbf{r}\|$.

$$\frac{d}{dt}\left[\frac{\mathbf{r}}{r}\right] = \frac{r\mathbf{r}' - \mathbf{r}(dr/dt)}{r^2} = \frac{r\mathbf{r}' - \mathbf{r}[(\mathbf{r} \cdot \mathbf{r}')/r]}{r^2} = \frac{r^2\mathbf{r}' - (\mathbf{r} \cdot \mathbf{r}')\mathbf{r}}{r^3} \quad \text{(using Exercise 87)}$$

$$= \frac{(x^2 + y^2 + z^2)(x'\mathbf{i} + y'\mathbf{j} + z'\mathbf{k}) - (xx' + yy' + zz')(x\mathbf{i} + y\mathbf{j} + z\mathbf{k})}{r^3}$$

$$= \frac{1}{r^3}[(x'y^2 + x'z^2 - xyy' - xzz')\mathbf{i} + (x^2y' + z^2y' - xx'y - zz'y)\mathbf{j} + (x^2z' + y^2z' - xx'z - yy'z)\mathbf{k}]$$

$$= \frac{1}{r^3}\begin{vmatrix} \mathbf{i} & \mathbf{j} & \mathbf{k} \\ yz' - y'z & -(xz' - x'z) & xy' - x'y \\ x & y & z \end{vmatrix} = \frac{1}{r^3}\{[\mathbf{r} \times \mathbf{r}'] \times \mathbf{r}\}$$

89. From Exercise 86, we have concluded that planetary motion is planar. Assume that the planet moves in the *xy*-plane with the sun at the origin. From Exercise 88, we have

$$\mathbf{r}' \times \mathbf{L} = GM\left(\frac{\mathbf{r}}{r} + \mathbf{e}\right).$$

Since $\mathbf{r}' \times \mathbf{L}$ and $\mathbf{r}$ are both perpendicular to $\mathbf{L}$, so is $\mathbf{e}$. Thus, $\mathbf{e}$ lies in the *xy*-plane. Situate the coordinate system so that $\mathbf{e}$ lies along the positive *x*-axis and θ is the angle between $\mathbf{e}$ and $\mathbf{r}$. Let $e = \|\mathbf{e}\|$. Then $\mathbf{r} \cdot \mathbf{e} = \|\mathbf{r}\|\,\|\mathbf{e}\|\cos\theta = re\cos\theta$. Also,

$$\|\mathbf{L}\|^2 = \mathbf{L} \cdot \mathbf{L} = (\mathbf{r} \times \mathbf{r}') \cdot \mathbf{L}$$

$$= \mathbf{r} \cdot (\mathbf{r}' \times \mathbf{L}) = \mathbf{r} \cdot \left[GM\left(\mathbf{e} + \frac{\mathbf{r}}{r}\right)\right] = GM\left[\mathbf{r} \cdot \mathbf{e} + \frac{\mathbf{r} \cdot \mathbf{r}}{r}\right] = GM[re\cos\theta + r]$$

Thus,

$$\frac{\|\mathbf{L}\|^2/GM}{1 + e\cos\theta} = r$$

and the planetary motion is a conic section. Since the planet returns to its initial position periodically, the conic is an ellipse.

91. $A = \dfrac{1}{2}\displaystyle\int_\alpha^\beta r^2\,d\theta$

Thus,

$$\frac{dA}{dt} = \frac{dA}{d\theta}\frac{d\theta}{dt} = \frac{1}{2}r^2\frac{d\theta}{dt} = \frac{1}{2}\|\mathbf{L}\|$$

and $\mathbf{r}$ sweeps out area at a constant rate.

Review Exercises for Chapter 11

1. $\mathbf{r}(t) = t\mathbf{i} + \csc t\,\mathbf{k}$

   (a) Domain: $t \neq n\pi$, n an integer

   (b) Continuous except at $t = n\pi$, n an integer

3. $\mathbf{r}(t) = \ln t\,\mathbf{i} + t\mathbf{j} + t\mathbf{k}$

   (a) Domain: $(0, \infty)$

   (b) Continuous for all $t > 0$

5. (a) $\mathbf{r}(0) = \mathbf{i}$

   (b) $\mathbf{r}(-2) = -3\mathbf{i} + 4\mathbf{j} + \frac{8}{3}\mathbf{k}$

   (c) $\mathbf{r}(c - 1) = (2(c - 1) + 1)\mathbf{i} + (c - 1)^2\mathbf{j} - \frac{1}{3}(c - 1)^3\mathbf{k}$

               $= (2c - 1)\mathbf{i} + (c - 1)^2\mathbf{j} - \frac{1}{3}(c - 1)^3\mathbf{k}$

   (d) $\mathbf{r}(1 + \Delta t) - \mathbf{r}(1) = ([2(1 + \Delta t) + 1]\mathbf{i} + [1 + \Delta t]^2\mathbf{j} - \frac{1}{3}[1 + \Delta t]^3\mathbf{k}) - (3\mathbf{i} + \mathbf{j} - \frac{1}{3}\mathbf{k})$

                        $= 2\Delta t\mathbf{i} + \Delta t(\Delta t + 2)\mathbf{j} - \frac{1}{3}(\Delta t^3 + 3\Delta t^2 + 3\Delta t)\mathbf{k}$

7. $\mathbf{r}(t) = \cos t\mathbf{i} + 2\sin^2 t\mathbf{j}$

   $x(t) = \cos t, \; y(t) = 2\sin^2 t$

   $x^2 + \dfrac{y}{2} = 1$

       $y = 2(1 - x^2)$

   $-1 \le x \le 1$

9. $\mathbf{r}(t) = \mathbf{i} + t\mathbf{j} + t^2\mathbf{k}$

   $x = 1$

   $y = t$

   $z = t^2 \implies z = y^2$

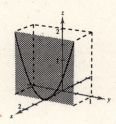

11. $\mathbf{r}(t) = \mathbf{i} + \sin t\mathbf{j} + \mathbf{k}$

   $x = 1, \; y = \sin t, \; z = 1$

t	0	$\dfrac{\pi}{2}$	π	$\dfrac{3\pi}{2}$
x	1	1	1	1
y	0	1	0	-1
z	1	1	1	1

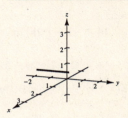

13. $\mathbf{r}(t) = t\mathbf{i} + \ln t\mathbf{j} + \frac{1}{2}t^2\mathbf{k}$

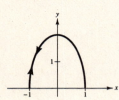

15. One possible answer is:

    $\mathbf{r}_1(t) = 4t\mathbf{i} + 3t\mathbf{j}, \qquad 0 \le t \le 1$

    $\mathbf{r}_2(t) = 4\mathbf{i} + (3 - t)\mathbf{j}, \qquad 0 \le t \le 3$

    $\mathbf{r}_3(t) = (4 - t)\mathbf{i}, \qquad 0 \le t \le 4$

17. The vector joining the points is $\langle 7, 4, -10 \rangle$. One path is

    $\mathbf{r}(t) = \langle -2 + 7t, -3 + 4t, 8 - 10t \rangle$.

19. $z = x^2 + y^2, \; x + y = 0, \; t = x$

   $x = t, \; y = -t, \; z = 2t^2$

    $\mathbf{r}(t) = t\mathbf{i} - t\mathbf{j} + 2t^2\mathbf{k}$

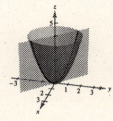

21. $\displaystyle\lim_{t \to 2^-} (t^2\mathbf{i} + \sqrt{4 - t^2}\mathbf{j} + \mathbf{k}) = 4\mathbf{i} + \mathbf{k}$

23. $\mathbf{r}(t) = 3t\mathbf{i} + (t - 1)\mathbf{j}, \ \mathbf{u}(t) = t\mathbf{i} + t^2\mathbf{j} + \frac{2}{3}t^3\mathbf{k}$

  (a) $\mathbf{r}'(t) = 3\mathbf{i} + \mathbf{j}$

  (b) $\mathbf{r}''(t) = \mathbf{0}$

  (c) $\mathbf{r}(t) \cdot \mathbf{u}(t) = 3t^2 + t^2(t - 1) = t^3 + 2t^2$

    $D_t[\mathbf{r}(t) \cdot \mathbf{u}(t)] = 3t^2 + 4t$

  (d) $\mathbf{u}(t) - 2\mathbf{r}(t) = -5t\mathbf{i} + (t^2 - 2t + 2)\mathbf{j} + \frac{2}{3}t^3\mathbf{k}$

    $D_t[\mathbf{u}(t) - 2\mathbf{r}(t)] = -5\mathbf{i} + (2t - 2)\mathbf{j} + 2t^2\mathbf{k}$

  (e) $\|\mathbf{r}(t)\| = \sqrt{10t^2 - 2t + 1}$

    $D_t[\|\mathbf{r}(t)\|] = \dfrac{10t - 1}{\sqrt{10t^2 - 2t + 1}}$

  (f) $\mathbf{r}(t) \times \mathbf{u}(t) = \frac{2}{3}(t^4 - t^3)\mathbf{i} - 2t^4\mathbf{j} + (3t^3 - t^2 + t)\mathbf{k}$

    $D_t[\mathbf{r}(t) \times \mathbf{u}(t)] = \left(\frac{8}{3}t^3 - 2t^2\right)\mathbf{i} - 8t^3\mathbf{j} + (9t^2 - 2t + 1)\mathbf{k}$

25. $x(t)$ and $y(t)$ are increasing functions at $t = t_0$, and $z(t)$ is a decreasing function at $t = t_0$.

27. $\int (\cos t\mathbf{i} + t \cos t\mathbf{j}) \, dt = \sin t\mathbf{i} + (t \sin t + \cos t)\mathbf{j} + \mathbf{C}$

29. $\int \|\cos t\mathbf{i} + \sin t\mathbf{j} + t\mathbf{k}\| \, dt = \int \sqrt{1 + t^2} \, dt = \frac{1}{2}\left[t\sqrt{1 + t^2} + \ln|t + \sqrt{1 + t^2}|\right] + \mathbf{C}$

31. $\mathbf{r}(t) = \int (2t\mathbf{i} + e^t\mathbf{j} + e^{-t}\mathbf{k}) \, dt = t^2\mathbf{i} + e^t\mathbf{j} - e^{-t}\mathbf{k} + \mathbf{C}$

  $\mathbf{r}(0) = \mathbf{j} - \mathbf{k} + \mathbf{C} = \mathbf{i} + 3\mathbf{j} - 5\mathbf{k} \Rightarrow \mathbf{C} = \mathbf{i} + 2\mathbf{j} - 4\mathbf{k}$

  $\mathbf{r}(t) = (t^2 + 1)\mathbf{i} + (e^t + 2)\mathbf{j} - (e^{-t} + 4)\mathbf{k}$

33. $\int_{-2}^{2} (3t\mathbf{i} + 2t^2\mathbf{j} - t^3\mathbf{k}) \, dt = \left[\frac{3t^2}{2}\mathbf{i} + \frac{2t^3}{3}\mathbf{j} - \frac{t^4}{4}\mathbf{k}\right]_{-2}^{2} = \frac{32}{3}\mathbf{j}$

35. $\int_{0}^{2} (e^{t/2}\mathbf{i} - 3t^2\mathbf{j} - \mathbf{k}) \, dt = \left[2e^{t/2}\mathbf{i} - t^3\mathbf{j} - t\mathbf{k}\right]_{0}^{2} = (2e - 2)\mathbf{i} - 8\mathbf{j} - 2\mathbf{k}$

37. $\mathbf{r}(t) = \langle \cos^3 t, \sin^3 t, 3t \rangle$

  $\mathbf{v}(t) = \mathbf{r}'(t) = \langle -3\cos^2 t \sin t, 3\sin^2 t \cos t, 3 \rangle$

    $\|\mathbf{v}(t)\| = \sqrt{9\cos^4 t \sin^2 t + 9\sin^4 t \cos^2 t + 9}$

        $= 3\sqrt{\cos^2 t \sin^2 t(\cos^2 t + \sin^2 t) + 1}$

        $= 3\sqrt{\cos^2 t \sin^2 t + 1}$

  $\mathbf{a}(t) = \mathbf{v}'(t) = \langle -6\cos t(-\sin^2 t) + (-3\cos^2 t)\cos t, \ 6\sin t \cos^2 t + 3\sin^2 t(-\sin t), 0 \rangle$

        $= \langle 3\cos t(2\sin^2 t - \cos^2 t), \ 3\sin t(2\cos^2 t - \sin^2 t), 0 \rangle$

39. $\mathbf{r}(t) = \left\langle \ln(t - 3), t^2, \frac{1}{2}t \right\rangle, \ t_0 = 4$

  $\mathbf{r}'(t) = \left\langle \dfrac{1}{t - 3}, 2t, \dfrac{1}{2} \right\rangle$

  $\mathbf{r}'(4) = \left\langle 1, 8, \dfrac{1}{2} \right\rangle$ direction numbers

  Since $\mathbf{r}(4) = \langle 0, 16, 2 \rangle$, the parametric equations are

  $x = t, y = 16 + 8t, z = 2 + \frac{1}{2}t$.

    $\mathbf{r}(t_0 + 0.1) = \mathbf{r}(4.1) \approx \langle 0.1, 16.8, 2.05 \rangle$

41. $\mathbf{r}(t) = \left\langle v_0 t \cos \theta, v_0 t \sin \theta - \frac{1}{2}gt^2 \right\rangle$

    $= \left\langle \dfrac{75\sqrt{3}}{2}t, \dfrac{75}{2}t - 16t^2 \right\rangle$

  $\dfrac{75}{2}t - 16t^2 \Rightarrow t = \dfrac{75}{32}$

  Range $= \dfrac{75\sqrt{3}}{2}\left(\dfrac{75}{32}\right) = \dfrac{5625}{64}\sqrt{3} \approx 152.2$ feet

  or, Range $= v_0 \cos \theta \left[\dfrac{v_0 \sin \theta}{\frac{1}{2}g}\right] = \dfrac{v_0^2 \sin 2\theta}{g}$.

    $= \dfrac{75^2 \sin(60°)}{32} \approx 152.2$ feet

43. Range $= x = \dfrac{v_0^2}{9.8}\sin 2\theta = 80 \Rightarrow v_0 = \sqrt{\dfrac{(80)(9.8)}{\sin 40°}} \approx 34.9$ m/sec (see Exercise 41.)

45. $\mathbf{r}(t) = 5t\mathbf{i}$

$\mathbf{v}(t) = 5\mathbf{i}$

$\|\mathbf{v}(t)\| = 5$

$\mathbf{a}(t) = \mathbf{0}$

$\mathbf{T}(t) = \mathbf{i}$

$\mathbf{N}(t)$ does not exist.

$\mathbf{a} \cdot \mathbf{T} = 0$

$\mathbf{a} \cdot \mathbf{N}$ does not exist.

(The curve is a line.)

47. $\mathbf{r}(t) = t\mathbf{i} + \sqrt{t}\mathbf{j}$

$\mathbf{v}(t) = \mathbf{i} + \dfrac{1}{2\sqrt{t}}\mathbf{j}$

$\|\mathbf{v}(t)\| = \dfrac{\sqrt{4t+1}}{2\sqrt{t}}$

$\mathbf{a}(t) = -\dfrac{1}{4t\sqrt{t}}\mathbf{j}$

$\mathbf{T}(t) = \dfrac{\mathbf{i} + (1/2\sqrt{t})\mathbf{j}}{(\sqrt{4t+1})/2\sqrt{t}} = \dfrac{2\sqrt{t}\mathbf{i} + \mathbf{j}}{\sqrt{4t+1}}$

$\mathbf{N}(t) = \dfrac{\mathbf{i} - 2\sqrt{t}\mathbf{j}}{\sqrt{4t+1}}$

$\mathbf{a} \cdot \mathbf{T} = \dfrac{-1}{4t\sqrt{t}\sqrt{4t+1}}$

$\mathbf{a} \cdot \mathbf{N} = \dfrac{1}{2t\sqrt{4t+1}}$

49. $\mathbf{r}(t) = e^t\mathbf{i} + e^{-t}\mathbf{j}$

$\mathbf{v}(t) = e^t\mathbf{i} - e^{-t}\mathbf{j}$

$\|\mathbf{v}(t)\| = \sqrt{e^{2t} + e^{-2t}}$

$\mathbf{a}(t) = e^t\mathbf{i} + e^{-t}\mathbf{j}$

$\mathbf{T}(t) = \dfrac{e^t\mathbf{i} - e^{-t}\mathbf{j}}{\sqrt{e^{2t} + e^{-2t}}}$

$\mathbf{N}(t) = \dfrac{e^{-t}\mathbf{i} + e^t\mathbf{j}}{\sqrt{e^{2t} + e^{-2t}}}$

$\mathbf{a} \cdot \mathbf{T} = \dfrac{e^{2t} - e^{-2t}}{\sqrt{e^{2t} + e^{-2t}}}$

$\mathbf{a} \cdot \mathbf{N} = \dfrac{2}{\sqrt{e^{2t} + e^{-2t}}}$

51. $\mathbf{r}(t) = t\mathbf{i} + t^2\mathbf{j} + \dfrac{1}{2}t^2\mathbf{k}$

$\mathbf{v}(t) = \mathbf{i} + 2t\mathbf{j} + t\mathbf{k}$

$\|\mathbf{v}\| = \sqrt{1 + 5t^2}$

$\mathbf{a}(t) = 2\mathbf{j} + \mathbf{k}$

$\mathbf{T}(t) = \dfrac{\mathbf{i} + 2t\mathbf{j} + t\mathbf{k}}{\sqrt{1 + 5t^2}}$

$\mathbf{N}(t) = \dfrac{-5t\mathbf{i} + 2\mathbf{j} + \mathbf{k}}{\sqrt{5}\sqrt{1 + 5t^2}}$

$\mathbf{a} \cdot \mathbf{T} = \dfrac{5t}{\sqrt{1 + 5t^2}}$

$\mathbf{a} \cdot \mathbf{N} = \dfrac{5}{\sqrt{5}\sqrt{1 + 5t^2}} = \dfrac{\sqrt{5}}{\sqrt{1 + 5t^2}}$

53. $\mathbf{r}(t) = 2\cos t\mathbf{i} + 2\sin t\mathbf{j} + t\mathbf{k}, x = 2\cos t, y = 2\sin t, z = t$

When $t = \dfrac{3\pi}{4}$, $x = -\sqrt{2}$, $y = \sqrt{2}$, $z = \dfrac{3\pi}{4}$.

$\mathbf{r}'(t) = -2\sin t\mathbf{i} + 2\cos t\mathbf{j} + \mathbf{k}$

Direction numbers when $t = \dfrac{3\pi}{4}$, $a = -\sqrt{2}$, $b = -\sqrt{2}$, $c = 1$

$x = -\sqrt{2}t - \sqrt{2}, y = -\sqrt{2}t + \sqrt{2}, z = t + \dfrac{3\pi}{4}$

55. $v = \sqrt{\dfrac{9.56 \times 10^4}{4600}} \approx 4.56$ mi/sec (see Exercise 56, Section 11.4.)

57. $\mathbf{r}(t) = 2t\mathbf{i} - 3t\mathbf{j}, 0 \le t \le 5$

$\mathbf{r}'(t) = 2\mathbf{i} - 3\mathbf{j}$

$s = \displaystyle\int_a^b \|\mathbf{r}'(t)\| \, dt = \int_0^5 \sqrt{4 + 9} \, dt$

$= \sqrt{13}t \Big]_0^5 = 5\sqrt{13}$

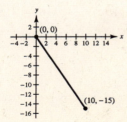

59. $\mathbf{r}(t) = 10\cos^3 t\mathbf{i} + 10\sin^3 t\mathbf{j}$

$\mathbf{r}'(t) = -30\cos^2 t\sin t\mathbf{i} + 30\sin^2 t\cos t\mathbf{j}$

$\|\mathbf{r}'(t)\| = 30\sqrt{\cos^4 t\sin^2 t + \sin^4 t\cos^2 t}$

$= 30|\cos t\sin t|$

$s = 4\int_0^{\pi/2} 30\cos t\cdot\sin t\,dt = \left[120\frac{\sin^2 t}{2}\right]_0^{\pi/2} = 60$

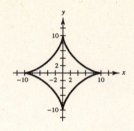

61. $\mathbf{r}(t) = -3t\mathbf{i} + 2t\mathbf{j} + 4t\mathbf{k},\ 0 \le t \le 3$

$\mathbf{r}'(t) = -3\mathbf{i} + 2\mathbf{j} + 4\mathbf{k}$

$s = \int_a^b \|\mathbf{r}'(t)\|\,dt = \int_0^3 \sqrt{9 + 4 + 16}\,dt = \int_0^3 \sqrt{29}\,dt = 3\sqrt{29}$

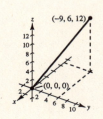

63. $\mathbf{r}(t) = \langle 8\cos t, 8\sin t, t\rangle,\ 0 \le t \le \dfrac{\pi}{2}$

$\mathbf{r}'(t) = \langle -8\sin t, 8\cos t, 1\rangle,\ \|\mathbf{r}'(t)\| = \sqrt{65}$

$s = \int_a^b \|\mathbf{r}'(t)\|\,dt = \int_0^{\pi/2} \sqrt{65}\,dt = \dfrac{\pi\sqrt{65}}{2}$

65. $\mathbf{r}(t) = \dfrac{1}{2}t\mathbf{i} + \sin t\mathbf{j} + \cos t\mathbf{k},\ \ 0 \le t \le \pi$

$\mathbf{r}'(t) = \dfrac{1}{2}\mathbf{i} + \cos t\mathbf{j} - \sin t\mathbf{k}$

$s = \int_0^\pi \|\mathbf{r}'(t)\|\,dt$

$= \int_0^\pi \sqrt{\dfrac{1}{4} + \cos^2 t + \sin^2 t}\,dt$

$= \dfrac{\sqrt{5}}{2}\int_0^\pi dt = \left[\dfrac{\sqrt{5}}{2}t\right]_0^\pi = \dfrac{\sqrt{5}}{2}\pi$

67. $\mathbf{r}(t) = 3t\mathbf{i} + 2t\mathbf{j}$

Line

$K = 0$

69. $\mathbf{r}(t) = 2t\mathbf{i} + \dfrac{1}{2}t^2\mathbf{j} + t^2\mathbf{k}$

$\mathbf{r}'(t) = 2\mathbf{i} + t\mathbf{j} + 2t\mathbf{k},\ \|\mathbf{r}'\| = \sqrt{5t^2 + 4}$

$\mathbf{r}''(t) = \mathbf{j} + 2\mathbf{k}$

$\mathbf{r}' \times \mathbf{r}'' = \begin{vmatrix} \mathbf{i} & \mathbf{j} & \mathbf{k} \\ 2 & t & 2t \\ 0 & 1 & 2 \end{vmatrix} = -4\mathbf{j} + 2\mathbf{k},\ \|\mathbf{r}' \times \mathbf{r}''\| = \sqrt{20}$

$K = \dfrac{\|\mathbf{r}' \times \mathbf{r}''\|}{\|\mathbf{r}'\|^3} = \dfrac{\sqrt{20}}{(5t^2 + 4)^{3/2}} = \dfrac{2\sqrt{5}}{(4 + 5t^2)^{3/2}}$

71. $y = \dfrac{1}{2}x^2 + 2$

$y' = x$

$y'' = 1$

$K = \dfrac{|y''|}{[1 + (y')^2]^{3/2}} = \dfrac{1}{(1 + x^2)^{3/2}}$

At $x = 4, K = \dfrac{1}{17^{3/2}}$ and $r = 17^{3/2} = 17\sqrt{17}$.

73. $y = \ln x$

$y' = \dfrac{1}{x}, y'' = -\dfrac{1}{x^2}$

$K = \dfrac{|y''|}{[1 + (y')^2]^{3/2}} = \dfrac{1/x^2}{[1 + (1/x)^2]^{3/2}}$

At $x = 1, K = \dfrac{1}{2^{3/2}} = \dfrac{1}{2\sqrt{2}} = \dfrac{\sqrt{2}}{4}$ and $r = 2\sqrt{2}$.

75. The curvature changes abruptly from zero to a nonzero constant at the points B and C.

Problem Solving for Chapter 11

1. $x(t) = \int_0^t \cos\left(\frac{\pi u^2}{2}\right) du, \, y(t) = \int_0^t \sin\left(\frac{\pi u^2}{2}\right) du$

$x'(t) = \cos\left(\frac{\pi t^2}{2}\right), \, y'(t) = \sin\left(\frac{\pi t^2}{2}\right)$

(a) $s = \int_0^a \sqrt{x'(t)^2 + y'(t)^2}\, dt = \int_0^a dt = a$

(b) $x''(t) = -\pi t \sin\left(\frac{\pi t^2}{2}\right), \, y''(t) = \pi t \cos\left(\frac{\pi t^2}{2}\right)$

$$K = \frac{\left|\pi t \cos^2\left(\frac{\pi t^2}{2}\right) + \pi t \sin^2\left(\frac{\pi t^2}{2}\right)\right|}{1} = \pi t$$

At $t = a, K = \pi a$.

(c) $K = \pi a = \pi(\text{length})$

3. Bomb: $\mathbf{r}_1(t) = \langle 5000 - 400t, 3200 - 16t^2 \rangle$

Projectile: $\mathbf{r}_2(t) = \langle (v_0 \cos \theta)t, (v_0 \sin \theta)t - 16t^2 \rangle$

At 1600 feet: Bomb:

$3200 - 16t^2 = 1600 \implies t = 10$ seconds.

Projectile will travel 5 seconds:

$5(v_0 \sin \theta) - 16(25) = 1600$

$v_0 \sin \theta = 400.$

Horizontal position:

At $t = 10$, bomb is at $5000 - 400(10) = 1000$.

At $t = 5$, projectile is at $5v_0 \cos \theta$.

Thus, $v_0 \cos \theta = 200$.

Combining, $\dfrac{v_0 \sin \theta}{v_0 \cos \theta} = \dfrac{400}{200} \implies \tan \theta = 2 \implies \theta \approx 63.4°$.

$v_0 = \dfrac{200}{\cos \theta} \approx 447.2$ ft/sec

5. $x'(\theta) = 1 - \cos \theta, \, y'(\theta) = \sin \theta, \, 0 \le \theta \le 2\pi$

$\sqrt{x'(\theta)^2 + y'(\theta)^2} = \sqrt{(1 - \cos \theta)^2 + \sin^2 \theta}$

$$= \sqrt{2 - 2\cos \theta} = \sqrt{4 \sin^2 \frac{\theta}{2}}$$

$s(t) = \int_\pi^t 2 \sin \frac{\theta}{2}\, d\theta = \left[-4 \cos \frac{\theta}{2}\right]_\pi^t = -4 \cos \frac{t}{2}$

$x''(\theta) = \sin \theta, \, y''(\theta) = \cos \theta$

$K = \dfrac{|(1 - \cos \theta)\cos \theta - \sin \theta \sin \theta|}{\left(2 \sin \frac{\theta}{2}\right)^3} = \dfrac{|\cos \theta - 1|}{8 \sin^3 \frac{\theta}{2}}$

$$= \frac{1}{4 \sin \frac{\theta}{2}}$$

Thus, $\rho = \dfrac{1}{K} = 4 \sin \dfrac{t}{2}$ and

$s^2 + \rho^2 = 16 \cos^2\left(\dfrac{t}{2}\right) + 16 \sin^2\left(\dfrac{t}{2}\right) = 16.$

7. $\|\mathbf{r}(t)\|^2 = \mathbf{r}(t) \cdot \mathbf{r}(t)$

$\dfrac{d}{dt}(\|\mathbf{r}(t)\|)^2 = 2\|\mathbf{r}(t)\|\dfrac{d}{dt}\|\mathbf{r}(t)\|$

$= \mathbf{r}(t) \cdot \mathbf{r}'(t) + \mathbf{r}'(t) \cdot \mathbf{r}(t) \implies \dfrac{d}{dt}\|\mathbf{r}(t)\| = \dfrac{\mathbf{r}(t) \cdot \mathbf{r}'(t)}{\|\mathbf{r}(t)\|}$

9. $\mathbf{r}(t) = 4 \cos t\mathbf{i} + 4 \sin t\mathbf{j} + 3t\mathbf{k}, \ t = \dfrac{\pi}{2}$

$\mathbf{r}'(t) = -4 \sin t\mathbf{i} + 4 \cos t\mathbf{j} + 3\mathbf{k}, \ \|\mathbf{r}'(t)\| = 5$

$\mathbf{r}''(t) = -4 \cos t\mathbf{i} - 4 \sin t\mathbf{j}$

$\mathbf{T} = -\dfrac{4}{5} \sin t\mathbf{i} + \dfrac{4}{5} \cos t\mathbf{j} + \dfrac{3}{5}\mathbf{k}$

$\mathbf{T}' = -\dfrac{4}{5} \cos t\mathbf{i} - \dfrac{4}{5} \sin t\mathbf{j}$

$\mathbf{N} = -\cos t\mathbf{i} - \sin t\mathbf{j}$

$\mathbf{B} = \mathbf{T} \times \mathbf{N} = \dfrac{3}{5} \sin t\mathbf{i} - \dfrac{3}{5} \cos t\mathbf{j} + \dfrac{4}{5}\mathbf{k}$

At $t = \dfrac{\pi}{2}, \ \mathbf{T}\!\left(\dfrac{\pi}{2}\right) = -\dfrac{4}{5}\mathbf{i} + \dfrac{3}{5}\mathbf{k}$

$\mathbf{N}\!\left(\dfrac{\pi}{2}\right) = -\mathbf{j}$

$\mathbf{B}\!\left(\dfrac{\pi}{2}\right) = \dfrac{3}{5}\mathbf{i} + \dfrac{4}{5}\mathbf{k}$

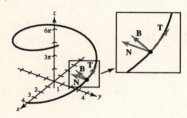

11. (a) $\|\mathbf{B}\| = \|\mathbf{T} \times \mathbf{N}\| = 1 \ \text{ constant length} \ \Longrightarrow \ \dfrac{d\mathbf{B}}{ds} \perp \mathbf{B}$

$\dfrac{d\mathbf{B}}{ds} = \dfrac{d}{ds}(\mathbf{T} \times \mathbf{N}) = (\mathbf{T} \times \mathbf{N}') + (\mathbf{T}' \times \mathbf{N})$

$\mathbf{T} \cdot \dfrac{d\mathbf{B}}{ds} = \mathbf{T} \cdot (\mathbf{T} \times \mathbf{N}') + \mathbf{T} \cdot (\mathbf{T}' \times \mathbf{N})$

$= (\mathbf{T} \times \mathbf{T}) \cdot \mathbf{N}' + \mathbf{T} \cdot \left(\mathbf{T}' \times \dfrac{\mathbf{T}'}{\|\mathbf{T}'\|}\right) = 0$

Hence, $\dfrac{d\mathbf{B}}{ds} \perp \mathbf{B}$ and $\dfrac{d\mathbf{B}}{ds} \perp \mathbf{T} \ \Longrightarrow \ \dfrac{d\mathbf{B}}{ds} = -\tau\mathbf{N}$

for some scalar τ.

(b) $\mathbf{B} = \mathbf{T} \times \mathbf{N}$. Using Exercise 10.4, number 64,

$\mathbf{B} \times \mathbf{N} = (\mathbf{T} \times \mathbf{N}) \times \mathbf{N} = -\mathbf{N} \times (\mathbf{T} \times \mathbf{N})$

$= -[(\mathbf{N} \cdot \mathbf{N})\mathbf{T} - (\mathbf{N} \cdot \mathbf{T})\mathbf{N}]$

$= -\mathbf{T}$

$\mathbf{B} \times \mathbf{T} = (\mathbf{T} \times \mathbf{N}) \times \mathbf{T} = -\mathbf{T} \times (\mathbf{T} \times \mathbf{N})$

$= -[(\mathbf{T} \cdot \mathbf{N})\mathbf{T} - (\mathbf{T} \cdot \mathbf{T})\mathbf{N}]$

$= \mathbf{N}.$

Now, $K\mathbf{N} = \left\|\dfrac{d\mathbf{T}}{ds}\right\| \dfrac{\mathbf{T}'(s)}{\|\mathbf{T}'(s)\|} = \mathbf{T}'(s) = \dfrac{d\mathbf{T}}{ds}.$

Finally,

$\mathbf{N}'(s) = \dfrac{d}{ds}(\mathbf{B} \times \mathbf{T}) = (\mathbf{B} \times \mathbf{T}') + (\mathbf{B}' \times \mathbf{T})$

$= (\mathbf{B} \times K\mathbf{N}) + (-\tau\mathbf{N} \times \mathbf{T})$

$= -K\mathbf{T} + \tau\mathbf{B}.$

13. $\mathbf{r}(t) = \langle t \cos \pi t, \ t \sin \pi t \rangle, \ 0 \le t \le 2$

(a)

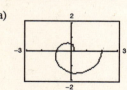

(b) Length $= \displaystyle\int_0^2 \|\mathbf{r}'(t)\| \, dt$

$= \displaystyle\int_0^2 \sqrt{\pi^2 t^2 + 1} \, dt \approx 6.766 \quad \text{(graphing utility)}$

(c) $K = \dfrac{\pi(\pi^2 t^2 + 2)}{[\pi^2 t^2 + 1]^{3/2}}$

$K(0) = 2\pi$

$K(1) = \dfrac{\pi(\pi^2 + 2)}{(\pi^2 + 1)^{3/2}} \approx 1.04$

$K(2) \approx 0.51$

(e) $\displaystyle\lim_{t \to \infty} K = 0$

(d)

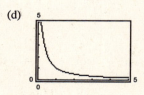

(f) As $t \to \infty$, the graph spirals outward and the curvature decreases.

C H A P T E R   1 2
Functions of Several Variables

CHAPTER 12
Functions of Several Variables

Section 12.1   Introduction to Functions of Several Variables

Solutions to Odd-Numbered Exercises

1. $x^2z + yz - xy = 10$

$z(x^2 + y) = 10 + xy$

$z = \dfrac{10 + xy}{x^2 + y}$

Yes, z is a function of x and y.

3. $\dfrac{x^2}{4} + \dfrac{y^2}{9} + z^2 = 1$

No, z is not a function of x and y. For example, $(x, y) = (0, 0)$ corresponds to both $z = \pm 1$.

5. $f(x, y) = \dfrac{x}{y}$

(a) $f(3, 2) = \dfrac{3}{2}$

(b) $f(-1, 4) = -\dfrac{1}{4}$

(c) $f(30, 5) = \dfrac{30}{5} = 6$

(d) $f(5, y) = \dfrac{5}{y}$

(e) $f(x, 2) = \dfrac{x}{2}$

(f) $f(5, t) = \dfrac{5}{t}$

7. $f(x, y) = xe^y$

(a) $f(5, 0) = 5e^0 = 5$

(b) $f(3, 2) = 3e^2$

(c) $f(2, -1) = 2e^{-1} = \dfrac{2}{e}$

(d) $f(5, y) = 5e^y$

(e) $f(x, 2) = xe^2$

(f) $f(t, t) = te^t$

9. $h(x, y, z) = \dfrac{xy}{z}$

(a) $h(2, 3, 9) = \dfrac{(2)(3)}{9} = \dfrac{2}{3}$

(b) $h(1, 0, 1) = \dfrac{(1)(0)}{1} = 0$

11. $f(x, y) = x \sin y$

(a) $f\left(2, \dfrac{\pi}{4}\right) = 2 \sin \dfrac{\pi}{4} = \sqrt{2}$

(b) $f(3, 1) = 3 \sin 1$

13. $g(x, y) = \displaystyle\int_x^y (2t - 3)\, dt$

(a) $g(0, 4) = \displaystyle\int_0^4 (2t - 3)\, dt = \left[t^2 - 3t \right]_0^4 = 4$

(b) $g(1, 4) = \displaystyle\int_1^4 (2t - 3)\, dt = \left[t^2 - 3t \right]_1^4 = 6$

15. $f(x, y) = x^2 - 2y$

(a) $\dfrac{f(x + \Delta x, y) - f(x, y)}{\Delta x} = \dfrac{[(x + \Delta x)^2 - 2y] - (x^2 - 2y)}{\Delta x}$

$= \dfrac{x^2 + 2x(\Delta x) + (\Delta x)^2 - 2y - x^2 + 2y}{\Delta x} = \dfrac{\Delta x(2x + \Delta x)}{\Delta x} = 2x + \Delta x, \ \Delta x \neq 0$

(b) $\dfrac{f(x, y + \Delta y) - f(x, y)}{\Delta y} = \dfrac{[x^2 - 2(y + \Delta y)] - (x^2 - 2y)}{\Delta y} = \dfrac{x^2 - 2y - 2\Delta y - x^2 + 2y}{\Delta y} = \dfrac{-2\Delta y}{\Delta y} = -2, \ \Delta y \neq 0$

17. $f(x, y) = \sqrt{4 - x^2 - y^2}$

Domain: $4 - x^2 - y^2 \geq 0$

$$x^2 + y^2 \leq 4$$

$$\{(x, y): x^2 + y^2 \leq 4\}$$

Range: $0 \leq z \leq 2$

19. $f(x, y) = \arcsin(x + y)$

Domain:
$\{(x, y): -1 \leq x + y \leq 1\}$

Range: $-\dfrac{\pi}{2} \leq z \leq \dfrac{\pi}{2}$

21. $f(x, y) = \ln(4 - x - y)$

Domain: $4 - x - y > 0$

$$x + y < 4$$

$$\{(x, y): y < -x + 4\}$$

Range: all real numbers

23. $z = \dfrac{x + y}{xy}$

Domain: $\{(x, y): x \neq 0 \text{ and } y \neq 0\}$

Range: all real numbers

25. $f(x, y) = e^{x/y}$

Domain: $\{(x, y): y \neq 0\}$

Range: $z > 0$

27. $g(x, y) = \dfrac{1}{xy}$

Domain: $\{(x, y): x \neq 0 \text{ and } y \neq 0\}$

Range: all real numbers except zero

29. $f(x, y) = \dfrac{-4x}{x^2 + y^2 + 1}$

   (a) View from the positive x-axis: $(20, 0, 0)$

   (c) View from the first octant: $(20, 15, 25)$

   (b) View where x is negative, y and z are positive: $(-15, 10, 20)$

   (d) View from the line $y = x$ in the xy-plane: $(20, 20, 0)$

31. $f(x, y) = 5$

Plane: $z = 5$

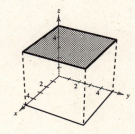

33. $f(x, y) = y^2$

Since the variable x is missing, the surface is a cylinder with rulings parallel to the x-axis. The generating curve is $z = y^2$. The domain is the entire xy-plane and the range is $z \geq 0$.

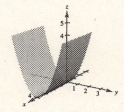

35. $z = 4 - x^2 - y^2$

Paraboloid

Domain: entire xy-plane

Range: $z \leq 4$

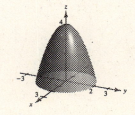

37. $f(x, y) = e^{-x}$

Since the variable y is missing, the surface is a cylinder with rulings parallel to the y-axis. The generating curve is $z = e^{-x}$. The domain is the entire xy-plane and the range is $z > 0$.

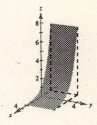

39. $z = y^2 - x^2 + 1$

Hyperbolic paraboloid

Domain: entire xy-plane

Range: $-\infty < z < \infty$

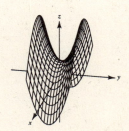

41. $f(x, y) = x^2 e^{(-xy/2)}$

43. $f(x, y) = x^2 + y^2$

(a)

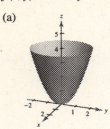

(b) g is a vertical translation of f two units upward

(c) g is a horizontal translation of f two units to the right. The vertex moves from $(0, 0, 0)$ to $(0, 2, 0)$.

(d) g is a reflection of f in the xy-plane followed by a vertical translation 4 units upward.

(e)

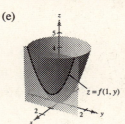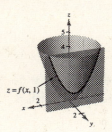

45. $z = e^{1-x^2-y^2}$

Level curves:

$$c = e^{1-x^2-y^2}$$

$$\ln c = 1 - x^2 - y^2$$

$$x^2 + y^2 = 1 - \ln c$$

Circles centered at $(0, 0)$

Matches (c)

47. $z = \ln|y - x^2|$

Level curves:

$$c = \ln|y - x^2|$$

$$\pm e^c = y - x^2$$

$$y = x^2 \pm e^c$$

Parabolas

Matches (b)

49. $z = x + y$

Level curves are parallel lines of the form $x + y = c$.

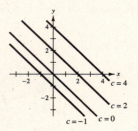

51. $f(x, y) = \sqrt{25 - x^2 - y^2}$

The level curves are of the form

$$c = \sqrt{25 - x^2 - y^2},$$
$$x^2 + y^2 = 25 - c^2.$$

Thus, the level curves are circles of radius 5 or less, centered at the origin.

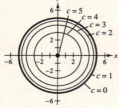

53. $f(x, y) = xy$

The level curves are hyperbolas of the form $xy = c$.

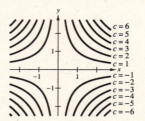

55. $f(x, y) = \dfrac{x}{x^2 + y^2}$

The level curves are of the form

$$c = \frac{x}{x^2 + y^2}$$

$$x^2 - \frac{x}{c} + y^2 = 0$$

$$\left(x - \frac{1}{2c}\right)^2 + y^2 = \left(\frac{1}{2c}\right)^2$$

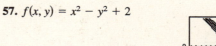

Thus, the level curves are circles passing through the origin and centered at $(1/2c, 0)$.

57. $f(x, y) = x^2 - y^2 + 2$

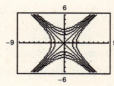

59. $g(x, y) = \dfrac{8}{1 + x^2 + y^2}$

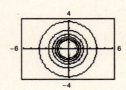

61. See Definition, page 838.

63. No, The following graphs are not hemispheres.

$$z = e^{-(x^2+y^2)}$$

$$z = x^2 + y^2$$

65. The surface is sloped like a saddle. The graph is not unique. Any vertical translation would have the same level curves.

One possible function is

$$f(x, y) = x^2 - y^2.$$

67. $V(I, R) = 1000\left[\dfrac{1 + 0.10(1 - R)}{1 + I}\right]^{10}$

Tax Rate	Inflation Rate		
	0	0.03	0.05
0	2593.74	1929.99	1592.33
0.28	2004.23	1491.34	1230.42
0.35	1877.14	1396.77	1152.40

69. $f(x, y, z) = x - 2y + 3z$

$c = 6$

$6 = x - 2y + 3z$

Plane

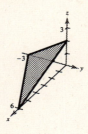

71. $f(x, y, z) = x^2 + y^2 + z^2$

$c = 9$

$9 = x^2 + y^2 + z^2$

Sphere

73. $f(x, y, z) = 4x^2 + 4y^2 - z^2$

$c = 0$

$0 = 4x^2 + 4y^2 - z^2$

Elliptic cone

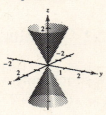

75. $N(d, L) = \left(\dfrac{d - 4}{4}\right)^2 L$

(a) $N(22, 12) = \left(\dfrac{22 - 4}{4}\right)^2 (12) = 243$ board-feet

(b) $N(30, 12) = \left(\dfrac{30 - 4}{4}\right)^2 (12) = 507$ board-feet

77. $T = 600 - 0.75x^2 - 0.75y^2$

The level curves are of the form

$$c = 600 - 0.75x^2 - 0.75y^2$$

$$x^2 + y^2 = \frac{600 - c}{0.75}.$$

The level curves are circles centered at the origin.

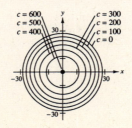

79. $C = 0.75xy + \quad 2(0.40)xz + 2(0.40)yz$

base + front & back + two ends

$$= 0.75xy + 0.80(xz + yz)$$

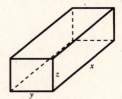

81. $PV = kT$, $20(2600) = k(300)$

(a) $k = \dfrac{20(2600)}{300} = \dfrac{520}{3}$

(b) $P = \dfrac{kT}{V} = \dfrac{520}{3}\left(\dfrac{T}{V}\right)$

The level curves are of the form: $c = \left(\dfrac{520}{3}\right)\left(\dfrac{T}{V}\right)$

$$V = \dfrac{520}{3c}T$$

Thus, the level curves are lines through the origin with slope $\dfrac{520}{3c}$.

83. (a) Highest pressure at C

(b) Lowest pressure at A

(c) Highest wind velocity at B

85. (a) The boundaries between colors represent level curves

(b) No, the colors represent intervals of different lengths, as indicated in the box

(c) You could use more colors, which means using smaller intervals

87. False. Let

$$f(x, y) = 2xy$$

$$f(1, 2) = f(2, 1), \text{ but } 1 \ne 2$$

89. False. Let

$$f(x, y) = 5.$$

Then, $f(2x, 2y) = 5 \ne 2^2 f(x, y)$.

Section 12.2   Limits and Continuity

1. Let $\varepsilon > 0$ be given. We need to find $\delta > 0$ such that $|f(x, y) - L| = |y - b| < \varepsilon$

whenever $0 < \sqrt{(x - a)^2 + (y - b)^2} < \delta$. Take $\delta = \varepsilon$.

Then if $0 < \sqrt{(x - a)^2 + (y - b)^2} < \delta = \varepsilon$, we have

$$\sqrt{(y - b)^2} < \varepsilon$$

$$|y - b| < \varepsilon.$$

3. $\displaystyle\lim_{(x, y)\to(a, b)} [f(x, y) - g(x, y)] = \lim_{(x, y)\to(a, b)} f(x, y) - \lim_{(x, y)\to(a, b)} g(x, y) = 5 - 3 = 2$

5. $\displaystyle\lim_{(x, y)\to(a, b)} [f(x, y)g(x, y)] = \left[\lim_{(x, y)\to(a, b)} f(x, y)\right]\left[\lim_{(x, y)\to(a, b)} g(x, y)\right] = 5(3) = 15$

7. $\displaystyle\lim_{(x, y)\to(2, 1)} (x + 3y^2) = 2 + 3(1)^2 = 5$

Continuous everywhere

9. $\displaystyle\lim_{(x, y)\to(2, 4)} \dfrac{x + y}{x - y} = \dfrac{2 + 4}{2 - 4} = -3$

Continuous for $x \ne y$

11. $\displaystyle\lim_{(x, y)\to(0, 1)} \dfrac{\arcsin(x/y)}{1 + xy} = \arcsin 0 = 0$

Continuous for $xy \ne -1, y \ne 0, |x/y| \le 1$

13. $\displaystyle\lim_{(x, y)\to(-1, 2)} e^{xy} = e^{-2} = \dfrac{1}{e^2}$

Continuous everywhere

15. $\displaystyle\lim_{(x, y, z)\to(1, 2, 5)} \sqrt{x + y + z} = \sqrt{8} = 2\sqrt{2}$

Continuous for $x + y + z \ge 0$

17. $\displaystyle\lim_{(x, y)\to(0, 0)} e^{xy} = 1$

Continuous everywhere

19. $\displaystyle\lim_{(x,\,y)\to(0,\,0)} \ln(x^2 + y^2) = \ln(0) = -\infty$

The limit does not exist.

Continuous except at $(0, 0)$

21. $f(x, y) = \dfrac{xy}{x^2 + y^2}$

Continuous except at $(0, 0)$

Path: $y = 0$

(x, y)	$(1, 0)$	$(0.5, 0)$	$(0.1, 0)$	$(0.01, 0)$	$(0.001, 0)$
$f(x, y)$	0	0	0	0	0

Path: $y = x$

(x, y)	$(1, 1)$	$(0.5, 0.5)$	$(0.1, 0.1)$	$(0.01, 0.01)$	$(0.001, 0.001)$
$f(x, y)$	$\frac{1}{2}$	$\frac{1}{2}$	$\frac{1}{2}$	$\frac{1}{2}$	$\frac{1}{2}$

The limit does not exist because along the path $y = 0$ the function equals 0, whereas along the path $y = x$ the function equals $\frac{1}{2}$.

23. $f(x, y) = -\dfrac{xy^2}{x^2 + y^4}$

Continuous except at $(0, 0)$

Path: $x = y^2$

(x, y)	$(1, 1)$	$(0.25, 0.5)$	$(0.01, 0.1)$	$(0.0001, 0.01)$	$(0.000001, 0.001)$
$f(x, y)$	$-\frac{1}{2}$	$-\frac{1}{2}$	$-\frac{1}{2}$	$-\frac{1}{2}$	$-\frac{1}{2}$

Path: $x = -y^2$

(x, y)	$(-1, 1)$	$(-0.25, 0.5)$	$(-0.01, 0.1)$	$(-0.0001, 0.01)$	$(-0.000001, 0.001)$
$f(x, y)$	$\frac{1}{2}$	$\frac{1}{2}$	$\frac{1}{2}$	$\frac{1}{2}$	$\frac{1}{2}$

The limit does not exist because along the path $x = y^2$ the function equals $-\frac{1}{2}$, whereas along the path $x = -y^2$ the function equals $\frac{1}{2}$.

25. $\displaystyle\lim_{(x,\,y)\to(0,\,0)} f(x, y) = \lim_{(x,\,y)\to(0,\,0)} \left(\frac{x^2 + 2xy^2 + y^2}{x^2 + y^2} \right)$

$\displaystyle = \lim_{(x,\,y)\to(0,\,0)} \left(1 + \frac{2xy^2}{x^2 + y^2} \right) = 1$

(same limit for g)

Thus, f is not continuous at $(0, 0)$, whereas g is continuous at $(0, 0)$.

27. $\displaystyle\lim_{(x,\,y)\to(0,\,0)} (\sin x + \sin y) = 0$

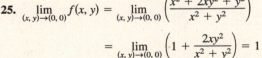

29. $\displaystyle\lim_{(x,\,y)\to(0,\,0)} \frac{x^2 y}{x^4 + 4y^2}$

Does not exist

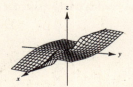

31. $f(x, y) = \dfrac{10xy}{2x^2 + 3y^2}$

The limit does not exist. Use the paths $x = 0$ and $x = y$.

33. $\displaystyle\lim_{(x, y)\to(0, 0)} \frac{\sin(x^2 + y^2)}{x^2 + y^2} = \lim_{r\to 0} \frac{\sin r^2}{r^2} = \lim_{r\to 0} \frac{2r \cos r^2}{2r} = \lim_{r\to 0} \cos r^2 = 1$

35. $\displaystyle\lim_{(x, y)\to(0, 0)} \frac{x^3 + y^3}{x^2 + y^2} = \lim_{r\to 0} \frac{r^3(\cos^3 \theta + \sin^3 \theta)}{r^2} = \lim_{r\to 0} r(\cos^3 \theta + \sin^3 \theta) = 0$

37. $f(x, y, z) = \dfrac{1}{\sqrt{x^2 + y^2 + z^2}}$

Continuous except at $(0, 0, 0)$

39. $f(x, y, z) = \dfrac{\sin z}{e^x + e^y}$

Continuous everywhere

41. $f(t) = t^2$

$g(x, y) = 3x - 2y$

$f(g(x, y)) = f(3x - 2y)$

$\qquad = (3x - 2y)^2$

$\qquad = 9x^2 - 12xy + 4y^2$

Continuous everywhere

43. $f(t) = \dfrac{1}{t}$

$g(x, y) = 3x - 2y$

$f(g(x, y)) = f(3x - 2y) = \dfrac{1}{3x - 2y}$

Continuous for $y \neq \dfrac{3x}{2}$

45. $f(x, y) = x^2 - 4y$

(a) $\displaystyle\lim_{\Delta x\to 0} \frac{f(x + \Delta x, y) - f(x, y)}{\Delta x} = \lim_{\Delta x\to 0} \frac{[(x + \Delta x)^2 - 4y] - (x^2 - 4y)}{\Delta x}$

$\qquad\qquad = \displaystyle\lim_{\Delta x\to 0} \frac{2x\Delta x - (\Delta x)^2}{\Delta x} = \lim_{\Delta x\to 0} (2x - \Delta x) = 2x$

(b) $\displaystyle\lim_{\Delta y\to 0} \frac{f(x, y + \Delta y) - f(x, y)}{\Delta y} = \lim_{\Delta y\to 0} \frac{[x^2 - 4(y + \Delta y)] - (x^2 - 4y)}{\Delta y}$

$\qquad\qquad = \displaystyle\lim_{\Delta y\to 0} \frac{-4\Delta y}{\Delta y} = \lim_{\Delta y\to 0} (-4) = -4$

47. $f(x, y) = 2x + xy - 3y$

(a) $\displaystyle\lim_{\Delta x\to 0} \frac{f(x + \Delta x, y) - f(x, y)}{\Delta x} = \lim_{\Delta x\to 0} \frac{[2(x + \Delta x) + (x + \Delta x)y - 3y] - (2x + xy - 3y)}{\Delta x}$

$\qquad\qquad = \displaystyle\lim_{\Delta x\to 0} \frac{2\Delta x + \Delta xy}{\Delta x} = \lim_{\Delta x\to 0} (2 + y) = 2 + y$

(b) $\displaystyle\lim_{\Delta y\to 0} \frac{f(x, y + \Delta y) - f(x, y)}{\Delta y} = \lim_{\Delta y\to 0} \frac{[2x + x(y + \Delta y) - 3(y + \Delta y)] - (2x + xy - 3y)}{\Delta y}$

$\qquad\qquad = \displaystyle\lim_{\Delta y\to 0} \frac{x\Delta y - 3\Delta y}{\Delta y} = \lim_{\Delta y\to 0} (x - 3) = x - 3$

49. See the definition on page 851.

Show that the value of $\displaystyle\lim_{(x, y)\to(x_0, y_0)} f(x, y)$ is not the same

for two different paths to (x_0, y_0).

51. No.

The existence of $f(2, 3)$ has no bearing on the existence of the limit as $(x, y)\to(2, 3)$.

53. Since $\displaystyle\lim_{(x, y)\to(a, b)} f(x, y) = L_1$, then for $\varepsilon/2 > 0$, there corresponds $\delta_1 > 0$ such that $|f(x, y) - L_1| < \varepsilon/2$ whenever

$$0 < \sqrt{(x - a)^2 + (y - b)^2} < \delta_1.$$

Since $\displaystyle\lim_{(x, y)\to(a, b)} g(x, y) = L_2$, then for $\varepsilon/2 > 0$, there corresponds $\delta_2 > 0$ such that $|g(x, y) - L_2| < \varepsilon/2$ whenever

$$0 < \sqrt{(x - a)^2 + (y - b)^2} < \delta_2.$$

Let δ be the smaller of δ_1 and δ_2. By the triangle inequality, whenever $\sqrt{(x - a)^2 + (y - b)^2} < \delta$, we have

$$|f(x, y) + g(x, y) - (L_1 + L_2)| = |(f(x, y) - L_1) + (g(x, y) - L_2)| \le |f(x, y) - L_1| + |g(x, y) - L_2| < \frac{\varepsilon}{2} + \frac{\varepsilon}{2} = \varepsilon.$$

Therefore, $\displaystyle\lim_{(x, y)\to(a, b)} [f(x, y) + g(x, y)] = L_1 + L_2$.

55. True. Assuming $f(x, 0)$ exists for $x \ne 0$.

57. False. Let

$$f(x, y) = \begin{cases} \ln(x^2 + y^2), & (x, y) \ne (0, 0) \\ 0, & x = 0, y = 0 \end{cases}$$

See Exercise 19.

Section 12.3   Partial Derivatives

1. $f_x(4, 1) < 0$

3. $f_y(4, 1) > 0$

5. $f(x, y) = 2x - 3y + 5$

$f_x(x, y) = 2$

$f_y(x, y) = -3$

7. $z = x\sqrt{y}$

$\dfrac{\partial z}{\partial x} = \sqrt{y}$

$\dfrac{\partial z}{\partial y} = \dfrac{x}{2\sqrt{y}}$

9. $z = x^2 - 5xy + 3y^2$

$\dfrac{\partial z}{\partial x} = 2x - 5y$

$\dfrac{\partial z}{\partial y} = -5x + 6y$

11. $z = x^2 e^{2y}$

$\dfrac{\partial z}{\partial x} = 2xe^{2y}$

$\dfrac{\partial z}{\partial y} = 2x^2 e^{2y}$

13. $z = \ln(x^2 + y^2)$

$\dfrac{\partial z}{\partial x} = \dfrac{2x}{x^2 + y^2}$

$\dfrac{\partial z}{\partial y} = \dfrac{2y}{x^2 + y^2}$

15. $z = \ln\left(\dfrac{x + y}{x - y}\right) = \ln(x + y) - \ln(x - y)$

$\dfrac{\partial z}{\partial x} = \dfrac{1}{x + y} - \dfrac{1}{x - y} = -\dfrac{2y}{x^2 - y^2}$

$\dfrac{\partial z}{\partial y} = \dfrac{1}{x + y} + \dfrac{1}{x - y} = \dfrac{2x}{x^2 - y^2}$

17. $z = \dfrac{x^2}{2y} + \dfrac{4y^2}{x}$

$\dfrac{\partial z}{\partial x} = \dfrac{2x}{2y} - \dfrac{4y^2}{x^2} = \dfrac{x^3 - 4y^3}{x^2 y}$

$\dfrac{\partial z}{\partial y} = -\dfrac{x^2}{2y^2} + \dfrac{8y}{x} = \dfrac{-x^3 + 16y^3}{2xy^2}$

19. $h(x, y) = e^{-(x^2 + y^2)}$

$h_x(x, y) = -2xe^{-(x^2 + y^2)}$

$h_y(x, y) = -2ye^{-(x^2 + y^2)}$

21. $f(x, y) = \sqrt{x^2 + y^2}$

$f_x(x, y) = \dfrac{1}{2}(x^2 + y^2)^{-1/2}(2x) = \dfrac{x}{\sqrt{x^2 + y^2}}$

$f_y(x, y) = \dfrac{1}{2}(x^2 + y^2)^{-1/2}(2y) = \dfrac{y}{\sqrt{x^2 + y^2}}$

23. $z = \tan(2x - y)$

$\dfrac{\partial z}{\partial x} = 2\sec^2(2x - y)$

$\dfrac{\partial z}{\partial y} = -\sec^2(2x - y)$

25. $z = e^y \sin xy$

$$\frac{\partial z}{\partial x} = ye^y \cos xy$$

$$\frac{\partial z}{\partial y} = e^y \sin xy + xe^y \cos xy$$

$$= e^y(x \cos xy + \sin xy)$$

27. $f(x, y) = \int_x^y (t^2 - 1)\, dt$

$$= \left[\frac{t^3}{3} - t\right]_x^y = \left(\frac{y^3}{3} - y\right) - \left(\frac{x^3}{3} - x\right)$$

$$f_x(x, y) = -x^2 + 1 = 1 - x^2$$

$$f_y(x, y) = y^2 - 1$$

[You could also use the Second Fundamental Theorem of Calculus.]

29. $f(x, y) = 2x + 3y$

$$\frac{\partial f}{\partial x} = \lim_{\Delta x \to 0} \frac{f(x + \Delta x, y) - f(x, y)}{\Delta x} = \lim_{\Delta x \to 0} \frac{2(x + \Delta x) + 3y - 2x - 3y}{\Delta x} = \lim_{\Delta x \to 0} \frac{2\Delta x}{\Delta x} = 2$$

$$\frac{\partial f}{\partial y} = \lim_{\Delta y \to 0} \frac{f(x, y + \Delta y) - f(x, y)}{\Delta y} = \lim_{\Delta y \to 0} \frac{2x + 3(y + \Delta y) - 2x - 3y}{\Delta y} = \lim_{\Delta y \to 0} \frac{3\Delta y}{\Delta y} = 3$$

31. $f(x, y) = \sqrt{x + y}$

$$\frac{\partial f}{\partial x} = \lim_{\Delta x \to 0} \frac{f(x + \Delta x, y) - f(x, y)}{\Delta x} = \lim_{\Delta x \to 0} \frac{\sqrt{x + \Delta x + y} - \sqrt{x + y}}{\Delta x}$$

$$= \lim_{\Delta x \to 0} \frac{\left(\sqrt{x + \Delta x + y} - \sqrt{x + y}\right)\left(\sqrt{x + \Delta x + y} + \sqrt{x + y}\right)}{\Delta x\left(\sqrt{x + \Delta x + y} + \sqrt{x + y}\right)}$$

$$= \lim_{\Delta x \to 0} \frac{1}{\sqrt{x + \Delta x + y} + \sqrt{x + y}} = \frac{1}{2\sqrt{x + y}}$$

$$\frac{\partial f}{\partial y} = \lim_{\Delta y \to 0} \frac{f(x, y + \Delta y) - f(x, y)}{\Delta y} = \lim_{\Delta y \to 0} \frac{\sqrt{x + y + \Delta y} - \sqrt{x + y}}{\Delta y}$$

$$= \lim_{\Delta y \to 0} \frac{\left(\sqrt{x + y + \Delta y} - \sqrt{x + y}\right)\left(\sqrt{x + y + \Delta y} + \sqrt{x + y}\right)}{\Delta y\left(\sqrt{x + y + \Delta y} + \sqrt{x + y}\right)}$$

$$= \lim_{\Delta y \to 0} \frac{1}{\sqrt{x + y + \Delta y} + \sqrt{x + y}} = \frac{1}{2\sqrt{x + y}}$$

33. $g(x, y) = 4 - x^2 - y^2$

$g_x(x, y) = -2x$

At $(1, 1)$: $g_x(1, 1) = -2$

$g_y(x, y) = -2y$

At $(1, 1)$: $g_y(1, 1) = -2$

35. $z = e^{-x} \cos y$

$$\frac{\partial z}{\partial x} = -e^{-x} \cos y$$

At $(0, 0)$: $\frac{\partial z}{\partial x} = -1$

$$\frac{\partial z}{\partial y} = -e^{-x} \sin y$$

At $(0, 0)$: $\frac{\partial z}{\partial y} = 0$

37. $f(x, y) = \arctan \dfrac{y}{x}$

$$f_x(x, y) = \frac{1}{1 + (y^2/x^2)}\left(-\frac{y}{x^2}\right) = \frac{-y}{x^2 + y^2}$$

At $(2, -2)$: $f_x(2, -2) = \dfrac{1}{4}$

$$f_y(x, y) = \frac{1}{1 + (y^2/x^2)}\left(\frac{1}{x}\right) = \frac{x}{x^2 + y^2}$$

At $(2, -2)$: $f_y(2, -2) = \dfrac{1}{4}$

39. $f(x, y) = \dfrac{xy}{x - y}$

$$f_x(x, y) = \frac{y(x - y) - xy}{(x - y)^2} = \frac{-y^2}{(x - y)^2}$$

At $(2, -2)$: $f_x(2, -2) = -\dfrac{1}{4}$

$$f_y(x, y) = \frac{x(x - y) + xy}{(x - y)^2} = \frac{x^2}{(x - y)^2}$$

At $(2, -2)$: $f_y(2, -2) = \dfrac{1}{4}$

41. $z = \sqrt{49 - x^2 - y^2}$, $x = 2$,

    $(2, 3, 6)$

    Intersecting curve: $z = \sqrt{45 - y^2}$

$$\frac{\partial z}{\partial y} = \frac{-y}{\sqrt{45 - y^2}}$$

At $(2, 3, 6)$: $\dfrac{\partial z}{\partial y} = \dfrac{-3}{\sqrt{45 - 9}} = -\dfrac{1}{2}$

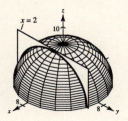

43. $z = 9x^2 - y^2$, $y = 3$, $(1, 3, 0)$

    Intersecting curve: $z = 9x^2 - 9$

$$\frac{\partial z}{\partial x} = 18x$$

At $(1, 3, 0)$: $\dfrac{\partial z}{\partial x} = 18(1) = 18$

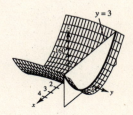

45. $f_x(x, y) = 2x + 4y - 4$, $f_y(x, y) = 4x + 2y + 16$

    $f_x = f_y = 0$: $2x + 4y = 4$

                    $4x + 2y = -16$

    Solving for x and y,

        $x = -6$ and $y = 4$.

47. $f_x(x, y) = -\dfrac{1}{x^2} + y$, $f_y(x, y) = -\dfrac{1}{y^2} + x$

    $f_x = f_y = 0$: $-\dfrac{1}{x^2} + y = 0$ and $-\dfrac{1}{y^2} + x = 0$

$$y = \frac{1}{x^2} \text{ and } x = \frac{1}{y^2}$$

    $y = y^4 \implies y = 1 = x$

    Points: $(1, 1)$

49. (a) The graph is that of f_y.

    (b) The graph is that of f_x.

51. $w = \sqrt{x^2 + y^2 + z^2}$

$$\frac{\partial w}{\partial x} = \frac{x}{\sqrt{x^2 + y^2 + z^2}}$$

$$\frac{\partial w}{\partial y} = \frac{y}{\sqrt{x^2 + y^2 + z^2}}$$

$$\frac{\partial w}{\partial z} = \frac{z}{\sqrt{x^2 + y^2 + z^2}}$$

53. $F(x, y, z) = \ln\sqrt{x^2 + y^2 + z^2}$

$$= \frac{1}{2}\ln(x^2 + y^2 + z^2)$$

$$F_x(x, y, z) = \frac{x}{x^2 + y^2 + z^2}$$

$$F_y(x, y, z) = \frac{y}{x^2 + y^2 + z^2}$$

$$F_z(x, y, z) = \frac{z}{x^2 + y^2 + z^2}$$

55. $H(x, y, z) = \sin(x + 2y + 3z)$

    $H_x(x, y, z) = \cos(x + 2y + 3z)$

    $H_y(x, y, z) = 2\cos(x + 2y + 3z)$

    $H_z(x, y, z) = 3\cos(x + 2y + 3z)$

57. $z = x^2 - 2xy + 3y^2$

$$\frac{\partial z}{\partial x} = 2x - 2y$$

$$\frac{\partial^2 z}{\partial x^2} = 2$$

$$\frac{\partial^2 z}{\partial y \partial x} = -2$$

$$\frac{\partial z}{\partial y} = -2x + 6y$$

$$\frac{\partial^2 z}{\partial y^2} = 6$$

$$\frac{\partial^2 z}{\partial x \partial y} = -2$$

59. $z = \sqrt{x^2 + y^2}$

$$\frac{\partial z}{\partial x} = \frac{x}{\sqrt{x^2 + y^2}}$$

$$\frac{\partial^2 z}{\partial x^2} = \frac{y^2}{(x^2 + y^2)^{3/2}}$$

$$\frac{\partial^2 z}{\partial y \partial x} = \frac{-xy}{(x^2 + y^2)^{3/2}}$$

$$\frac{\partial z}{\partial y} = \frac{y}{\sqrt{x^2 + y^2}}$$

$$\frac{\partial^2 z}{\partial y^2} = \frac{x^2}{(x^2 + y^2)^{3/2}}$$

$$\frac{\partial^2 z}{\partial x \partial y} = \frac{-xy}{(x^2 + y^2)^{3/2}}$$

61. $z = e^x \tan y$

$$\frac{\partial z}{\partial x} = e^x \tan y$$

$$\frac{\partial^2 z}{\partial x^2} = e^x \tan y$$

$$\frac{\partial^2 z}{\partial y \partial x} = e^x \sec^2 y$$

$$\frac{\partial z}{\partial y} = e^x \sec^2 y$$

$$\frac{\partial^2 z}{\partial y^2} = 2e^x \sec^2 y \tan y$$

$$\frac{\partial^2 z}{\partial x \partial y} = e^x \sec^2 y$$

63. $z = \arctan \dfrac{y}{x}$

$$\frac{\partial z}{\partial x} = \frac{1}{1 + (y^2/x^2)}\left(-\frac{y}{x^2}\right) = \frac{-y}{x^2 + y^2}$$

$$\frac{\partial^2 z}{\partial x^2} = \frac{2xy}{(x^2 + y^2)^2}$$

$$\frac{\partial^2 z}{\partial y \partial x} = \frac{-(x^2 + y^2) + y(2y)}{(x^2 + y^2)^2} = \frac{y^2 - x^2}{(x^2 + y^2)^2}$$

$$\frac{\partial z}{\partial y} = \frac{1}{1 + (y^2/x^2)}\left(\frac{1}{x}\right) = \frac{x}{x^2 + y^2}$$

$$\frac{\partial^2 z}{\partial y^2} = \frac{-2xy}{(x^2 + y^2)^2}$$

$$\frac{\partial^2 z}{\partial x \partial y} = \frac{(x^2 + y^2) - x(2x)}{(x^2 + y^2)^2} = \frac{y^2 - x^2}{(x^2 + y^2)^2}$$

65. $z = x \sec y$

$$\frac{\partial z}{\partial x} = \sec y$$

$$\frac{\partial^2 z}{\partial x^2} = 0$$

$$\frac{\partial^2 z}{\partial y \partial x} = \sec y \tan y$$

$$\frac{\partial z}{\partial y} = x \sec y \tan y$$

$$\frac{\partial^2 z}{\partial y^2} = x \sec y(\sec^2 y + \tan^2 y)$$

$$\frac{\partial^2 z}{\partial x \partial y} = \sec y \tan y$$

Therefore, $\dfrac{\partial^2 z}{\partial y \partial x} = \dfrac{\partial^2 z}{\partial x \partial y}$.

There are no points for which $z_x = 0 = z_y$, because

$$\frac{\partial z}{\partial x} = \sec y \neq 0.$$

67. $z = \ln\left(\dfrac{x}{x^2 + y^2}\right) = \ln x - \ln(x^2 + y^2)$

$$\frac{\partial z}{\partial x} = \frac{1}{x} - \frac{2x}{x^2 + y^2} = \frac{y^2 - x^2}{x(x^2 + y^2)}$$

$$\frac{\partial^2 z}{\partial x^2} = \frac{x^4 - 4x^2y^2 - y^4}{x^2(x^2 + y^2)^2}$$

$$\frac{\partial^2 z}{\partial y \partial x} = \frac{4xy}{(x^2 + y^2)^2}$$

$$\frac{\partial z}{\partial y} = -\frac{2y}{x^2 + y^2}$$

$$\frac{\partial^2 z}{\partial y^2} = \frac{2(y^2 - x^2)}{(x^2 + y^2)^2}$$

$$\frac{\partial^2 z}{\partial x \partial y} = \frac{4xy}{(x^2 + y^2)^2}$$

There are no points for which $z_x = z_y = 0$.

69. $f(x, y, z) = xyz$

$f_x(x, y, z) = yz$

$f_y(x, y, z) = xz$

$f_{yy}(x, y, z) = 0$

$f_{xy}(x, y, z) = z$

$f_{yx}(x, y, z) = z$

$f_{yyx}(x, y, z) = 0$

$f_{xyy}(x, y, z) = 0$

$f_{yxy}(x, y, z) = 0$

Therefore, $f_{xyy} = f_{yxy} = f_{yyx} = 0$.

71. $f(x, y, z) = e^{-x} \sin yz$

$f_x(x, y, z) = -e^{-x} \sin yz$

$f_y(x, y, z) = ze^{-x} \cos yz$

$f_{yy}(x, y, z) = -z^2e^{-x} \sin yz$

$f_{xy}(x, y, z) = -ze^{-x} \cos yz$

$f_{yx}(x, y, z) = -ze^{-x} \cos yz$

$f_{yyx}(x, y, z) = z^2e^{-x} \sin yz$

$f_{xyy}(x, y, z) = z^2e^{-x} \sin yz$

$f_{yxy}(x, y, z) = z^2e^{-x} \sin yz$

Therefore, $f_{xyy} = f_{yxy} = f_{yyx}$.

73. $z = 5xy$

$$\frac{\partial z}{\partial x} = 5y$$

$$\frac{\partial^2 z}{\partial x^2} = 0$$

$$\frac{\partial z}{\partial y} = 5x$$

$$\frac{\partial^2 z}{\partial y^2} = 0$$

Therefore, $\dfrac{\partial^2 z}{\partial x^2} + \dfrac{\partial^2 z}{\partial y^2} = 0 + 0 = 0.$

75. $z = e^x \sin y$

$$\frac{\partial z}{\partial x} = e^x \sin y$$

$$\frac{\partial^2 z}{\partial x^2} = e^x \sin y$$

$$\frac{\partial z}{\partial y} = e^x \cos y$$

$$\frac{\partial^2 z}{\partial y^2} = -e^x \sin y$$

Therefore, $\dfrac{\partial^2 z}{\partial x^2} + \dfrac{\partial^2 z}{\partial y^2} = e^x \sin y - e^x \sin y = 0.$

77. $z = \sin(x - ct)$

$$\frac{\partial z}{\partial t} = -c \cos(x - ct)$$

$$\frac{\partial^2 z}{\partial t^2} = -c^2 \sin(x - ct)$$

$$\frac{\partial z}{\partial x} = \cos(x - ct)$$

$$\frac{\partial^2 z}{\partial x^2} = -\sin(x - ct)$$

Therefore, $\dfrac{\partial^2 z}{\partial t^2} = c^2 \dfrac{\partial^2 z}{\partial x^2}.$

79. $z = e^{-t} \cos \dfrac{x}{c}$

$$\frac{\partial z}{\partial t} = -e^{-t} \cos \frac{x}{c}$$

$$\frac{\partial z}{\partial x} = -\frac{1}{c} e^{-t} \sin \frac{x}{c}$$

$$\frac{\partial^2 z}{\partial x^2} = -\frac{1}{c^2} e^{-t} \cos \frac{x}{c}$$

Therefore, $\dfrac{\partial z}{\partial t} = c^2 \dfrac{\partial^2 z}{\partial x^2}.$

81. See the definition on page 859.

83.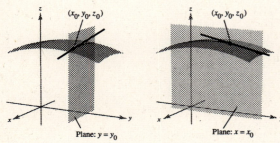

Plane: $y = y_0$ Plane: $x = x_0$

$\dfrac{\partial f}{\partial x}$ denotes the slope of the surface in the x-direction.

$\dfrac{\partial f}{\partial y}$ denotes the slope of the surface in the y-direction.

85. The plane $z = x + y = f(x, y)$ satisfies

$$\frac{\partial f}{\partial x} > 0 \text{ and } \frac{\partial f}{\partial y} > 0.$$

87. (a) $C = 32\sqrt{xy} + 175x + 205y + 1050$

$$\frac{\partial C}{\partial x} = 16\sqrt{\frac{y}{x}} + 175$$

$$\frac{\partial C}{\partial x}\bigg]_{(80,\, 20)} = 16\sqrt{\frac{1}{4}} + 175 = 183$$

$$\frac{\partial C}{\partial y} = 16\sqrt{\frac{x}{y}} + 205$$

$$\frac{\partial C}{\partial y}\bigg]_{(80,\, 20)} = 16\sqrt{4} + 205 = 237$$

(b) The fireplace-insert stove results in the cost increasing at a faster rate because

$$\frac{\partial C}{\partial y} > \frac{\partial C}{\partial x}.$$

89. An increase in either price will cause a decrease in demand.

91. $T = 500 - 0.6x^2 - 1.5y^2$

$$\frac{\partial T}{\partial x} = -1.2x, \frac{\partial T}{\partial x}(2, 3) = -2.4°/m$$

$$\frac{\partial T}{\partial y} = -3y = \frac{\partial T}{\partial y}(2, 3) = -9°/m$$

93.
$$PV = mRT$$

$$T = \frac{PV}{mR} \implies \frac{\partial T}{\partial P} = \frac{V}{mR}$$

$$P = \frac{mRT}{V} \implies \frac{\partial P}{\partial V} = -\frac{mRT}{V^2}$$

$$V = \frac{mRT}{P} \implies \frac{\partial V}{\partial T} = \frac{mR}{P}$$

$$\frac{\partial T}{\partial P} \cdot \frac{\partial P}{\partial V} \cdot \frac{\partial V}{\partial T} = \left(\frac{V}{mR}\right)\left(-\frac{mRT}{V^2}\right)\left(\frac{mR}{P}\right)$$

$$= -\frac{mRT}{VP} = -\frac{mRT}{mRT} = -1$$

95. (a) $\frac{\partial z}{\partial x} = -1.83$

$$\frac{\partial z}{\partial y} = -1.09$$

(b) As the consumption of skim milk (x) increases, the consumption of whole milk (z) decreases.

Similarly, as the consumption of reduced-fat milk (y) increases, the consumption of whole milk (z) decreases.

97. $f(x, y) = \begin{cases} \dfrac{xy(x^2 - y^2)}{x^2 + y^2}, & (x, y) \neq (0, 0) \\ 0, & (x, y) = (0, 0) \end{cases}$

(a) $f_x(x, y) = \dfrac{(x^2 + y^2)(3x^2y - y^3) - (x^3y - xy^3)(2x)}{(x^2 + y^2)^2} = \dfrac{y(x^4 + 4x^2y^2 - y^4)}{(x^2 + y^2)^2}$

$f_y(x, y) = \dfrac{(x^2 + y^2)(x^3 - 3xy^2) - (x^3y - xy^3)(2y)}{(x^2 + y^2)^2} = \dfrac{x(x^4 - 4x^2y^2 - y^4)}{(x^2 + y^2)^2}$

(b) $f_x(0, 0) = \lim_{\Delta x \to 0} \dfrac{f(\Delta x, 0) - f(0, 0)}{\Delta x} = \lim_{\Delta x \to 0} \dfrac{0/[(\Delta x)^2] - 0}{\Delta x} = 0$

$f_y(0, 0) = \lim_{\Delta y \to 0} \dfrac{f(0, \Delta y) - f(0, 0)}{\Delta y} = \lim_{\Delta y \to 0} \dfrac{0/[(\Delta y)^2] - 0}{\Delta y} = 0$

(c) $f_{xy}(0, 0) = \dfrac{\partial}{\partial y}\left(\dfrac{\partial f}{\partial x}\right)\Big|_{(0, 0)} = \lim_{\Delta y \to 0} \dfrac{f_x(0, \Delta y) - f_x(0, 0)}{\Delta y} = \lim_{\Delta y \to 0} \dfrac{\Delta y(-(\Delta y)^4)}{((\Delta y)^2)^2(\Delta y)} = \lim_{\Delta y \to 0}(-1) = -1$

$f_{yx}(0, 0) = \dfrac{\partial}{\partial x}\left(\dfrac{\partial f}{\partial y}\right)\Big|_{(0, 0)} = \lim_{\Delta x \to 0} \dfrac{f_y(\Delta x, 0) - f_y(0, 0)}{\Delta x} = \lim_{\Delta x \to 0} \dfrac{\Delta x((\Delta x)^4)}{((\Delta x)^2)^2(\Delta x)} = \lim_{\Delta x \to 0} 1 = 1$

(d) f_{yx} or f_{xy} or both are not continuous at $(0, 0)$.

99. True

101. True

Section 12.4   Differentials

1. $z = 3x^2y^3$

$dz = 6xy^3\, dx + 9x^2y^2\, dy$

3. $z = \dfrac{-1}{x^2 + y^2}$

$dz = \dfrac{2x}{(x^2 + y^2)^2}\, dx + \dfrac{2y}{(x^2 + y^2)^2}\, dy$

$= \dfrac{2}{(x^2 + y^2)^2}(x\, dx + y\, dy)$

5. $z = x \cos y - y \cos x$

$dz = (\cos y + y \sin x)\,dx + (-x \sin y - \cos x)\,dy = (\cos y + y \sin x)\,dx - (x \sin y + \cos x)\,dy$

7. $z = e^x \sin y$

$dz = (e^x \sin y)\,dx + (e^x \cos y)\,dy$

9. $w = 2z^3 y \sin x$

$dw = 2z^3 y \cos x\,dx + 2z^3 \sin x\,dy + 6z^2 y \sin x\,dz$

11. (a) $f(1, 2) = 4$

$f(1.05, 2.1) = 3.4875$

$\Delta z = f(1.05, 2.1) - f(1, 2) = -0.5125$

(b) $dz = -2x\,dx - 2y\,dy$

$= -2(0.05) - 4(0.1) = -0.5$

13. (a) $f(1, 2) = \sin 2$

$f(1.05, 2.1) = 1.05 \sin 2.1$

$\Delta z = f(1.05, 2.1) - f(1, 2) \approx -0.00293$

(b) $dz = \sin y\,dx + x \cos y\,dy$

$= (\sin 2)(0.05) + (\cos 2)(0.1) \approx 0.00385$

15. (a) $f(1, 2) = -5$

$f(1.05, 2.1) = -5.25$

$\Delta z = -0.25$

(b) $dz = 3\,dx - 4\,dy$

$= 3(0.05) - 4(0.1) = -0.25$

17. Let $z = \sqrt{x^2 + y^2},\ x = 5,\ y = 3,\ dx = 0.05,\ dy = 0.1$. Then: $dz = \dfrac{x}{\sqrt{x^2 + y^2}}\,dx + \dfrac{y}{\sqrt{x^2 + y^2}}\,dy$

$\sqrt{(5.05)^2 + (3.1)^2} - \sqrt{5^2 + 3^2} \approx \dfrac{5}{\sqrt{5^2 + 3^2}}(0.05) + \dfrac{3}{\sqrt{5^2 + 3^2}}(0.1) = \dfrac{0.55}{\sqrt{34}} \approx 0.094$

19. Let $z = (1 - x^2)/y^2,\ x = 3,\ y = 6,\ dx = 0.05,\ dy = -0.05$. Then: $dz = -\dfrac{2x}{y^2}\,dx + \dfrac{-2(1 - x^2)}{y^3}\,dy$

$\dfrac{1 - (3.05)^2}{(5.95)^2} - \dfrac{1 - 3^2}{6^2} \approx -\dfrac{2(3)}{6^2}(0.05) - \dfrac{2(1 - 3^2)}{6^3}(-0.05) \approx -0.012$

21. See the definition on page 869.

23. The tangent plane to the surface $z = f(x, y)$ at the point P is a linear approximation of z.

25. $A = lh$

$dA = l\,dh + h\,dl$

$\Delta A = (l + dl)(h + dl) - lh$

$= h\,dl + l\,dh + dl\,dh$

$\Delta A - dA = dl\,dh$

27. $V = \dfrac{\pi r^2 h}{3}$

$r = 3$

$h = 6$

$dV = \dfrac{2\pi rh}{3}\,dr + \dfrac{\pi r^2}{3}\,dh = \dfrac{\pi r}{3}(2h\,dr + r\,dh)$

$\Delta V = \dfrac{\pi}{3}[(r + \Delta r)^2(h + \Delta h) - r^2 h]$

$= \dfrac{\pi}{3}[(3 + \Delta r)^2(6 + \Delta h) - 54]$

Δr	Δh	dV	ΔV	$\Delta V - dV$
0.1	0.1	4.7124	4.8391	0.1267
0.1	-0.1	2.8274	2.8264	-0.0010
0.001	0.002	0.0565	0.0566	0.0001
-0.0001	0.0002	-0.0019	-0.0019	0.0000

29. (a) $dz = -1.83 \, dx - 1.09 \, dy$

(b) $dz = \dfrac{\partial z}{\partial x} dx + \dfrac{\partial z}{\partial y} dy$

$= -1.83(\pm 0.25) + (-1.09)(\pm 0.25)$

$= \pm 0.73$

Maximum propagated error: ± 0.73

Relative error: $\dfrac{dz}{z} = \dfrac{\pm 0.73}{(-1.83)(7.2) - 1.09(8.5) + 28.7} = \dfrac{\pm 0.73}{6.259} \approx \pm 0.1166 = 11.67\%$

31. $V = \pi r^2 h \implies dV = (2\pi rh) \, dr + (\pi r^2) \, dh$

$\dfrac{dV}{V} = 2\dfrac{dr}{r} + \dfrac{dh}{h}$

$= 2(0.04) + (0.02) = 0.10 = 10\%$

33. $A = \frac{1}{2} ab \sin C$

$dA = \frac{1}{2}[(b \sin C) \, da + (a \sin C) \, db + (ab \cos C) \, dC]$

$= \frac{1}{2}[4(\sin 45°)(\pm \frac{1}{16}) + 3(\sin 45°)(\pm \frac{1}{16}) + 12(\cos 45°)(\pm 0.02)] \approx \pm 0.24 \text{ in.}^2$

35. (a) $V = \dfrac{1}{2} bhl$

$= \left(18 \sin \dfrac{\theta}{2}\right)\left(18 \cos \dfrac{\theta}{2}\right)(16)(12)$

$= 31{,}104 \sin \theta \text{ in.}^3$

$= 18 \sin \theta \text{ ft}^3$

V is maximum when $\sin \theta = 1$ or $\theta = \pi/2$.

(b) $V = \dfrac{s^2}{2}(\sin \theta)l$

$dV = s(\sin \theta)l \, ds + \dfrac{s^2}{2} l(\cos \theta) \, d\theta + \dfrac{s^2}{2}(\sin \theta) \, dl$

$= 18\left(\sin \dfrac{\pi}{2}\right)(16)(12)\left(\dfrac{1}{2}\right) + \dfrac{18^2}{2}(16)(12)\left(\cos \dfrac{\pi}{2}\right)\left(\dfrac{\pi}{90}\right) + \dfrac{18^2}{2}\left(\sin \dfrac{\pi}{2}\right)\left(\dfrac{1}{2}\right)$

$= 1809 \text{ in}^3 \approx 1.047 \text{ ft}^3$

37. $P = \dfrac{E^2}{R}$

$dP = \dfrac{2E}{R} dE - \dfrac{E^2}{R^2} dR$

$\dfrac{dP}{P} = \dfrac{\dfrac{2E}{R} dE - \dfrac{E^2}{R^2} dR}{P} = 2\dfrac{dE}{E} - \dfrac{dR}{R} = 2(0.02) - (-0.03) = 0.07 = 7\%$

39. $L = 0.00021 \left(\ln \dfrac{2h}{r} - 0.75 \right)$

$dL = 0.00021 \left[\dfrac{dh}{h} - \dfrac{dr}{r} \right] = 0.00021 \left[\dfrac{(\pm 1/100)}{100} - \dfrac{(\pm 1/16)}{2} \right] \approx (\pm 6.6) \times 10^{-6}$

$L = 0.00021(\ln 100 - 0.75) \pm dL \approx 8.096 \times 10^{-4} \pm dL = 8.096 \times 10^{-4} \pm 6.6 \times 10^{-6}$ micro–henrys

41. $z = f(x, y) = x^2 - 2x + y$

$\Delta z = f(x + \Delta x, y + \Delta y) - f(x, y)$

$\quad = (x^2 + 2x(\Delta x) + (\Delta x)^2 - 2x - 2(\Delta x) + y + (\Delta y)) - (x^2 - 2x + y)$

$\quad = 2x(\Delta x) + (\Delta x)^2 - 2(\Delta x) + (\Delta y)$

$\quad = (2x - 2)\,\Delta x + \Delta y + \Delta x(\Delta x) + 0(\Delta y)$

$\quad = f_x(x, y)\,\Delta x + f_y(x, y)\,\Delta y + \epsilon_1 \Delta x + \epsilon_2 \Delta y$ where $\epsilon_1 = \Delta x$ and $\epsilon_2 = 0$.

As $(\Delta x, \Delta y) \to (0, 0)$, $\epsilon_1 \to 0$ and $\epsilon_2 \to 0$.

43. $z = f(x, y) = x^2 y$

$\Delta z = f(x + \Delta x, \, y + \Delta y) - f(x, y)$

$\quad = (x^2 + 2x(\Delta x) + (\Delta x)^2)(y + \Delta y) - x^2 y$

$\quad = 2xy(\Delta x) + y(\Delta x)^2 + x^2 \Delta y + 2x(\Delta x)(\Delta y) + (\Delta x)^2 \, \Delta y$

$\quad = 2xy(\Delta x) + x^2 \Delta y + (y\Delta x)\,\Delta x + [2x\Delta x + (\Delta x)^2]\,\Delta y$

$\quad = f_x(x, y)\,\Delta x + f_y(x, y)\,\Delta y + \epsilon_1 \Delta x + \epsilon_2 \Delta y$ where $\epsilon_1 = y(\Delta x)$ and $\epsilon_2 = 2x\Delta x + (\Delta x)^2$.

As $(\Delta x, \Delta y) \to (0, 0)$, $\epsilon_1 \to 0$ and $\epsilon_2 \to 0$.

45. $f(x, y) = \begin{cases} \dfrac{3x^2 y}{x^4 + y^2}, & (x, y) \neq (0, 0) \\[2mm] 0, & (x, y) = (0, 0) \end{cases}$

(a) $f_x(0, 0) = \lim\limits_{\Delta x \to 0} \dfrac{f(\Delta x, 0) - f(0, 0)}{\Delta x} = \lim\limits_{\Delta x \to 0} \dfrac{\dfrac{0}{(\Delta x)^4} - 0}{\Delta x} = 0$

$f_y(0, 0) = \lim\limits_{\Delta y \to 0} \dfrac{f(0, \Delta y) - f(0, 0)}{\Delta y} = \lim\limits_{\Delta y \to 0} \dfrac{\dfrac{0}{(\Delta y)^2} - 0}{\Delta y} = 0$

Thus, the partial derivatives exist at $(0, 0)$.

(b) Along the line $y = x$: $\lim\limits_{(x, y) \to (0, 0)} f(x, y) = \lim\limits_{x \to 0} \dfrac{3x^3}{x^4 + x^2} = \lim\limits_{x \to 0} \dfrac{3x}{x^2 + 1} = 0$

Along the curve $y = x^2$: $\lim\limits_{(x, y) \to (0, 0)} f(x, y) = \dfrac{3x^4}{2x^4} = \dfrac{3}{2}$

f is not continuous at $(0, 0)$. Therefore, f is not differentiable at $(0, 0)$. (See Theroem 12.5)

47. Essay. For example, we can use the equation $F = ma$:

$dF = \dfrac{\partial F}{\partial m}\, dm + \dfrac{\partial F}{\partial a}\, da = a\, dm + m\, da.$

Section 12.5   Chain Rules for Functions of Several Variables

1. $w = x^2 + y^2$

$x = e^t$

$y = e^{-t}$

$\dfrac{dw}{dt} = \dfrac{\partial w}{\partial x}\dfrac{dx}{dt} + \dfrac{\partial w}{\partial y}\dfrac{dy}{dt} = 2xe^t + 2y(-e^{-t}) = 2(e^{2t} - e^{-2t})$

3. $w = x \sec y$

$x = e^t$

$y = \pi - t$

$\dfrac{dw}{dt} = \dfrac{\partial w}{\partial x}\dfrac{dx}{dt} + \dfrac{\partial w}{\partial y}\dfrac{dy}{dt} = (\sec y)(e^t) + (x \sec y \tan y)(-1)$

$= e^t \sec(\pi - t)[1 - \tan(\pi - t)]$

$= -e^t (\sec t + \sec t \tan t)$

5. $w = xy,\ x = 2 \sin t,\ y = \cos t$

(a) $\dfrac{dw}{dt} = \dfrac{\partial w}{\partial x}\dfrac{dx}{dt} + \dfrac{\partial w}{\partial y}\dfrac{dy}{dt} = 2y \cos t + x(-\sin t) = 2y \cos t - x \sin t$

$= 2(\cos^2 t - \sin^2 t) = 2 \cos 2t$

(b) $w = 2 \sin t \cos t = \sin 2t,\ \dfrac{dw}{dt} = 2 \cos 2t$

7. $w = x^2 + y^2 + z^2$

$x = e^t \cos t$

$y = e^t \sin t$

$z = e^t$

(a) $\dfrac{dw}{dt} = \dfrac{\partial w}{\partial x}\dfrac{dx}{dt} + \dfrac{\partial w}{\partial y}\dfrac{dy}{dt} + \dfrac{\partial w}{\partial z}\dfrac{dz}{dt} = 2x(-e^t \sin t + e^t \cos t) + 2y(e^t \cos t + e^t \sin t) + 2ze^t = 4e^{2t}$

(b) $w = (e^t \cos t)^2 + (e^t \sin t)^2 + (e^t)^2 = 2e^{2t},\ \dfrac{dw}{dt} = 4e^{2t}$

9. $w = xy + xz + yz,\ x = t - 1,\ y = t^2 - 1,\ z = t$

(a) $\dfrac{dw}{dt} = \dfrac{\partial w}{\partial x}\dfrac{dx}{dt} + \dfrac{\partial w}{\partial y}\dfrac{dy}{dt} + \dfrac{\partial w}{\partial z}\dfrac{dz}{dt} = (y + z) + (x + z)(2t) + (x + y)$

$= (t^2 - 1 + t) + (t - 1 + t)(2t) + (t - 1 + t^2 - 1) = 3(2t^2 - 1)$

(b) $w = (t - 1)(t^2 - 1) + (t - 1)t + (t^2 - 1)t$

$\dfrac{dw}{dt} = 2t(t - 1) + (t^2 - 1) + 2t - 1 + 3t^2 - 1 = 3(2t^2 - 1)$

11. Distance $= f(t) = \sqrt{(x_1 - x_2)^2 + (y_1 - y_2)^2} = \sqrt{(10 \cos 2t - 7 \cos t)^2 + (6 \sin 2t - 4 \sin t)^2}$

$f'(t) = \dfrac{1}{2}[(10 \cos 2t - 7 \cos t)^2 + (6 \sin 2t - 4 \sin t)^2]^{-1/2}$

$[[2(10 \cos 2t - 7 \cos t)(-20 \sin 2t + 7 \sin t)] + [2(6 \sin 2t - 4 \sin t)(12 \cos 2t - 4 \cos t)]]$

$f'\left(\dfrac{\pi}{2}\right) = \dfrac{1}{2}[(-10)^2 + 4^2]^{-1/2}[[2(-10)(7)] + (2(-4)(-12)]$

$= \dfrac{1}{2}(116)^{-1/2}(-44) = \dfrac{-22}{2\sqrt{29}} = \dfrac{-11\sqrt{29}}{29} \approx -2.04$

13. $w = \arctan(2xy), \; x = \cos t, \; y = \sin t, \; t = 0$

$$\frac{dw}{dt} = \frac{\partial w}{\partial x}\frac{dx}{dt} + \frac{\partial w}{\partial y}\frac{dy}{dt}$$

$$= \frac{2y}{1 + (4x^2y^2)}(-\sin t) + \frac{2x}{1 + (4x^2y^2)}(\cos t)$$

$$= \frac{2\sin t}{1 + 4\cos^2 t \sin^2 t}(-\sin t) + \frac{2\cos t}{1 + 4\cos^2 t \sin^2 t}(\cos t)$$

$$= \frac{2\cos^2 t - 2\sin^2 t}{1 + 4\cos^2 t \sin^2 t}$$

$$\frac{d^2w}{dt^2} = \frac{(1 + 4\cos^2 t \sin^2 t)(-8\cos t \sin t) - (2\cos^2 t - 2\sin^2 t)(8\cos^3 t \sin t - 8\sin^3 t \cos t)}{(1 + 4\cos^2 t \sin^2 t)^2}$$

$$= \frac{-8\cos t \sin t(1 + 2\sin^4 t + 2\cos^4 t)}{(1 + 4\cos^2 t \sin^2 t)^2}$$

At $t = 0, \dfrac{d^2w}{dt^2} = 0.$

15. $w = x^2 + y^2$

$x = s + t$

$y = s - t$

$\dfrac{\partial w}{\partial s} = 2x + 2y = 2(x + y) = 4s$

$\dfrac{\partial w}{\partial t} = 2x + 2y(-1) = 2(x - y) = 4t$

When $s = 2$ and $t = -1,$

$\dfrac{\partial w}{\partial s} = 8$ and $\dfrac{\partial w}{\partial t} = -4.$

17. $w = x^2 - y^2$

$x = s \cos t$

$y = s \sin t$

$\dfrac{\partial w}{\partial s} = 2x \cos t - 2y \sin t$

$\quad = 2s \cos^2 t - 2s \sin^2 t = 2s \cos 2t$

$\dfrac{\partial w}{\partial t} = 2x(-s \sin t) - 2y(s \cos t) = -2s^2 \sin 2t$

When $s = 3$ and $t = \dfrac{\pi}{4}, \dfrac{\partial w}{\partial s} = 0$ and $\dfrac{\partial w}{\partial t} = -18.$

19. $w = x^2 - 2xy + y^2, \; x = r + \theta, \; y = r - \theta$

(a) $\dfrac{\partial w}{\partial r} = (2x - 2y)(1) + (-2x + 2y)(1) = 0$

$\dfrac{\partial w}{\partial \theta} = (2x - 2y)(1) + (-2x + 2y)(-1)$

$\quad = 4x - 4y = 4(x - y)$

$\quad = 4[(r + \theta) - (r - \theta)] = 8\theta$

(b) $w = (r + \theta)^2 - 2(r + \theta)(r - \theta) + (r - \theta)^2$

$\quad = (r^2 + 2r\theta + \theta^2) - 2(r^2 - \theta^2) + (r^2 - 2r\theta + \theta^2)$

$\quad = 4\theta^2$

$\dfrac{\partial w}{\partial r} = 0$

$\dfrac{\partial w}{\partial \theta} = 8\theta$

21. $w = \arctan \dfrac{y}{x}, \ x = r \cos \theta, \ y = r \sin \theta$

(a) $\dfrac{\partial w}{\partial r} = \dfrac{-y}{x^2 + y^2} \cos \theta + \dfrac{x}{x^2 + y^2} \sin \theta = \dfrac{-r \sin \theta \cos \theta}{r^2} + \dfrac{r \cos \theta \sin \theta}{r^2} = 0$

$\dfrac{\partial w}{\partial \theta} = \dfrac{-y}{x^2 + y^2}(-r \sin \theta) + \dfrac{x}{x^2 + y^2}(r \cos \theta) = \dfrac{-(r \sin \theta)(-r \sin \theta)}{r^2} + \dfrac{(r \cos \theta)(r \cos \theta)}{r^2} = 1$

(b) $w = \arctan \dfrac{r \sin \theta}{r \cos \theta} = \arctan(\tan \theta) = \theta$

$\dfrac{\partial w}{\partial r} = 0$

$\dfrac{\partial w}{\partial \theta} = 1$

23. $w = xyz, \ x = s + t, \ y = s - t, \ z = st^2$

$\dfrac{\partial w}{\partial s} = yz(1) + xz(1) + xy(t^2)$

$= (s - t)st^2 + (s + t)st^2 + (s + t)(s - t)t^2$

$= 2s^2t^2 + s^2t^2 - t^4 = 3s^2t^2 - t^4 = t^2(3s^2 - t^2)$

$\dfrac{\partial w}{\partial t} = yz(1) + xz(-1) + xy(2st)$

$= (s - t)st^2 - (s + t)st^2 + (s + t)(s - t)(2st)$

$= -2st^3 + 2s^3t - 2st^3 = 2s^3t - 4st^3 = 2st(s^2 - 2t^2)$

25. $w = ze^{x/y}, \ x = s - t, \ y = s + t, \ z = st$

$\dfrac{\partial w}{\partial s} = \dfrac{z}{y}e^{x/y}(1) + \dfrac{-zx}{y^2}e^{x/y}(1) + e^{x/y}(t)$

$= e^{(s-t)/(s+t)}\left[\dfrac{st}{s + t} - \dfrac{(s - t)st}{(s + t)^2} + t \right]$

$= e^{(s-t)/(s+t)}\left[\dfrac{st(s + t) - s^2t + st^2 + t(s + t)^2}{(s + t)^2} \right]$

$= e^{(s-t)/(s+t)}\dfrac{t(s^2 + 4st + t^2)}{(s + t)^2}$

$\dfrac{\partial w}{\partial t} = \dfrac{z}{y}e^{x/y}(-1) + \dfrac{-zx}{y^2}e^{x/y}(1) + e^{x/y}(s)$

$= e^{(s-t)/(s+t)}\left[-\dfrac{st}{s + t} - \dfrac{st(s - t)}{(s + t)^2} + s \right]$

$= e^{(s-t)/(s+t)}\left[\dfrac{-st(s + t) - st(s - t) + s(s + t)^2}{(s + t)^2} \right]$

$= e^{(s-t)/(s+t)}\dfrac{s(s^2 + t^2)}{(s + t)^2}$

27. $x^2 - 3xy + y^2 - 2x + y - 5 = 0$

$\dfrac{dy}{dx} = -\dfrac{F_x(x, y)}{F_y(x, y)} = -\dfrac{2x - 3y - 2}{-3x + 2y + 1}$

$= \dfrac{3y - 2x + 2}{2y - 3x + 1}$

29. $\ln \sqrt{x^2 + y^2} + xy = 4$

$\dfrac{1}{2}\ln(x^2 + y^2) + xy - 4 = 0$

$\dfrac{dy}{dx} = -\dfrac{F_x(x, y)}{F_y(x, y)} = -\dfrac{\dfrac{x}{x^2 + y^2} + y}{\dfrac{y}{x^2 + y^2} + x} = -\dfrac{x + x^2y + y^3}{y + xy^2 + x^3}$

31. $F(x, y, z) = x^2 + y^2 + z^2 - 25$

$F_x = 2x$

$F_y = 2y$

$F_z = 2z$

$\dfrac{\partial z}{\partial x} = -\dfrac{F_x}{F_z} = -\dfrac{x}{z}$

$\dfrac{\partial z}{\partial y} = -\dfrac{F_y}{F_z} = -\dfrac{y}{z}$

33. $F(x, y, z) = \tan(x + y) + \tan(y + z) - 1$

$F_x = \sec^2(x + y)$

$F_y = \sec^2(x + y) + \sec^2(y + z)$

$F_z = \sec^2(y + z)$

$\dfrac{\partial z}{\partial x} = -\dfrac{F_x}{F_z} = -\dfrac{\sec^2(x + y)}{\sec^2(y + z)}$

$\dfrac{\partial z}{\partial y} = -\dfrac{F_y}{F_z} = -\dfrac{\sec^2(x + y) + \sec^2(y + z)}{\sec^2(y + z)}$

$= -\left(\dfrac{\sec^2(x + y)}{\sec^2(y + z)} + 1 \right)$

35. $F(x, y, z) = x^2 + 2yz + z^2 - 1$

$F_x = 2x$

$F_y = 2z$

$F_z = 2y + 2z$

(i) $\dfrac{\partial z}{\partial x} = -\dfrac{F_x}{F_z} = -\dfrac{2x}{2y + 2z} = -\dfrac{x}{y + z}$

(ii) $\dfrac{\partial z}{\partial y} = -\dfrac{F_y}{F_z} = -\dfrac{2z}{2y + 2z} = -\dfrac{z}{y + z}$

37. $F(x, y, z) = e^{xz} + xy$

$\dfrac{\partial z}{\partial x} = -\dfrac{F_x(x, y, z)}{F_z(x, y, z)} = -\dfrac{ze^{xz} + y}{xe^{xz}}$

$\dfrac{\partial z}{\partial y} = -\dfrac{F_y(x, y, z)}{F_z(x, y, z)} = \dfrac{-x}{xe^{xz}} = \dfrac{-1}{e^{xz}} = -e^{-xz}$

39. $F(x, y, z, w) = xyz + xzw - yzw + w^2 - 5$

$F_x = yz + zw$

$F_y = xz - zw$

$F_z = xy + xw - yw$

$F_w = xz - yz + 2w$

$\dfrac{\partial w}{\partial x} = -\dfrac{F_x}{F_w} = -\dfrac{z(y + w)}{xz - yz + 2w}$

$\dfrac{\partial w}{\partial y} = -\dfrac{F_y}{F_w} = -\dfrac{z(x - w)}{xz - yz + 2w}$

$\dfrac{\partial w}{\partial z} = -\dfrac{F_z}{F_w} = -\dfrac{xy + xw - yw}{xz - yz + 2w}$

41. $F(x, y, z, w) = \cos xy + \sin yz + wz - 20$

$\dfrac{\partial w}{\partial x} = \dfrac{-F_x}{F_w} = \dfrac{y \sin xy}{z}$

$\dfrac{\partial w}{\partial y} = \dfrac{-F_y}{F_w} = \dfrac{x \sin xy - z \cos yz}{z}$

$\dfrac{\partial w}{\partial z} = \dfrac{-F_z}{F_w} = -\dfrac{y \cos zy + w}{z}$

43. $f(x, y) = \dfrac{xy}{\sqrt{x^2 + y^2}}$

$f(tx, ty) = \dfrac{(tx)(ty)}{\sqrt{(tx)^2 + (ty)^2}} = t\left(\dfrac{xy}{\sqrt{x^2 + y^2}}\right) = tf(x, y)$

Degree: 1

$xf_x(x, y) + yf_y(x, y) = x\left(\dfrac{y^3}{(x^2 + y^2)^{3/2}}\right) + y\left(\dfrac{x^3}{(x^2 + y^2)^{3/2}}\right)$

$= \dfrac{xy}{\sqrt{x^2 + y^2}} = 1f(x, y)$

45. $f(x, y) = e^{x/y}$

$f(tx, ty) = e^{tx/ty} = e^{x/y} = f(x, y)$

Degree: 0

$xf_x(x, y) + yf_y(x, y) = x\left(\dfrac{1}{y}e^{x/y}\right) + y\left(-\dfrac{x}{y^2}e^{x/y}\right) = 0$

47. $\dfrac{dw}{dt} = \dfrac{\partial w}{\partial x}\dfrac{dx}{dt} + \dfrac{\partial w}{\partial y}\dfrac{dy}{dt}$ (Page 876)

49. $w = f(x, y)$ is the explicit form of a function of two variables, as in $z = x^2 + y^2$.
The implicit form is of the form $F(x, y, z) = 0$, as in $z - x^2 - y^2 = 0$.

51. $A = \dfrac{1}{2}bh = \left(x \sin \dfrac{\theta}{2}\right)\left(x \cos \dfrac{\theta}{2}\right) = \dfrac{x^2}{2} \sin \theta$

$\dfrac{dA}{dt} = \dfrac{\partial A}{\partial x}\dfrac{dx}{dt} + \dfrac{\partial A}{\partial \theta}\dfrac{d\theta}{dt} = x \sin \theta \dfrac{dx}{dt} + \dfrac{x^2}{2} \cos \theta \dfrac{d\theta}{dt}$

$= 6\left(\sin \dfrac{\pi}{4}\right)\left(\dfrac{1}{2}\right) + \dfrac{6^2}{2}\left(\cos \dfrac{\pi}{4}\right)\left(\dfrac{\pi}{90}\right) = \dfrac{3\sqrt{2}}{2} + \dfrac{\pi\sqrt{2}}{10} \text{ m}^2/\text{hr}$

$\approx 2.566 \text{ m}^2/\text{hr}$

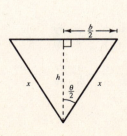

53. (a) $V = \dfrac{1}{3}\pi r^2 h$

$\dfrac{dV}{dt} = \dfrac{1}{3}\pi\left(2rh\dfrac{dr}{dt} + r^2\dfrac{dh}{dt}\right) = \dfrac{1}{3}\pi[2(12)(36)(6) + (12)^2(-4)] = 1536\pi$ in.^{3}/min

(b) $S = \pi r\sqrt{r^2 + h^2} + \pi r^2$ (Surface area includes base.)

$\dfrac{dS}{dt} = \pi\left[\left(\sqrt{r^2 + h^2} + \dfrac{r^2}{\sqrt{r^2 + h^2}} + 2r\right)\dfrac{dr}{dt} + \dfrac{rh}{\sqrt{r^2 + h^2}}\dfrac{dh}{dt}\right]$

$\qquad = \pi\left[\left(\sqrt{12^2 + 36^2} + \dfrac{144}{\sqrt{12^2 + 36^2}} + 2(12)\right)(6) + \dfrac{36(12)}{\sqrt{12^2 + 36^2}}(-4)\right]$

$\qquad = \pi\left[\left(12\sqrt{10} + \dfrac{12}{\sqrt{10}}\right)(6) + 144 + \dfrac{36}{\sqrt{10}}(-4)\right]$

$\qquad = \dfrac{648\pi}{\sqrt{10}} + 144\pi$ in.^{2}/min $= \dfrac{36\pi}{5}\left(20 + 9\sqrt{10}\right)$ in.^{2}/min

55. $I = \dfrac{1}{2}m(r_1{}^2 + r_2{}^2)$

$\dfrac{dI}{dt} = \dfrac{1}{2}m\left[2r_1\dfrac{dr_1}{dt} + 2r_2\dfrac{dr_2}{dt}\right] = m[(6)(2) + (8)(2)] = 28m$ cm^2/sec

57. (a) $\qquad\qquad\qquad \tan\phi = \dfrac{2}{x}$

$\qquad\qquad\qquad \tan(\theta + \phi) = \dfrac{4}{x}$

$\qquad\qquad\qquad \dfrac{\tan\theta + \tan\phi}{1 - \tan\theta\tan\phi} = \dfrac{4}{x}$

$\qquad\qquad\qquad \dfrac{\tan\theta + (2/x)}{1 - (2/x)\tan\theta} = \dfrac{4}{x}$

$\qquad\qquad\qquad x\tan\theta + 2 = 4 - \dfrac{8}{x}\tan\theta$

$\qquad x^2\tan\theta - 2x + 8\tan\theta = 0$

(b) $F(x, \theta) = (x^2 + 8)\tan\theta - 2x = 0$

$\dfrac{d\theta}{dx} = -\dfrac{F_x}{F_\theta} = -\dfrac{2x\tan\theta - 2}{\sec^2\theta(x^2 + 8)} = \dfrac{2\cos^2\theta - 2x\sin\theta\cos\theta}{x^2 + 8}$

(c) $\dfrac{d\theta}{dx} = 0 \Rightarrow 2\cos^2\theta = 2x\sin\theta\cos\theta \Rightarrow \cos\theta = x\sin\theta \Rightarrow \tan\theta = \dfrac{1}{x}$

Thus, $x^2\left(\dfrac{1}{x}\right) - 2x + 8\left(\dfrac{1}{x}\right) = 0 \Rightarrow \dfrac{8}{x} = x \Rightarrow x = 2\sqrt{2}$ ft.

59. $\qquad w = f(x, y)$

$\qquad\quad x = u - v$

$\qquad\quad y = v - u$

$\qquad \dfrac{\partial w}{\partial u} = \dfrac{\partial w}{\partial x}\dfrac{dx}{du} + \dfrac{\partial w}{\partial y}\dfrac{dy}{du} = \dfrac{\partial w}{\partial x} - \dfrac{\partial w}{\partial y}$

$\qquad \dfrac{\partial w}{\partial v} = \dfrac{\partial w}{\partial x}\dfrac{dx}{dv} + \dfrac{\partial w}{\partial y}\dfrac{dy}{dv} = -\dfrac{\partial w}{\partial x} + \dfrac{\partial w}{\partial y}$

$\dfrac{\partial w}{\partial u} + \dfrac{\partial w}{\partial v} = 0$

61. $w = f(x, y)$, $x = r\cos\theta$, $y = r\sin\theta$

$$\frac{\partial w}{\partial r} = \frac{\partial w}{\partial x}\cos\theta + \frac{\partial w}{\partial y}\sin\theta$$

$$\frac{\partial w}{\partial\theta} = \frac{\partial w}{\partial x}(-r\sin\theta) + \frac{\partial w}{\partial y}(r\cos\theta)$$

(a)
$$r\cos\theta\frac{\partial w}{\partial r} = \frac{\partial w}{\partial x}r\cos^2\theta + \frac{\partial w}{\partial y}r\sin\theta\cos\theta$$

$$-\sin\theta\frac{\partial w}{\partial\theta} = \frac{\partial w}{\partial x}(r\sin^2\theta) - \frac{\partial w}{\partial y}r\sin\theta\cos\theta$$

$$r\cos\theta\frac{\partial w}{\partial r} - \sin\theta\frac{\partial w}{\partial\theta} = \frac{\partial w}{\partial x}(r\cos^2\theta + r\sin^2\theta)$$

$$r\frac{\partial w}{\partial x} = \frac{\partial w}{\partial r}(r\cos\theta) - \frac{\partial w}{\partial\theta}\sin\theta$$

$$\frac{\partial w}{\partial x} = \frac{\partial w}{\partial r}\cos\theta - \frac{\partial w}{\partial\theta}\frac{\sin\theta}{r} \qquad \text{(First Formula)}$$

$$r\sin\theta\frac{\partial w}{\partial r} = \frac{\partial w}{\partial x}r\sin\theta\cos\theta + \frac{\partial w}{\partial y}r\sin^2\theta$$

$$\cos\theta\frac{\partial w}{\partial\theta} = \frac{\partial w}{\partial x}(-r\sin\theta\cos\theta) + \frac{\partial w}{\partial y}(r\cos^2\theta)$$

$$r\sin\theta\frac{\partial w}{\partial r} + \cos\theta\frac{\partial w}{\partial\theta} = \frac{\partial w}{\partial y}(r\sin^2\theta + r\cos^2\theta)$$

$$r\frac{\partial w}{\partial y} = \frac{\partial w}{\partial r}r\sin\theta + \frac{\partial w}{\partial\theta}\cos\theta$$

$$\frac{\partial w}{\partial y} = \frac{\partial w}{\partial r}\sin\theta + \frac{\partial w}{\partial\theta}\frac{\cos\theta}{r} \qquad \text{(Second Formula)}$$

(b) $\left(\dfrac{\partial w}{\partial r}\right)^2 + \dfrac{1}{r^2}\left(\dfrac{\partial w}{\partial\theta}\right)^2 = \left(\dfrac{\partial w}{\partial x}\right)^2\cos^2\theta + 2\dfrac{\partial w}{\partial x}\dfrac{\partial w}{\partial y}\sin\theta\cos\theta + \left(\dfrac{\partial w}{\partial y}\right)^2\sin^2\theta + \left(\dfrac{\partial w}{\partial x}\right)^2\sin^2\theta -$

$$2\frac{\partial w}{\partial x}\frac{\partial w}{\partial y}\sin\theta\cos\theta + \left(\frac{\partial w}{\partial y}\right)^2\cos^2\theta = \left(\frac{\partial w}{\partial x}\right)^2 + \left(\frac{\partial w}{\partial y}\right)^2$$

63. Given $\dfrac{\partial u}{\partial x} = \dfrac{\partial v}{\partial y}$ and $\dfrac{\partial u}{\partial y} = -\dfrac{\partial v}{\partial x}$, $x = r\cos\theta$ and $y = r\sin\theta$.

$$\frac{\partial u}{\partial r} = \frac{\partial u}{\partial x}\cos\theta + \frac{\partial u}{\partial y}\sin\theta = \frac{\partial v}{\partial y}\cos\theta - \frac{\partial v}{\partial x}\sin\theta$$

$$\frac{\partial v}{\partial\theta} = \frac{\partial v}{\partial x}(-r\sin\theta) + \frac{\partial v}{\partial y}(r\cos\theta) = r\left[\frac{\partial v}{\partial y}\cos\theta - \frac{\partial v}{\partial x}\sin\theta\right]$$

Therefore, $\dfrac{\partial u}{\partial r} = \dfrac{1}{r}\dfrac{\partial v}{\partial\theta}$.

$$\frac{\partial v}{\partial r} = \frac{\partial v}{\partial x}\cos\theta + \frac{\partial v}{\partial y}\sin\theta = -\frac{\partial u}{\partial y}\cos\theta + \frac{\partial u}{\partial x}\sin\theta$$

$$\frac{\partial u}{\partial\theta} = \frac{\partial u}{\partial x}(-r\sin\theta) + \frac{\partial u}{\partial y}(r\cos\theta) = -r\left[-\frac{\partial u}{\partial y}\cos\theta + \frac{\partial u}{\partial x}\sin\theta\right]$$

Therefore, $\dfrac{\partial v}{\partial r} = -\dfrac{1}{r}\dfrac{\partial u}{\partial\theta}$.

Section 12.6   Directional Derivatives and Gradients

1. $f(x, y) = 3x - 4xy + 5y$

$$\mathbf{v} = \frac{1}{2}(\mathbf{i} + \sqrt{3}\mathbf{j})$$

$$\nabla f(x, y) = (3 - 4y)\mathbf{i} + (-4x + 5)\mathbf{j}$$

$$\nabla f(1, 2) = -5\mathbf{i} + \mathbf{j}$$

$$\mathbf{u} = \frac{\mathbf{v}}{\|\mathbf{v}\|} = \frac{1}{2}\mathbf{i} + \frac{\sqrt{3}}{2}\mathbf{j}$$

$$D_{\mathbf{u}} f(1, 2) = \nabla f(1, 2) \cdot \mathbf{u} = \frac{1}{2}(-5 + \sqrt{3})$$

3. $f(x, y) = xy$

$$\mathbf{v} = \mathbf{i} + \mathbf{j}$$

$$\nabla f(x, y) = y\mathbf{i} + x\mathbf{j}$$

$$\nabla f(2, 3) = 3\mathbf{i} + 2\mathbf{j}$$

$$\mathbf{u} = \frac{\mathbf{v}}{\|\mathbf{v}\|} = \frac{\sqrt{2}}{2}\mathbf{i} + \frac{\sqrt{2}}{2}\mathbf{j}$$

$$D_{\mathbf{u}} f(2, 3) = \nabla f(2, 3) \cdot \mathbf{u} = \frac{5\sqrt{2}}{2}$$

5. $g(x, y) = \sqrt{x^2 + y^2}$

$$\mathbf{v} = 3\mathbf{i} - 4\mathbf{j}$$

$$\nabla g = \frac{x}{\sqrt{x^2 + y^2}}\mathbf{i} + \frac{y}{\sqrt{x^2 + y^2}}\mathbf{j}$$

$$\nabla g(3, 4) = \frac{3}{5}\mathbf{i} + \frac{4}{5}\mathbf{j}$$

$$\mathbf{u} = \frac{\mathbf{v}}{\|\mathbf{v}\|} = \frac{3}{5}\mathbf{i} - \frac{4}{5}\mathbf{j}$$

$$D_{\mathbf{u}} g(3, 4) = \nabla g(3, 4) \cdot \mathbf{u} = -\frac{7}{25}$$

7. $h(x, y) = e^x \sin y$

$$\mathbf{v} = -\mathbf{i}$$

$$\nabla h = e^x \sin y\mathbf{i} + e^x \cos y\mathbf{j}$$

$$\nabla h\left(1, \frac{\pi}{2}\right) = e\mathbf{i}$$

$$\mathbf{u} = \frac{\mathbf{v}}{\|\mathbf{v}\|} = -\mathbf{i}$$

$$D_{\mathbf{u}} h\left(1, \frac{\pi}{2}\right) = \nabla h\left(1, \frac{\pi}{2}\right) \cdot \mathbf{u} = -e$$

9. $f(x, y, z) = xy + yz + xz$

$$\mathbf{v} = 2\mathbf{i} + \mathbf{j} - \mathbf{k}$$

$$\nabla f(x, y, z) = (y + z)\mathbf{i} + (x + z)\mathbf{j} + (x + y)\mathbf{k}$$

$$\nabla f(1, 1, 1) = 2\mathbf{i} + 2\mathbf{j} + 2\mathbf{k}$$

$$\mathbf{u} = \frac{\mathbf{v}}{\|\mathbf{v}\|} = \frac{\sqrt{6}}{3}\mathbf{i} + \frac{\sqrt{6}}{6}\mathbf{j} - \frac{\sqrt{6}}{6}\mathbf{k}$$

$$D_{\mathbf{u}} f(1, 1, 1) = \nabla f(1, 1, 1) \cdot \mathbf{u} = \frac{2\sqrt{6}}{3}$$

11. $h(x, y, z) = x \arctan yz$

$$\mathbf{v} = \langle 1, 2, -1 \rangle$$

$$\nabla h(x, y, z) = \arctan yz\mathbf{i} + \frac{xz}{1 + (yz)^2}\mathbf{j} + \frac{xy}{1 + (yz)^2}\mathbf{k}$$

$$\nabla h(4, 1, 1) = \frac{\pi}{4}\mathbf{i} + 2\mathbf{j} + 2\mathbf{k}$$

$$\mathbf{u} = \frac{\mathbf{v}}{\|\mathbf{v}\|} = \left\langle \frac{1}{\sqrt{6}}, \frac{2}{\sqrt{6}}, -\frac{1}{\sqrt{6}} \right\rangle$$

$$D_{\mathbf{u}} h(4, 1, 1) = \nabla h(4, 1, 1) \cdot \mathbf{u} = \frac{\pi + 8}{4\sqrt{6}} = \frac{(\pi + 8)\sqrt{6}}{24}$$

13. $f(x, y) = x^2 + y^2$

$$\mathbf{u} = \frac{1}{\sqrt{2}}\mathbf{i} + \frac{1}{\sqrt{2}}\mathbf{j}$$

$$\nabla f = 2x\mathbf{i} + 2y\mathbf{j}$$

$$D_{\mathbf{u}} f = \nabla f \cdot \mathbf{u} = \frac{2}{\sqrt{2}}x + \frac{2}{\sqrt{2}}y = \sqrt{2}(x + y)$$

15. $f(x, y) = \sin(2x - y)$

$$\mathbf{u} = \frac{1}{2}\mathbf{i} - \frac{\sqrt{3}}{2}\mathbf{j}$$

$$\nabla f = 2\cos(2x - y)\mathbf{i} - \cos(2x - y)\mathbf{j}$$

$$D_{\mathbf{u}} f = \nabla f \cdot \mathbf{u} = \cos(2x - y) + \frac{\sqrt{3}}{2}\cos(2x - y)$$

$$= \left(\frac{2 + \sqrt{3}}{2}\right)\cos(2x - y)$$

17. $f(x, y) = x^2 + 4y^2$

$\mathbf{v} = -2\mathbf{i} - 2\mathbf{j}$

$\nabla f = 2x\mathbf{i} + 8y\mathbf{j}$

$\mathbf{u} = \dfrac{\mathbf{v}}{\|\mathbf{v}\|} = -\dfrac{1}{\sqrt{2}}\mathbf{i} - \dfrac{1}{\sqrt{2}}\mathbf{j}$

$D_{\mathbf{u}} f = -\dfrac{2}{\sqrt{2}}x - \dfrac{8}{\sqrt{2}}y = -\sqrt{2}(x + 4y)$

At $P = (3, 1)$, $D_{\mathbf{u}} f = -7\sqrt{2}$.

19. $h(x, y, z) = \ln(x + y + z)$

$\mathbf{v} = 3\mathbf{i} + 3\mathbf{j} + \mathbf{k}$

$\nabla h = \dfrac{1}{x + y + z}(\mathbf{i} + \mathbf{j} + \mathbf{k})$

At $(1, 0, 0)$, $\nabla h = \mathbf{i} + \mathbf{j} + \mathbf{k}$.

$\mathbf{u} = \dfrac{\mathbf{v}}{\|\mathbf{v}\|} = \dfrac{1}{\sqrt{19}}(3\mathbf{i} + 3\mathbf{j} + \mathbf{k})$

$D_{\mathbf{u}} h = \nabla h \cdot \mathbf{u} = \dfrac{7}{\sqrt{19}} = \dfrac{7\sqrt{19}}{19}$

21. $f(x, y) = 3x - 5y^2 + 10$

$\nabla f(x, y) = 3\mathbf{i} - 10y\mathbf{j}$

$\nabla f(2, 1) = 3\mathbf{i} - 10\mathbf{j}$

23. $z = \cos(x^2 + y^2)$

$\nabla z(x, y) = -2x \sin(x^2 + y^2)\mathbf{i} - 2y \sin(x^2 + y^2)\mathbf{j}$

$\nabla z(3, -4) = -6 \sin 25\mathbf{i} + 8 \sin 25\mathbf{j} \approx 0.7941\mathbf{i} - 1.0588\mathbf{j}$

25. $w = 3x^2y - 5yz + z^2$

$\nabla w(x, y, z) = 6xy\mathbf{i} + (3x^2 - 5z)\mathbf{j} + (2z - 5y)\mathbf{k}$

$\nabla w(1, 1, -2) = 6\mathbf{i} + 13\mathbf{j} - 9\mathbf{k}$

27. $\overrightarrow{PQ} = 2\mathbf{i} + 4\mathbf{j}$, $\mathbf{u} = \dfrac{1}{\sqrt{5}}\mathbf{i} + \dfrac{2}{\sqrt{5}}\mathbf{j}$

$\nabla g(x, y) = 2x\mathbf{i} + 2y\mathbf{j}$, $\nabla g(1, 2) = 2\mathbf{i} + 4\mathbf{j}$

$D_{\mathbf{u}} g = \nabla g \cdot \mathbf{u} = \dfrac{2}{\sqrt{5}} + \dfrac{8}{\sqrt{5}} = \dfrac{10}{\sqrt{5}} = 2\sqrt{5}$

29. $\overrightarrow{PQ} = 2\mathbf{i} + \mathbf{j}$, $\mathbf{u} = \dfrac{2}{\sqrt{5}}\mathbf{i} + \dfrac{1}{\sqrt{5}}\mathbf{j}$

$\nabla f(x, y) = -e^{-x} \cos y\mathbf{i} - e^{-x} \sin y\mathbf{j}$

$\nabla f(0, 0) = -\mathbf{i}$

$D_{\mathbf{u}} f = \nabla f \cdot \mathbf{u} = -\dfrac{2}{\sqrt{5}} = -\dfrac{2\sqrt{5}}{5}$

31. $h(x, y) = x \tan y$

$\nabla h(x, y) = \tan y\mathbf{i} + x \sec^2 y\mathbf{j}$

$\nabla h\left(2, \dfrac{\pi}{4}\right) = \mathbf{i} + 4\mathbf{j}$

$\left\| \nabla h\left(2, \dfrac{\pi}{4}\right) \right\| = \sqrt{17}$

33. $g(x, y) = \ln \sqrt[3]{x^2 + y^2} = \dfrac{1}{3} \ln(x^2 + y^2)$

$\nabla g(x, y) = \dfrac{1}{3}\left[\dfrac{2x}{x^2 + y^2}\mathbf{i} + \dfrac{2y}{x^2 + y^2}\mathbf{j} \right]$

$\nabla g(1, 2) = \dfrac{1}{3}\left(\dfrac{2}{5}\mathbf{i} + \dfrac{4}{5}\mathbf{j} \right) = \dfrac{2}{15}(\mathbf{i} + 2\mathbf{j})$

$\|\nabla g(1, 2)\| = \dfrac{2\sqrt{5}}{15}$

35. $f(x, y, z) = \sqrt{x^2 + y^2 + z^2}$

$\nabla f(x, y, z) = \dfrac{1}{\sqrt{x^2 + y^2 + z^2}}(x\mathbf{i} + y\mathbf{j} + z\mathbf{k})$

$\nabla f(1, 4, 2) = \dfrac{1}{\sqrt{21}}(\mathbf{i} + 4\mathbf{j} + 2\mathbf{k})$

$\|\nabla f(1, 4, 2)\| = 1$

37. $f(x, y, z) = xe^{yz}$

$\nabla f(x, y, z) = e^{yz}\mathbf{i} + xze^{yz}\mathbf{j} + xye^{yz}\mathbf{k}$

$\nabla f(2, 0, -4) = \mathbf{i} - 8\mathbf{j}$

$\|\nabla f(2, 0, -4)\| = \sqrt{65}$

For Exercises 39–45, $f(x, y) = 3 - \frac{x}{3} - \frac{y}{2}$ and $D_\theta f(x, y) = -\left(\frac{1}{3}\right)\cos\theta - \left(\frac{1}{2}\right)\sin\theta$.

39. $f(x, y) = 3 - \frac{x}{3} - \frac{y}{2}$

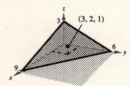

41. (a) $D_{4\pi/3} f(3, 2) = -\left(\frac{1}{3}\right)\left(-\frac{1}{2}\right) - \left(\frac{1}{2}\right)\left(-\frac{\sqrt{3}}{2}\right)$

$$= \frac{2 + 3\sqrt{3}}{12}$$

(b) $D_{-\pi/6} f(3, 2) = -\left(\frac{1}{3}\right)\left(\frac{\sqrt{3}}{2}\right) - \left(\frac{1}{2}\right)\left(-\frac{1}{2}\right)$

$$= \frac{3 - 2\sqrt{3}}{12}$$

43. (a) $\mathbf{v} = -3\mathbf{i} + 4\mathbf{j}$

$\|\mathbf{v}\| = \sqrt{9 + 16} = 5$

$\mathbf{u} = -\frac{3}{5}\mathbf{i} + \frac{4}{5}\mathbf{j}$

$D_{\mathbf{u}} f = \nabla f \cdot \mathbf{u} = \frac{1}{5} - \frac{2}{5} = -\frac{1}{5}$

(b) $\mathbf{v} = \mathbf{i} + 3\mathbf{j}$

$\|\mathbf{v}\| = \sqrt{10}$

$\mathbf{u} = \frac{1}{\sqrt{10}}\mathbf{i} + \frac{3}{\sqrt{10}}\mathbf{j}$

$D_{\mathbf{u}} f = \nabla f \cdot \mathbf{u} = \frac{-11}{6\sqrt{10}} = -\frac{11\sqrt{10}}{60}$

45. $\|\nabla f\| = \sqrt{\frac{1}{9} + \frac{1}{4}} = \frac{1}{6}\sqrt{13}$

For Exercises 47 and 49, $f(x, y) = 9 - x^2 - y^2$ and $D_\theta f(x, y) = -2x\cos\theta - 2y\sin\theta = -2(x\cos\theta + y\sin\theta)$.

47. $f(x, y) = 9 - x^2 - y^2$

49. $\nabla f(1, 2) = -2\mathbf{i} - 4\mathbf{j}$

$\|\nabla f(1, 2)\| = \sqrt{4 + 16} = \sqrt{20} = 2\sqrt{5}$

51. (a) In the direction of the vector $-4\mathbf{i} + \mathbf{j}$.

(b) $\nabla f = \frac{1}{10}(2x - 3y)\mathbf{i} + \frac{1}{10}(-3x + 2y)\mathbf{j}$

$\nabla f(1, 2) = \frac{1}{10}(-4)\mathbf{i} + \frac{1}{10}(1)\mathbf{j} = -\frac{2}{5}\mathbf{i} + \frac{1}{10}\mathbf{j}$

(Same direction as in part (a).)

(c) $-\nabla f = \frac{2}{5}\mathbf{i} - \frac{1}{10}\mathbf{j}$, the direction opposite that of the gradient.

53. $f(x, y) = x^2 - y^2$, $(4, -3, 7)$

(a)

—CONTINUED–

53. —CONTINUED—

(b) $D_{\mathbf{u}} f(x, y) = \nabla f(x, y) \cdot \mathbf{u} = 2x \cos \theta - 2y \sin \theta$

$D_{\mathbf{u}} f(4, -3) = 8 \cos \theta + 6 \sin \theta$

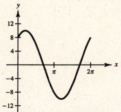

Generated by Mathematica

(c) Zeros: $\theta \approx 2.21, 5.36$

These are the angles θ for which $D_{\mathbf{u}} f(4, 3)$ equals zero.

(d) $g(\theta) = D_{\mathbf{u}} f(4, -3) = 8 \cos \theta + 6 \sin \theta$

$g'(\theta) = -8 \sin \theta + 6 \cos \theta$

Critical numbers: $\theta \approx 0.64, 3.79$

These are the angles for which $D_{\mathbf{u}} f(4, -3)$ is a maximum (0.64) and minimum (3.79).

(e) $\|\nabla f(4, -3)\| = \|2(4)\mathbf{i} - 2(-3)\mathbf{j}\| = \sqrt{64 + 36} = 10$, the maximum value of $D_{\mathbf{u}} f(4, -3)$, at $\theta \approx 0.64$.

(f) $f(x, y) = x^2 - y^2 = 7$

$\nabla f(4, -3) = 8\mathbf{i} + 6\mathbf{j}$ is perpendicular to the level curve at $(4, -3)$.

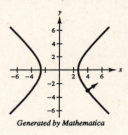

Generated by Mathematica

55. $f(x, y) = x^2 + y^2$

$c = 25, \ P = (3, 4)$

$\nabla f(x, y) = 2x\mathbf{i} + 2y\mathbf{j}$

$x^2 + y^2 = 25$

$\nabla f(3, 4) = 6\mathbf{i} + 8\mathbf{j}$

57. $f(x, y) = \dfrac{x}{x^2 + y^2}$

$c = \dfrac{1}{2}, \ P = (1, 1)$

$\nabla f(x, y) = \dfrac{y^2 - x^2}{(x^2 + y^2)^2}\mathbf{i} - \dfrac{2xy}{(x^2 + y^2)^2}\mathbf{j}$

$\dfrac{x}{x^2 + y^2} = \dfrac{1}{2}$

$x^2 + y^2 - 2x = 0$

$\nabla f(1, 1) = -\dfrac{1}{2}\mathbf{j}$

59. $4x^2 - y = 6$

$f(x, y) = 4x^2 - y$

$\nabla f(x, y) = 8x\mathbf{i} - \mathbf{j}$

$\nabla f(2, 10) = 16\mathbf{i} - \mathbf{j}$

$\dfrac{\nabla f(2, 10)}{\|\nabla f(2, 10)\|} = \dfrac{1}{\sqrt{257}}(16\mathbf{i} - \mathbf{j})$

$\qquad = \dfrac{\sqrt{257}}{257}(16\mathbf{i} - \mathbf{j})$

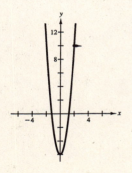

61. $9x^2 + 4y^2 = 40$

$f(x, y) = 9x^2 + 4y^2$

$\nabla f(x, y) = 18x\mathbf{i} + 8y\mathbf{j}$

$\nabla f(2, -1) = 36\mathbf{i} - 8\mathbf{j}$

$\dfrac{\nabla f(2, -1)}{\|\nabla f(2, -1)\|} = \dfrac{1}{\sqrt{85}}(9\mathbf{i} - 2\mathbf{j})$

$\qquad = \dfrac{\sqrt{85}}{85}(9\mathbf{i} - 2\mathbf{j})$

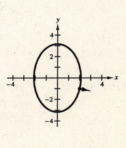

63. $T = \dfrac{x}{x^2 + y^2}$

$\nabla T = \dfrac{y^2 - x^2}{(x^2 + y^2)^2}\mathbf{i} - \dfrac{2xy}{(x^2 + y^2)^2}\mathbf{j}$

$\nabla T(3, 4) = \dfrac{7}{625}\mathbf{i} - \dfrac{24}{625}\mathbf{j} = \dfrac{1}{625}(7\mathbf{i} - 24\mathbf{j})$

65. See the definition, page 885.

67. Let $f(x, y)$ be a function of two variables and
$\mathbf{u} = \cos\theta\mathbf{i} + \sin\theta\mathbf{j}$ a unit vector.

(a) If $\theta = 0°$, then $D_{\mathbf{u}}f = \dfrac{\partial f}{\partial x}$.

(b) If $\theta = 90°$, then $D_{\mathbf{u}}f = \dfrac{\partial f}{\partial y}$.

69.

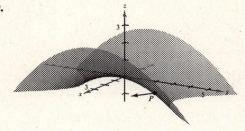

71.

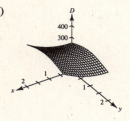

73. $T(x, y) = 400 - 2x^2 - y^2,$ $\qquad P = (10, 10)$

$\dfrac{dx}{dt} = -4x$ $\qquad\qquad \dfrac{dy}{dt} = -2y$

$x(t) = C_1 e^{-4t}$ $\qquad\qquad y(t) = C_2 e^{-2t}$

$10 = x(0) = C_1$ $\qquad\qquad 10 = y(0) = C_2$

$x(t) = 10e^{-4t}$ $\qquad\qquad y(t) = 10e^{-2t}$

$x = \dfrac{y^2}{10}$ $\qquad\qquad y^2(t) = 100e^{-4t}$

$y^2 = 10x$

75. (a)

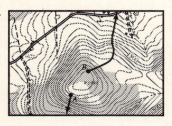

(b) The graph of $-D = -250 - 30x^2 - 50\sin(\pi y/2)$
would model the ocean floor.

(c) $D(1, 0.5) = 250 + 30(1) + 50\sin\dfrac{\pi}{4} \approx 315.4$ ft

(d) $\dfrac{\partial D}{\partial x} = 60x$ and $\dfrac{\partial D}{\partial x}(1, 0.5) = 60$

(e) $\dfrac{\partial D}{\partial y} = 25\pi\cos\dfrac{\pi y}{2}$ and $\dfrac{\partial D}{\partial y}(1, 0.5) = 25\pi\cos\dfrac{\pi}{4} \approx 55.5$

(f) $\nabla D = 60x\mathbf{i} + 25\pi\cos\left(\dfrac{\pi y}{2}\right)\mathbf{j}$

$\nabla D(1, 0.5) = 60\mathbf{i} + 55.5\mathbf{j}$

77. True

79. True

81. Let $f(x, y, z) = e^x\cos y + \dfrac{z^2}{2} + C$. Then $\nabla f(x, y, z) = e^x\cos y\mathbf{i} - e^x\sin y\mathbf{j} + z\mathbf{k}$.

Section 12.7   Tangent Planes and Normal Lines

1. $F(x, y, z) = 3x - 5y + 3z - 15 = 0$

$\qquad 3x - 5y + 3z = 15$ Plane

3. $F(x, y, z) = 4x^2 + 9y^2 - 4z^2 = 0$

$\qquad 4x^2 + 9y^2 = 4z^2$ Elliptic cone

5. $F(x, y, z) = x + y + z - 4$

$\qquad \nabla F = \mathbf{i} + \mathbf{j} + \mathbf{k}$

$\qquad \mathbf{n} = \dfrac{\nabla F}{\|\nabla F\|} = \dfrac{1}{\sqrt{3}}(\mathbf{i} + \mathbf{j} + \mathbf{k})$

$\qquad\qquad = \dfrac{\sqrt{3}}{3}(\mathbf{i} + \mathbf{j} + \mathbf{k})$

7. $F(x, y, z) = \sqrt{x^2 + y^2} - z$

$\qquad \nabla F(x, y, z) = \dfrac{x}{\sqrt{x^2 + y^2}}\mathbf{i} + \dfrac{y}{\sqrt{x^2 + y^2}}\mathbf{j} - \mathbf{k}$

$\qquad \nabla F(3, 4, 5) = \dfrac{3}{5}\mathbf{i} + \dfrac{4}{5}\mathbf{j} - \mathbf{k}$

$\qquad \mathbf{n} = \dfrac{\nabla F}{\|\nabla F\|} = \dfrac{5}{5\sqrt{2}}\left(\dfrac{3}{5}\mathbf{i} + \dfrac{4}{5}\mathbf{j} - \mathbf{k}\right)$

$\qquad\qquad = \dfrac{1}{5\sqrt{2}}(3\mathbf{i} + 4\mathbf{j} - 5\mathbf{k})$

$\qquad\qquad = \dfrac{\sqrt{2}}{10}(3\mathbf{i} + 4\mathbf{j} - 5\mathbf{k})$

9. $F(x, y, z) = x^2y^4 - z$

$\qquad \nabla F(x, y, z) = 2xy^4\mathbf{i} + 4x^2y^3\mathbf{j} - \mathbf{k}$

$\qquad \nabla F(1, 2, 16) = 32\mathbf{i} + 32\mathbf{j} - \mathbf{k}$

$\qquad \mathbf{n} = \dfrac{\nabla F}{\|\nabla F\|} = \dfrac{1}{\sqrt{2049}}(32\mathbf{i} + 32\mathbf{j} - \mathbf{k})$

$\qquad\qquad = \dfrac{\sqrt{2049}}{2049}(32\mathbf{i} + 32\mathbf{j} - \mathbf{k})$

11. $F(x, y, z) = \ln\left(\dfrac{x}{y - z}\right) = \ln x - \ln(y - z)$

$\qquad \nabla F(x, y, z) = \dfrac{1}{x}\mathbf{i} - \dfrac{1}{y - z}\mathbf{j} + \dfrac{1}{y - z}\mathbf{k}$

$\qquad \nabla F(1, 4, 3) = \mathbf{i} - \mathbf{j} + \mathbf{k}$

$\qquad \mathbf{n} = \dfrac{\nabla F}{\|\nabla F\|} = \dfrac{1}{\sqrt{3}}(\mathbf{i} - \mathbf{j} + \mathbf{k})$

$\qquad\qquad = \dfrac{\sqrt{3}}{3}(\mathbf{i} - \mathbf{j} + \mathbf{k})$

13. $F(x, y, z) = -x \sin y + z - 4$

$\qquad \nabla F(x, y, z) = -\sin y\,\mathbf{i} - x\cos y\,\mathbf{j} + \mathbf{k}$

$\qquad \nabla F\left(6, \dfrac{\pi}{6}, 7\right) = -\dfrac{1}{2}\mathbf{i} - 3\sqrt{3}\mathbf{j} + \mathbf{k}$

$\qquad \mathbf{n} = \dfrac{\nabla F}{\|\nabla F\|} = \dfrac{2}{\sqrt{113}}\left(-\dfrac{1}{2}\mathbf{i} - 3\sqrt{3}\mathbf{j} + \mathbf{k}\right)$

$\qquad\qquad = \dfrac{1}{\sqrt{113}}(-\mathbf{i} - 6\sqrt{3}\mathbf{j} + 2\mathbf{k})$

$\qquad\qquad = \dfrac{\sqrt{113}}{113}(-\mathbf{i} - 6\sqrt{3}\mathbf{j} + 2\mathbf{k})$

15. $f(x, y) = 25 - x^2 - y^2,\ (3, 1, 15)$

$\qquad F(x, y, z) = 25 - x^2 - y^2 - z$

$\qquad F_x(x, y, z) = -2x \qquad F_y(x, y, z) = -2y \qquad F_z(x, y, z) = -1$

$\qquad F_x(3, 1, 15) = -6 \qquad F_y(3, 1, 15) = -2 \qquad F_z(3, 1, 15) = -1$

$\qquad -6(x - 3) - 2(y - 1) - (z - 15) = 0$

$\qquad\qquad\qquad 0 = 6x + 2y + z - 35$

$\qquad\qquad 6x + 2y + z = 35$

17. $f(x, y) = \sqrt{x^2 + y^2}, \ (3, 4, 5)$

$F(x, y, z) = \sqrt{x^2 + y^2} - z$

$F_x(x, y, z) = \dfrac{x}{\sqrt{x^2 + y^2}}$ $F_y(x, y, z) = \dfrac{y}{\sqrt{x^2 + y^2}}$ $F_z(x, y, z) = -1$

$F_x(3, 4, 5) = \dfrac{3}{5}$ $F_y(3, 4, 5) = \dfrac{4}{5}$ $F_z(3, 4, 5) = -1$

$\dfrac{3}{5}(x - 3) + \dfrac{4}{5}(y - 4) - (z - 5) = 0$

$3(x - 3) + 4(y - 4) - 5(z - 5) = 0$

$\qquad\qquad 3x + 4y - 5z = 0$

19. $g(x, y) = x^2 - y^2, \ (5, 4, 9)$

$G(x, y, z) = x^2 - y^2 - z$

$G_x(x, y, z) = 2x$ $G_y(x, y, z) = -2y$ $G_z(x, y, z) = -1$

$G_x(5, 4, 9) = 10$ $G_y(5, 4, 9) = -8$ $G_z(5, 4, 9) = -1$

$10(x - 5) - 8(y - 4) - (z - 9) = 0$

$\qquad\qquad 10x - 8y - z = 9$

21. $z = e^x(\sin y + 1), \ \left(0, \dfrac{\pi}{2}, 2\right)$

$F(x, y, z) = e^x(\sin y + 1) - z$

$F_x(x, y, z) = e^x(\sin y + 1)$ $F_y(x, y, z) = e^x \cos y$ $F_z(x, y, z) = -1$

$F_x\left(0, \dfrac{\pi}{2}, 2\right) = 2$ $F_y\left(0, \dfrac{\pi}{2}, 2\right) = 0$ $F_z\left(0, \dfrac{\pi}{2}, 2\right) = -1$

$2x - z = -2$

23. $h(x, y) = \ln \sqrt{x^2 + y^2}, \ (3, 4, \ln 5)$

$H(x, y, z) = \ln \sqrt{x^2 + y^2} - z = \dfrac{1}{2}\ln(x^2 + y^2) - z$

$H_x(x, y, z) = \dfrac{x}{x^2 + y^2}$ $H_y(x, y, z) = \dfrac{y}{x^2 + y^2}$ $H_z(x, y, z) = -1$

$H_x(3, 4, \ln 5) = \dfrac{3}{25}$ $H_y(3, 4, \ln 5) = \dfrac{4}{25}$ $H_z(3, 4, \ln 5) = -1$

$\dfrac{3}{25}(x - 3) + \dfrac{4}{25}(y - 4) - (z - \ln 5) = 0$

$3(x - 3) + 4(y - 4) - 25(z - \ln 5) = 0$

$\qquad\qquad 3x + 4y - 25z = 25(1 - \ln 5)$

25. $x^2 + 4y^2 + z^2 = 36, \ (2, -2, 4)$

$F(x, y, z) = x^2 + 4y^2 + z^2 - 36$

$F_x(x, y, z) = 2x$ $F_y(x, y, z) = 8y$ $F_z(x, y, z) = 2z$

$F_x(2, -2, 4) = 4$ $F_y(2, -2, 4) = -16$ $F_z(2, -2, 4) = 8$

$4(x - 2) - 16(y + 2) + 8(z - 4) = 0$

$(x - 2) - 4(y + 2) + 2(z - 4) = 0$

$\qquad\qquad x - 4y + 2z = 18$

27. $xy^2 + 3x - z^2 = 4$, $(2, 1, -2)$

$F(x, y, z) = xy^2 + 3x - z^2 - 4$

$F_x(x, y, z) = y^2 + 3$ $\qquad$ $F_y(x, y, z) = 2xy$ $\qquad$ $F_z(x, y, z) = -2z$

$F_x(2, 1, -2) = 4$ $\qquad$ $F_y(2, 1, -2) = 4$ $\qquad$ $F_z(2, 1, -2) = 4$

$4(x - 2) + 4(y - 1) + 4(z + 2) = 0$

$$x + y + z = 1$$

29. $x^2 + y^2 + z = 9$, $(1, 2, 4)$

$F(x, y, z) = x^2 + y^2 + z - 9$

$F_x(x, y, z) = 2x$ $\qquad$ $F_y(x, y, z) = 2y$ $\qquad$ $F_z(x, y, z) = 1$

$F_x(1, 2, 4) = 2$ $\qquad$ $F_y(1, 2, 4) = 4$ $\qquad$ $F_z(1, 2, 4) = 1$

Direction numbers: 2, 4, 1

Plane: $2(x - 1) + 4(y - 2) + (z - 4) = 0$, $2x + 4y + z = 14$

Line: $\dfrac{x - 1}{2} = \dfrac{y - 2}{4} = \dfrac{z - 4}{1}$

31. $xy - z = 0$, $(-2, -3, 6)$

$F(x, y, z) = xy - z$

$F_x(x, y, z) = y$ $\qquad$ $F_y(x, y, z) = x$ $\qquad$ $F_z(x, y, z) = -1$

$F_x(-2, -3, 6) = -3$ $\qquad$ $F_y(-2, -3, 6) = -2$ $\qquad$ $F_z(-2, -3, 6) = -1$

Direction numbers: 3, 2, 1

Plane: $3(x + 2) + 2(y + 3) + (z - 6) = 0$, $3x + 2y + z = -6$

Line: $\dfrac{x + 2}{3} = \dfrac{y + 3}{2} = \dfrac{z - 6}{1}$

33. $z = \arctan \dfrac{y}{x}$, $\left(1, 1, \dfrac{\pi}{4}\right)$

$F(x, y, z) = \arctan \dfrac{y}{x} - z$

$F_x(x, y, z) = \dfrac{-y}{x^2 + y^2}$ $\qquad$ $F_y(x, y, z) = \dfrac{x}{x^2 + y^2}$ $\qquad$ $F_z(x, y, z) = -1$

$F_x\left(1, 1, \dfrac{\pi}{4}\right) = -\dfrac{1}{2}$ $\qquad$ $F_y\left(1, 1, \dfrac{\pi}{4}\right) = \dfrac{1}{2}$ $\qquad$ $F_z\left(1, 1, \dfrac{\pi}{4}\right) = -1$

Direction numbers: 1, -1, 2

Plane: $(x - 1) - (y - 1) + 2\left(z - \dfrac{\pi}{4}\right) = 0$, $x - y + 2z = \dfrac{\pi}{2}$

Line: $\dfrac{x - 1}{1} = \dfrac{y - 1}{-1} = \dfrac{z - (\pi/4)}{2}$

35. $z = f(x, y) = \dfrac{4xy}{(x^2 + 1)(y^2 + 1)}$, $-2 \le x \le 2$, $0 \le y \le 3$

(a) Let $F(x, y, z) = \dfrac{4xy}{(x^2 + 1)(y^2 + 1)} - z$

$$\nabla F(x, y, z) = \frac{4y}{y^2 + 1}\left(\frac{x^2 + 1 - 2x^2}{(x^2 + 1)^2}\right)\mathbf{i} + \frac{4x}{x^2 + 1}\left(\frac{y^2 + 1 - 2y^2}{(y^2 + 1)^2}\right)\mathbf{j} - \mathbf{k}$$

$$= \frac{4y(1 - x^2)}{(y^2 + 1)(x^2 + 1)^2}\mathbf{i} + \frac{4x(1 - y^2)}{(x^2 + 1)(y^2 + 1)^2}\mathbf{j} - \mathbf{k}$$

$\nabla F(1, 1, 1) = -\mathbf{k}$.

Direction numbers: $0, 0, -1$.

Line: $x = 1$, $y = 1$, $z = 1 - t$

Tangent plane: $0(x - 1) + 0(y - 1) - 1(z - 1) = 0 \implies z = 1$

(b) $\nabla F\left(-1, 2, -\dfrac{4}{5}\right) = 0\mathbf{i} + \dfrac{-4(-3)}{(2)(5)^2}\mathbf{j} - \mathbf{k} = \dfrac{6}{25}\mathbf{j} - \mathbf{k}$

Line: $x = -1$, $y = 2 + \dfrac{6}{25}t$, $z = -\dfrac{4}{5} - t$

Plane: $0(x + 1) + \dfrac{6}{25}(y - 2) - 1\left(z + \dfrac{4}{5}\right) = 0$

$$6y - 12 - 25z - 20 = 0$$

$$6y - 25z - 32 = 0$$

(c)

(d) At $(1, 1, 1)$, the tangent plane is parallel to the xy-plane, implying that the surface is level there. At $\left(-1, 2, -\dfrac{4}{5}\right)$, the function does not change in the x-direction.

37. $F_x(x_0, y_0, z_0)(x - x_0) + F_y(x_0, y_0, z_0)(y - y_0) + F_z(x_0, y_0, z_0)(z - z_0) = 0$

(Theorem 12.13)

39. $F(x, y, z) = x^2 + y^2 - 5$ $G(x, y, z) = x - z$

$\nabla F(x, y, z) = 2x\mathbf{i} + 2y\mathbf{j}$ $\nabla G(x, y, z) = \mathbf{i} - \mathbf{k}$

$\nabla F(2, 1, 2) = 4\mathbf{i} + 2\mathbf{j}$ $\nabla G(2, 1, 2) = \mathbf{i} - \mathbf{k}$

(a) $\nabla F \times \nabla G = \begin{vmatrix} \mathbf{i} & \mathbf{j} & \mathbf{k} \\ 4 & 2 & 0 \\ 1 & 0 & -1 \end{vmatrix} = -2\mathbf{i} + 4\mathbf{j} - 2\mathbf{k} = -2(\mathbf{i} - 2\mathbf{j} + \mathbf{k})$

Direction numbers: $1, -2, 1$. $\dfrac{x - 2}{1} = \dfrac{y - 1}{-2} = \dfrac{z - 2}{1}$

(b) $\cos \theta = \dfrac{|\nabla F \cdot \nabla G|}{\|\nabla F\| \, \|\nabla G\|} = \dfrac{4}{\sqrt{20}\sqrt{2}} = \dfrac{2}{\sqrt{10}} = \dfrac{\sqrt{10}}{5}$; not orthogonal

41. $F(x, y, z) = x^2 + z^2 - 25$ $G(x, y, z) = y^2 + z^2 - 25$

$\nabla F = 2x\mathbf{i} + 2z\mathbf{k}$ $\nabla G = 2y\mathbf{j} + 2z\mathbf{k}$

$\nabla F(3, 3, 4) = 6\mathbf{i} + 8\mathbf{k}$ $\nabla G(3, 3, 4) = 6\mathbf{j} + 8\mathbf{k}$

—CONTINUED—

41. —CONTINUED—

(a) $\nabla F \times \nabla G = \begin{vmatrix} \mathbf{i} & \mathbf{j} & \mathbf{k} \\ 6 & 0 & 8 \\ 0 & 6 & 8 \end{vmatrix} = -48\mathbf{i} - 48\mathbf{j} + 36\mathbf{k} = -12(4\mathbf{i} + 4\mathbf{j} - 3\mathbf{k})$

Direction numbers: 4, 4, −3. $\dfrac{x-3}{4} = \dfrac{y-3}{4} = \dfrac{z-4}{-3}$

(b) $\cos\theta = \dfrac{|\nabla F \cdot \nabla G|}{\|\nabla F\| \, \|\nabla G\|} = \dfrac{64}{(10)(10)} = \dfrac{16}{25}$; not orthogonal

43. $\quad F(x, y, z) = x^2 + y^2 + z^2 - 6 \qquad\qquad G(x, y, z) = x - y - z$

$\nabla F(x, y, z) = 2x\mathbf{i} + 2y\mathbf{j} + 2z\mathbf{k} \qquad \nabla G(x, y, z) = \mathbf{i} - \mathbf{j} - \mathbf{k}$

$\nabla F(2, 1, 1) = 4\mathbf{i} + 2\mathbf{j} + 2\mathbf{k} \qquad\quad \nabla G(2, 1, 1) = \mathbf{i} - \mathbf{j} - \mathbf{k}$

(a) $\nabla F \times \nabla G = \begin{vmatrix} \mathbf{i} & \mathbf{j} & \mathbf{k} \\ 4 & 2 & 2 \\ 1 & -1 & -1 \end{vmatrix} = 6\mathbf{j} - 6\mathbf{k} = 6(\mathbf{j} - \mathbf{k})$
$\qquad$ (b) $\cos\theta = \dfrac{|\nabla F \cdot \nabla G|}{\|\nabla F\| \, \|\nabla G\|} = 0$; orthogonal

Direction numbers: 0, 1, −1. $x = 2, \dfrac{y-1}{1} = \dfrac{z-1}{-1}$

45. $f(x, y) = 6 - x^2 - \dfrac{y^2}{4}, \; g(x, y) = 2x + y$

(a) $F(x, y, z) = z + x^2 + \dfrac{y^2}{4} - 6 \qquad G(x, y, z) = z - 2x - y$

$\nabla F(x, y, z) = 2x\mathbf{i} + \dfrac{1}{2}y\mathbf{j} + \mathbf{k} \qquad \nabla G(x, y, z) = -2\mathbf{i} - \mathbf{j} + \mathbf{k}$

$\nabla F(1, 2, 4) = 2\mathbf{i} + \mathbf{j} + \mathbf{k} \qquad\quad \nabla G(1, 2, 4) = -2\mathbf{i} - \mathbf{j} + \mathbf{k}$

The cross product of these gradients is parallel to the curve of intersection.

$$\nabla F(1, 2, 4) \times \nabla G(1, 2, 4) = \begin{vmatrix} \mathbf{i} & \mathbf{j} & \mathbf{k} \\ 2 & 1 & 1 \\ -2 & -1 & 1 \end{vmatrix} = 2\mathbf{i} - 4\mathbf{j}$$

Using direction numbers 1, −2, 0, you get $x = 1 + t, \; y = 2 - 2t, \; z = 4$.

$$\cos\theta = \dfrac{\nabla F \cdot \nabla G}{\|\nabla F\| \, \|\nabla G\|} = \dfrac{-4 - 1 + 1}{\sqrt{6}\sqrt{6}} = \dfrac{-4}{6} \implies \theta \approx 48.2°$$

(b)

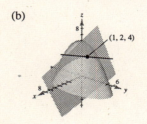

47. $F(x, y, z) = 3x^2 + 2y^2 - z - 15, \; (2, 2, 5)$

$\nabla F(x, y, z) = 6x\mathbf{i} + 4y\mathbf{j} - \mathbf{k}$

$\nabla F(2, 2, 5) = 12\mathbf{i} + 8\mathbf{j} - \mathbf{k}$

$\cos\theta = \dfrac{|\nabla F(2, 2, 5) \cdot \mathbf{k}|}{\|\nabla F(2, 2, 5)\|} = \dfrac{1}{\sqrt{209}}$

$\theta = \arccos\left(\dfrac{1}{\sqrt{209}}\right) \approx 86.03°$

49. $F(x, y, z) = x^2 - y^2 + z, \; (1, 2, 3)$

$\nabla F(x, y, z) = 2x\mathbf{i} - 2y\mathbf{j} + \mathbf{k}$

$\nabla F(1, 2, 3) = 2\mathbf{i} - 4\mathbf{j} + \mathbf{k}$

$\cos\theta = \dfrac{|\nabla F(1, 2, 3) \cdot \mathbf{k}|}{\|\nabla F(1, 2, 3)\|} = \dfrac{1}{\sqrt{21}}$

$\theta = \arccos\dfrac{1}{\sqrt{21}} \approx 77.40°$

51. $F(x, y, z) = 3 - x^2 - y^2 + 6y - z$

$\nabla F(x, y, z) = -2x\mathbf{i} + (-2y + 6)\mathbf{j} - \mathbf{k}$

$-2x = 0, \ x = 0$

$-2y + 6 = 0, \ y = 3$

$z = 3 - 0^2 - 3^2 + 6(3) = 12$

$(0, 3, 12)$ (vertex of paraboloid)

53. $T(x, y, z) = 400 - 2x^2 - y^2 - 4z^2, \ (4, 3, 10)$

$$\frac{dx}{dt} = -4kx \qquad \frac{dy}{dt} = -2ky \qquad \frac{dz}{dt} = -8kz$$

$x(t) = C_1 e^{-4kt} \qquad y(t) = C_2 e^{-2kt} \qquad z(t) = C_3 e^{-8kt}$

$x(0) = C_1 = 4 \qquad y(0) = C_2 = 3 \qquad z(0) = C_3 = 10$

$x = 4e^{-4kt} \qquad y = 3e^{-2kt} \qquad z = 10e^{-8kt}$

55. $F(x, y, z) = \dfrac{x^2}{a^2} + \dfrac{y^2}{b^2} + \dfrac{z^2}{c^2} - 1$

$F_x(x, y, z) = \dfrac{2x}{a^2}$

$F_y(x, y, z) = \dfrac{2y}{b^2}$

$F_z(x, y, z) = \dfrac{2z}{c^2}$

Plane: $\dfrac{2x_0}{a^2}(x - x_0) + \dfrac{2y_0}{b^2}(y - y_0) + \dfrac{2z_0}{c^2}(z - z_0) = 0$

$\dfrac{x_0 x}{a^2} + \dfrac{y_0 y}{b^2} + \dfrac{z_0 z}{c^2} = \dfrac{x_0^2}{a^2} + \dfrac{y_0^2}{b^2} + \dfrac{z_0^2}{c^2} = 1$

57. $F(x, y, z) = a^2 x^2 + b^2 y^2 - z^2$

$F_x(x, y, z) = 2a^2 x$

$F_y(x, y, z) = 2b^2 y$

$F_z(x, y, z) = -2z$

Plane: $2a^2 x_0(x - x_0) + 2b^2 y_0(y - y_0) - 2z_0(z - z_0) = 0$

$\qquad a^2 x_0 x + b^2 y_0 y - z_0 z = a^2 x_0^2 + b^2 y_0^2 - z_0^2 = 0$

Hence, the plane passes through the origin.

59. $f(x, y) = e^{x-y}$

$f_x(x, y) = e^{x-y}, \qquad f_y(x, y) = -e^{x-y}$

$f_{xx}(x, y) = e^{x-y}, \qquad f_{yy}(x, y) = e^{x-y}, \qquad f_{xy}(x, y) = -e^{x-y}$

(a) $P_1(x, y) \approx f(0, 0) + f_x(0, 0)x + f_y(0, 0)y = 1 + x - y$

(b) $P_2(x, y) \approx f(0, 0) + f_x(0, 0)x + f_y(0,0)y + \frac{1}{2}f_{xx}(0, 0)x^2 + f_{xy}(0, 0)xy + \frac{1}{2}f_{yy}(0, 0)y^2$

$\qquad = 1 + x - y + \frac{1}{2}x^2 - xy + \frac{1}{2}y^2$

(c) If $x = 0$, $P_2(0, y) = 1 - y + \frac{1}{2}y^2$. This is the second–degree Taylor polynomial for e^{-y}.

If $y = 0$, $P_2(x, 0) = 1 + x + \frac{1}{2}x^2$. This is the second–degree Taylor polynomial for e^x.

(d)

x	y	$f(x, y)$	$P_1(x, y)$	$P_2(x, y)$
0	0	1	1	1
0	0.1	0.9048	0.9000	0.9050
0.2	0.1	1.1052	1.1000	1.1050
0.2	0.5	0.7408	0.7000	0.7450
1	0.5	1.6487	1.5000	1.6250

(e)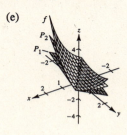

61. Given $w = F(x, y, z)$ where F is differentiable at

(x_0, y_0, z_0) and $\nabla F(x_0, y_0, z_0) \neq \mathbf{0}$,

the level surface of F at (x_0, y_0, z_0) is of the form $F(x, y, z) = C$ for some constant C. Let

$G(x, y, z) = F(x, y, z) - C = 0$.

Then $\nabla G(x_0, y_0, z_0) = \nabla F(x_0, y_0, z_0)$ where $\nabla G(x_0, y_0, z_0)$ is normal to $F(x_0, y_0, z_0) - C = 0$.

Therefore, $\nabla F(x_0, y_0 z_0)$ is normal to $F(x_0, y_0, z_0) = C$.

Section 12.8   Extrema of Functions of Two Variables

1. $g(x, y) = (x - 1)^2 + (y - 3)^2 \geq 0$

Relative minimum: $(1, 3, 0)$

$g_x = 2(x - 1) = 0 \Rightarrow x = 1$

$g_y = 2(y - 3) = 0 \Rightarrow y = 3$

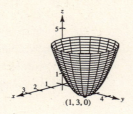

3. $f(x, y) = \sqrt{x^2 + y^2 + 1} \geq 1$

Relative minimum: $(0, 0, 1)$

Check: $f_x = \dfrac{x}{\sqrt{x^2 + y^2 + 1}} = 0 \Rightarrow x = 0$

$f_y = \dfrac{y}{\sqrt{x^2 + y^2 + 1}} = 0 \Rightarrow y = 0$

$f_{xx} = \dfrac{y^2 + 1}{(x^2 + y^2 + 1)^{3/2}}$, $f_{yy} = \dfrac{x^2 + 1}{(x^2 + y^2 + 1)^{3/2}}$, $f_{xy} = \dfrac{-xy}{(x^2 + y^2 + 1)^{3/2}}$

At the critical point $(0, 0)$, $f_{xx} > 0$ and $f_{xx} f_{yy} - (f_{xy})^2 > 0$. Therefore, $(0, 0, 1)$ is a relative minimum.

5. $f(x, y) = x^2 + y^2 + 2x - 6y + 6 = (x + 1)^2 + (y - 3)^2 - 4 \geq -4$

Relative minimum: $(-1, 3, -4)$

Check: $f_x = 2x + 2 = 0 \Rightarrow x = -1$

$f_y = 2y - 6 = 0 \Rightarrow y = 3$

$f_{xx} = 2$, $f_{yy} = 2$, $f_{xy} = 0$

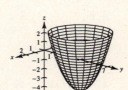

At the critical point $(-1, 3)$, $f_{xx} > 0$ and $f_{xx} f_{yy} - (f_{xy})^2 > 0$. Therefore, $(-1, 3, -4)$ is a relative minimum.

7. $f(x, y) = 2x^2 + 2xy + y^2 + 2x - 3$

$\left. \begin{array}{l} f_x = 4x + 2y + 2 = 0 \\ f_y = 2x + 2y = 0 \end{array} \right\}$ Solving simultaneously yields $x = -1$ and $y = 1$.

$f_{xx} = 4$, $f_{yy} = 2$, $f_{xy} = 2$

At the critical point $(-1, 1)$, $f_{xx} > 0$ and $f_{xx} f_{yy} - (f_{xy})^2 > 0$. Therefore, $(-1, 1, -4)$ is a relative minimum.

9. $f(x, y) = -5x^2 + 4xy - y^2 + 16x + 10$

$\left. \begin{array}{l} f_x = -10x + 4y + 16 = 0 \\ f_y = 4x - 2y = 0 \end{array} \right\}$ Solving simultaneously yields $x = 8$ and $y = 16$.

$f_{xx} = -10$, $f_{yy} = -2$, $f_{xy} = 4$

At the critical point $(8, 16)$, $f_{xx} < 0$ and $f_{xx} f_{yy} - (f_{xy})^2 > 0$. Therefore, $(8, 16, 74)$ is a relative maximum.

11. $f(x, y) = 2x^2 + 3y^2 - 4x - 12y + 13$

$f_x = 4x - 4 = 4(x - 1) = 0$ when $x = 1$.

$f_y = 6y - 12 = 6(y - 2) = 0$ when $y = 2$.

$f_{xx} = 4$, $f_{yy} = 6$, $f_{xy} = 0$

At the critical point $(1, 2)$, $f_{xx} > 0$ and $f_{xx} f_{yy} - (f_{xy})^2 > 0$. Therefore, $(1, 2, -1)$ is a relative minimum.

13. $f(x, y) = 2\sqrt{x^2 + y^2} + 3$

$\left. \begin{array}{l} f_x = \dfrac{2x}{\sqrt{x^2 + y^2}} = 0 \\ f_y = \dfrac{2y}{\sqrt{x^2 + y^2}} = 0 \end{array} \right\}$ $x = 0, y = 0$

Since $f(x, y) \geq 3$ for all (x, y), $(0, 0, 3)$ is relative minimum.

15. $g(x, y) = 4 - |x| - |y|$

$(0, 0)$ is the only critical point. Since $g(x, y) \le 4$ for all (x, y), $(0, 0, 4)$ is relative maximum.

17. $z = \dfrac{-4x}{x^2 + y^2 + 1}$

Relative minimum: $(1, 0, -2)$

Relative maximum: $(-1, 0, 2)$

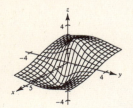

19. $z = (x^2 + 4y^2)e^{1-x^2-y^2}$

Relative minimum: $(0, 0, 0)$

Relative maxima: $(0, \pm 1, 4)$

Saddle points: $(\pm 1, 0, 1)$

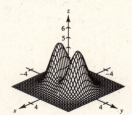

21. $h(x, y) = x^2 - y^2 - 2x - 4y - 4$

$h_x = 2x - 2 = 2(x - 1) = 0$ when $x = 1$.

$h_y = -2y - 4 = -2(y + 2) = 0$ when $y = -2$.

$h_{xx} = 2$, $h_{yy} = -2$, $h_{xy} = 0$

At the critical point $(1, -2)$, $h_{xx} h_{yy} - (h_{xy})^2 < 0$. Therefore, $(1, -2, -1)$ is a saddle point.

23. $h(x, y) = x^2 - 3xy - y^2$

$\left.\begin{array}{l} h_x = 2x - 3y = 0 \\ h_y = -3x - 2y = 0 \end{array}\right\}$ Solving simultaneously yields $x = 0$ and $y = 0$.

$h_{xx} = 2$, $h_{yy} = -2$, $h_{xy} = -3$

At the critical point $(0, 0)$, $h_{xx} h_{yy} - (h_{xy})^2 < 0$. Therefore, $(0, 0, 0)$ is a saddle point.

25. $f(x, y) = x^3 - 3xy + y^3$

$\left.\begin{array}{l} f_x = 3(x^2 - y) = 0 \\ f_y = 3(-x + y^2) = 0 \end{array}\right\}$ Solving by substitution yields two critical points $(0, 0)$ and $(1, 1)$.

$f_{xx} = 6x$, $f_{yy} = 6y$, $f_{xy} = -3$

At the critical point $(0, 0)$, $f_{xx} f_{yy} - (f_{xy})^2 < 0$. Therefore, $(0, 0, 0)$ is a saddle point. At the critical point $(1, 1)$, $f_{xx} = 6 > 0$ and $f_{xx} f_{yy} - (f_{xy})^2 > 0$. Therefore, $(1, 1, -1)$ is a relative minimum.

27. $f(x, y) = e^{-x} \sin y$

$\left.\begin{array}{l} f_x = -e^{-x} \sin y = 0 \\ f_y = e^{-x} \cos y = 0 \end{array}\right\}$ Since $e^{-x} > 0$ for all x and $\sin y$ and $\cos y$ are never both zero for a given value of y, there are no critical points.

29. $z = \dfrac{(x - y)^4}{x^2 + y^2} \ge 0$. $z = 0$ if $x = y \ne 0$.

Relative minimum at all points (x, x), $x \ne 0$.

31. $f_{xx} f_{yy} - (f_{xy})^2 = (9)(4) - 6^2 = 0$

Insufficient information.

33. $f_{xx} f_{yy} - (f_{xy})^2 = (-9)(6) - 10^2 < 0$

f has a saddle point at (x_0, y_0).

35. (a) The function f defined on a region R containing (x_0, y_0) has a relative minimum at (x_0, y_0) if $f(x, y) \geq f(x_0, y_0)$ for all (x, y) in R.

(b) The function f defined on a region R containing (x_0, y_0) has a relative maximum at (x_0, y_0) if $f(x, y) \leq f(x_0, y_0)$ for all (x, y) in R.

(c) A saddle point is a critical point which is not a relative extremum.

(d) See definition page 906.

37. 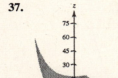   No extrema

39. 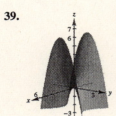   Saddle point

41. In this case, the point A will be a saddle point. The function could be

$$f(x, y) = xy.$$

43. $d = f_{xx} f_{yy} - f_{xy}^2 = (2)(8) - f_{xy}^2 = 16 - f_{xy}^2 > 0$

$\Rightarrow f_{xy}^2 < 16 \Rightarrow -4 < f_{xy} < 4$

45. $f(x, y) = x^3 + y^3$

$\left. \begin{matrix} f_x = 3x^2 = 0 \\ f_y = 3y^2 = 0 \end{matrix} \right\}$ Solving yields $x = y = 0$

$f_{xx} = 6x, \; f_{yy} = 6y, \; f_{xy} = 0$

At $(0, 0), f_{xx} f_{yy} - (f_{xy})^2 = 0$ and the test fails. $(0, 0, 0)$ is a saddle point.

47. $f(x, y) = (x - 1)^2(y + 4)^2 \geq 0$

$\left. \begin{matrix} f_x = 2(x - 1)(y + 4)^2 = 0 \\ f_y = 2(x - 1)^2(y + 4) = 0 \end{matrix} \right\}$ Solving yields the critical points $(1, a)$ and $(b, -4)$.

$f_{xx} = 2(y + 4)^2, f_{yy} = 2(x - 1)^2, f_{xy} = 4(x - 1)(y + 4)$

At both $(1, a)$ and $(b, -4), f_{xx} f_{yy} - (f_{xy})^2 = 0$ and the test fails.

Absolute minima: $(1, a, 0)$ and $(b, -4, 0)$

49. $f(x, y) = x^{2/3} + y^{2/3} \geq 0$

$\left. \begin{matrix} f_x = \dfrac{2}{3\sqrt[3]{x}} \\[2mm] f_y = \dfrac{2}{3\sqrt[3]{y}} \end{matrix} \right\}$ f_x and f_y are undefined at $x = 0, y = 0$. The critical point is $(0, 0)$.

$f_{xx} = -\dfrac{2}{9x\sqrt[3]{x}}, f_{yy} = -\dfrac{2}{9y\sqrt[3]{y}}, f_{xy} = 0$

At $(0, 0), f_{xx} f_{yy} - (f_{xy})^2$ is undefined and the test fails.

Absolute minimum: 0 at $(0, 0)$

51. $f(x, y, z) = x^2 + (y - 3)^2 + (z + 1)^2 \geq 0$

$\left. \begin{matrix} f_x = 2x = 0 \\ f_y = 2(y - 3) = 0 \\ f_z = 2(z + 1) = 0 \end{matrix} \right\}$ Solving yields the critical point $(0, 3, -1)$.

Absolute minimum: 0 at $(0, 3, -1)$

53. $f(x, y) = 12 - 3x - 2y$ has no critical points. On the line $y = x + 1, 0 \le x \le 1$,

$$f(x, y) = f(x) = 12 - 3x - 2(x + 1) = -5x + 10$$

and the maximum is 10, the minimum is 5. On the line $y = -2x + 4, 1 \le x \le 2$,

$$f(x, y) = f(x) = 12 - 3x - 2(-2x + 4) = x + 4$$

and the maximum is 6, the minimum is 5. On the line $y = -\frac{1}{2}x + 1, 0 \le x \le 2$,

$$f(x, y) = f(x) = 12 - 3x - 2\left(-\frac{1}{2}x + 1\right) = -2x + 10$$

and the maximum is 10, the minimum is 6.

Absolute maximum: 10 at $(0, 1)$

Absolute minimum: 5 at $(1, 2)$

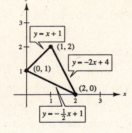

55. $f(x, y) = 3x^2 + 2y^2 - 4y$

$\left. \begin{array}{l} f_x = 6x = 0 \quad\;\; \Longrightarrow x = 0 \\ f_y = 4y - 4 = 0 \Longrightarrow y = 1 \end{array} \right\}$ $f(0, 1) = -2$

On the line $y = 4, -2 \le x \le 2$,

$$f(x, y) = f(x) = 3x^2 + 32 - 16 = 3x^2 + 16$$

and the maximum is 28, the minimum is 16. On the curve $y = x^2, -2 \le x \le 2$,

$$f(x, y) = f(x) = 3x^2 + 2(x^2)^2 - 4x^2 = 2x^4 - x^2 = x^2(2x^2 - 1)$$

and the maximum is 28, the minimum is $-\frac{1}{8}$.

Absolute maximum: 28 at $(\pm 2, 4)$

Absolute minimum: -2 at $(0, 1)$

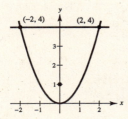

57. $f(x, y) = x^2 + xy, R = \{(x, y): |x| \le 2, |y| \le 1\}$

$\left. \begin{array}{l} f_x = 2x + y = 0 \\ f_y = x = 0 \end{array} \right\} x = y = 0$

$f(0, 0) = 0$

Along $y = 1, -2 \le x \le 2, f = x^2 + x, f' = 2x + 1 = 0 \Longrightarrow x = -\frac{1}{2}$.

Thus, $f(-2, 1) = 2, f\left(-\frac{1}{2}, 1\right) = -\frac{1}{4}$ and $f(2, 1) = 6$.

Along $y = -1, -2 \le x \le 2, f = x^2 - x, f' = 2x - 1 = 0 \Longrightarrow x = \frac{1}{2}$.

Thus, $f(-2, -1) = 6, f\left(\frac{1}{2}, -1\right) = -\frac{1}{4}, f(2, -1) = 2$.

Along $x = 2, -1 \le y \le 1, f = 4 + 2y \Longrightarrow f' = 2 \ne 0$.

Along $x = -2, -1 \le y \le 1, f = 4 - 2y \Longrightarrow f' = -2 \ne 0$.

Thus, the maxima are $f(2, 1) = 6$ and $f(-2, -1) = 6$ and the minima are $f\left(-\frac{1}{2}, 1\right) = -\frac{1}{4}$ and $f\left(\frac{1}{2}, -1\right) = -\frac{1}{4}$.

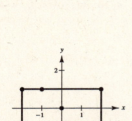

59. $f(x, y) = x^2 + 2xy + y^2, R = \{(x, y): x^2 + y^2 \le 8\}$

$\left. \begin{array}{l} f_x = 2x + 2y = 0 \\ f_y = 2x + 2y = 0 \end{array} \right\} y = -x$

$f(x, -x) = x^2 - 2x^2 + x^2 = 0$

On the boundary $x^2 + y^2 = 8$, we have $y^2 = 8 - x^2$ and $y = \pm\sqrt{8 - x^2}$. Thus,

$$f = x^2 \pm 2x\sqrt{8 - x^2} + (8 - x^2) = 8 \pm 2x\sqrt{8 - x^2}$$

$$f' = \pm[(8 - x^2)^{-1/2}(-2x^2) + 2(8 - x^2)^{1/2}] = \pm\frac{16 - 4x^2}{\sqrt{8 - x^2}}.$$

Then, $f' = 0$ implies $16 = 4x^2$ or $x = \pm 2$.

$$f(2, 2) = f(-2, -2) = 16 \quad \text{and} \quad f(2, -2) = f(-2, 2) = 0$$

Thus, the maxima are $f(2, 2) = 16$ and $f(-2, -2) = 16$, and the minima are $f(x, -x) = 0, |x| \le 2$.

61. $f(x, y) = \dfrac{4xy}{(x^2 + 1)(y^2 + 1)}, R = \{(x, y): 0 \le x \le 1, 0 \le y \le 1\}$

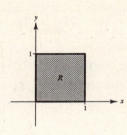

$f_x = \dfrac{4(1 - x^2)y}{(y^2 + 1)(x^2 + 1)^2} = 0 \implies x = 1 \text{ or } y = 0$

$f_y = \dfrac{4(1 - y^2)x}{(x^2 + 1)(y^2 + 1)^2} \implies x = 0 \text{ or } y = 1$

For $x = 0, y = 0$, also, and $f(0, 0) = 0$.

For $x = 1, y = 1, f(1, 1) = 1$.

The absolute maximum is $1 = f(1, 1)$.

The absolute minimum is $0 = f(0, 0)$. (In fact, $f(0, y) = f(x, 0) = 0$)

63. False

Let $f(x, y) = 1 - |x| - |y|$.

$(0, 0, 1)$ is a relative maximum, but $f_x(0, 0)$ and $f_y(0, 0)$ do not exist.

Section 12.9 Applications of Extrema of Functions of Two Variables

1. A point on the plane is given by $(x, y, 12 - 2x - 3y)$. The square of the distance from the origin to this point is

$$S = x^2 + y^2 + (12 - 2x - 3y)^2$$

$$S_x = 2x + 2(12 - 2x - 3y)(-2)$$

$$S_y = 2y + 2(12 - 2x - 3y)(-3)$$

From the equations $S_x = 0$ and $S_y = 0$, we obtain the system

$$5x + 6y = 24$$

$$3x + 5y = 18.$$

Solving simultaneously, we have $x = \frac{12}{7}, y = \frac{18}{7}$

$z = 12 - \frac{24}{7} - \frac{54}{7} = \frac{6}{7}$. Therefore, the distance from the origin to $\left(\frac{12}{7}, \frac{18}{7}, \frac{6}{7}\right)$ is

$$\sqrt{\left(\frac{12}{7}\right)^2 + \left(\frac{18}{7}\right)^2 + \left(\frac{6}{7}\right)^2} = \frac{6\sqrt{14}}{7}.$$

3. A point on the paraboloid is given by $(x, y, x^2 + y^2)$. The square of the distance from $(5, 5, 0)$ to a point on the paraboloid is given by

$$S = (x - 5)^2 + (y - 5)^2 + (x^2 + y^2)^2$$

$$S_x = 2(x - 5) + 4x(x^2 + y^2) = 0$$

$$S_y = 2(y - 5) + 4y(x^2 + y^2) = 0.$$

From the equations $S_x = 0$ and $S_y = 0$, we obtain the system

$$2x^3 + 2xy^2 + x - 5 = 0$$

$$2y^3 + 2x^2y + y - 5 = 0$$

Multiply the first equation by y and the second equation by x, and subtract to obtain $x = y$. Then, we have $x = 1$, $y = 1, z = 2$ and the distance is

$$\sqrt{(1 - 5)^2 + (1 - 5)^2 + (2 - 0)^2} = 6.$$

5. Let x, y and z be the numbers. Since $x + y + z = 30, z = 30 - x - y$.

$$P = xyz = 30xy - x^2y - xy^2$$

$$\left.\begin{array}{l} P_x = 30y - 2xy - y^2 = y(30 - 2x - y) = 0 \\ P_y = 30x - x^2 - 2xy = x(30 - x - 2y) = 0 \end{array}\right\} \begin{array}{l} 2x + y = 30 \\ x + 2y = 30 \end{array}$$

Solving simultaneously yields $x = 10, y = 10$, and $z = 10$.

7. Let x, y, and z be the numbers and let $S = x^2 + y^2 + z^2$. Since $x + y + z = 30$, we have

$$S = x^2 + y^2 + (30 - x - y)^2$$

$$\left.\begin{array}{l} S_x = 2x + 2(30 - x - y)(-1) = 0 \\ S_y = 2y + 2(30 - x - y)(-1) = 0 \end{array}\right\} \begin{array}{l} 2x + y = 30 \\ x + 2y = 30. \end{array}$$

Solving simultaneously yields $x = 10, y = 10$, and $z = 10$.

9. Let x, y, and z be the length, width, and height, respectively. Then the sum of the length and girth is given by $x + (2y + 2z) = 108$ or $x = 108 - 2y - 2z$. The volume is given by

$$V = xyz = 108zy - 2zy^2 - 2yz^2$$

$$V_y = 108z - 4yz - 2z^2 = z(108 - 4y - 2z) = 0$$

$$V_z = 108y - 2y^2 - 4yz = y(108 - 2y - 4z) = 0.$$

Solving the system $4y + 2z = 108$ and $2y + 4z = 108$, we obtain the solution $x = 36$ inches, $y = 18$ inches, and $z = 18$ inches.

11. Let $a + b + c = k$. Then

$$V = \frac{4\pi\,abc}{3} = \frac{4}{3}\pi\,ab(k - a - b)$$

$$= \frac{4}{3}\pi(kab - a^2b - ab^2)$$

$$V_a = \frac{4\pi}{3}(kb - 2ab - b^2) = 0 \Big] \; kb - 2ab - b^2 = 0$$

$$V_b = \frac{4\pi}{3}(ka - a^2 - 2ab) = 0 \Big\} \; ka - a^2 - 2ab = 0.$$

Solving this system simultaneously yields $a = b$ and substitution yields $b = k/3$. Therefore, the solution is $a = b = c = k/3$.

13. Let x, y, and z be the length, width, and height, respectively and let V_0 be the given volume.

Then $V_0 = xyz$ and $z = V_0/xy$. The surface area is

$$S = 2xy + 2yz + 2xz = 2\left(xy + \frac{V_0}{x} + \frac{V_0}{y}\right)$$

$$S_x = 2\left(y - \frac{V_0}{x^2}\right) = 0 \Big] \; x^2y - V_0 = 0$$

$$S_y = 2\left(x - \frac{V_0}{y^2}\right) = 0 \Big] \; xy^2 - V_0 = 0.$$

Solving simultaneously yields $x = \sqrt[3]{V_0}$, $y = \sqrt[3]{V_0}$, and $z = \sqrt[3]{V_0}$.

15. The distance from P to Q is $\sqrt{x^2 + 4}$. The distance from Q to R is $\sqrt{(y - x)^2 + 1}$. The distance from R to S is $10 - y$.

$$C = 3k\sqrt{x^2 + 4} + 2k\sqrt{(y - x)^2 + 1} + k(10 - y)$$

$$C_x = 3k\left(\frac{x}{\sqrt{x^2 + 4}}\right) + 2k\left(\frac{-(y - x)}{\sqrt{(y - x)^2 + 1}}\right) = 0$$

$$C_y = 2k\left(\frac{y - x}{\sqrt{(y - x)^2 + 1}}\right) - k = 0 \implies \frac{y - x}{\sqrt{(y - x)^2 + 1}} = \frac{1}{2}$$

$$3k\left(\frac{x}{\sqrt{x^2 + 4}}\right) + 2k\left(-\frac{1}{2}\right) = 0$$

$$\frac{x}{\sqrt{x^2 + 4}} = \frac{1}{3}$$

$$3x = \sqrt{x^2 + 4}$$

$$9x^2 = x^2 + 4$$

$$x^2 = \frac{1}{2}$$

$$x = \frac{\sqrt{2}}{2}$$

$$2(y - x) = \sqrt{(y - x)^2 + 1}$$

$$4(y - x)^2 = (y - x)^2 + 1$$

$$(y - x)^2 = \frac{1}{3}$$

$$y = \frac{1}{\sqrt{3}} + \frac{1}{\sqrt{2}} = \frac{2\sqrt{3} + 3\sqrt{2}}{6}$$

Therefore, $x = \dfrac{\sqrt{2}}{2} \approx 0.707$ km and $y = \dfrac{2\sqrt{3} + 3\sqrt{2}}{6} \approx 1.284$ kms.

17. Let h be the height of the trough and r the length of the slanted sides. We observe that the area of a trapezoidal cross section is given by

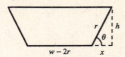

$$A = h\left[\frac{(w - 2r) + [(w - 2r) + 2x]}{2}\right] = (w - 2r + x)h$$

where $x = r \cos \theta$ and $h = r \sin \theta$. Substituting these expressions for x and h, we have

$$A(r, \theta) = (w - 2r + r \cos \theta)(r \sin \theta) = wr \sin \theta - 2r^2 \sin \theta + r^2 \sin \theta \cos \theta$$

Now

$$A_r(r, \theta) = w \sin \theta - 4r \sin \theta + 2r \sin \theta \cos \theta = \sin \theta(w - 4r + 2r \cos \theta) = 0 \implies w = r(4 - 2 \cos \theta)$$

$$A_\theta(r, \theta) = wr \cos \theta - 2r^2 \cos \theta + r^2 \cos 2\theta = 0.$$

Substituting the expression for w from $A_r(r, \theta) = 0$ into the equation $A_\theta(r, \theta) = 0$, we have

$$r^2(4 - 2 \cos \theta)\cos \theta - 2r^2 \cos \theta + r^2(2 \cos^2 \theta - 1) = 0$$

$$r^2(2 \cos \theta - 1) = 0 \text{ or } \cos \theta = \frac{1}{2}.$$

Therefore, the first partial derivatives are zero when $\theta = \pi/3$ and $r = w/3$. (Ignore the solution $r = \theta = 0$.) Thus, the trapezoid of maximum area occurs when each edge of width $w/3$ is turned up 60° from the horizontal.

19. $R(x_1, x_2) = -5x_1{}^2 - 8x_2{}^2 - 2x_1x_2 + 42x_1 + 102x_2$

$R_{x_1} = -10x_1 - 2x_2 + 42 = 0, \ 5x_1 + x_2 = 21$

$R_{x_2} = -16x_2 - 2x_1 + 102 = 0, \ x_1 + 8x_2 = 51$

Solving this system yields $x_1 = 3$ and $x_2 = 6$.

$R_{x_1x_1} = -10$

$R_{x_1x_2} = -2$

$R_{x_2x_2} = -16$

$R_{x_1x_1} < 0$ and $R_{x_1x_1}R_{x_2x_2} - (R_{x_1x_2})^2 > 0$

Thus, revenue is maximized when $x_1 = 3$ and $x_2 = 6$.

21. $P(x_1, x_2) = 15(x_1 + x_2) - C_1 - C_2$

$$= 15x_1 + 15x_2 - (0.02x_1{}^2 + 4x_1 + 500) - (0.05x_2{}^2 + 4x_2 + 275)$$

$$= -0.02x_1{}^2 - 0.05x_2{}^2 + 11x_1 + 11x_2 - 775$$

$P_{x_1} = -0.04x_1 + 11 = 0, \ x_1 = 275$

$P_{x_2} = -0.10x_2 + 11 = 0, \ x_2 = 110$

$P_{x_1x_1} = -0.04$

$P_{x_1x_2} = 0$

$P_{x_2x_2} = -0.10$

$P_{x_1x_1} < 0$ and $P_{x_1x_1}P_{x_2x_2} - (P_{x_1x_2})^2 > 0$

Therefore, profit is maximized when $x_1 = 275$ and $x_2 = 110$.

23. (a) $S(x, y) = d_1 + d_2 + d_3$

$$= \sqrt{(x - 0)^2 + (y - 0)^2} + \sqrt{(x + 2)^2 + (y - 2)^2} + \sqrt{(x - 4)^2 + (y - 2)^2}$$

$$= \sqrt{x^2 + y^2} + \sqrt{(x + 2)^2 + (y - 2)^2} + \sqrt{(x - 4)^2 + (y - 2)^2}$$

From the graph we see that the surface has a minimum.

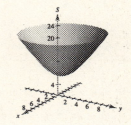

(b) $S_x(x, y) = \dfrac{x}{\sqrt{x^2 + y^2}} + \dfrac{x + 2}{\sqrt{(x + 2)^2 + (y - 2)^2}} + \dfrac{x - 4}{\sqrt{(x - 4)^2 + (y - 2)^2}}$

$S_y(x, y) = \dfrac{y}{\sqrt{x^2 + y^2}} + \dfrac{y - 2}{\sqrt{(x + 2)^2 + (y - 2)^2}} + \dfrac{y - 2}{\sqrt{(x - 4)^2 + (y - 2)^2}}$

(c) $-\nabla S(1, 1) = -S_x(1, 1)\mathbf{i} - S_y(1, 1)\mathbf{j} = -\dfrac{1}{\sqrt{2}}\mathbf{i} - \left(\dfrac{1}{\sqrt{2}} - \dfrac{2}{\sqrt{10}}\right)\mathbf{j}$

$$\tan \theta = \dfrac{\left(2/\sqrt{10}\right) - \left(1/\sqrt{2}\right)}{-1/\sqrt{2}} = 1 - \dfrac{2}{\sqrt{5}} \implies \theta \approx 186.027°$$

(d) $(x_2, y_2) = (x_1 - S_x(x_1, y_1)t, y_1 - S_y(x_1, y_1)t) = \left(1 - \dfrac{1}{\sqrt{2}}t, 1 + \left(\dfrac{2}{\sqrt{10}} - \dfrac{1}{\sqrt{2}}\right)t\right)$

$$S\left(1 - \dfrac{1}{\sqrt{2}}t, 1 + \left(\dfrac{2}{\sqrt{10}} - \dfrac{1}{\sqrt{2}}\right)t\right) = \sqrt{2 + \left(\dfrac{2\sqrt{10}}{5} - 2\sqrt{2}\right)t + \left(1 - \dfrac{2\sqrt{5}}{5} + \dfrac{2}{5}\right)t^2}$$

$$+ \sqrt{10 - \left(\dfrac{2\sqrt{10}}{5} + 2\sqrt{2}\right)t + \left(1 - \dfrac{2\sqrt{5}}{5} + \dfrac{2}{5}\right)t^2}$$

$$+ \sqrt{10 - \left(\dfrac{2\sqrt{10}}{5} - 4\sqrt{2}\right)t + \left(1 - \dfrac{2\sqrt{5}}{5} + \dfrac{2}{5}\right)t^2}$$

Using a computer algebra system, we find that the minimum occurs when $t \approx 1.344$. Thus, $(x_2, y_2) \approx (0.05, 0.90)$.

(e) $(x_3, y_3) = (x_2 - S_x(x_2, y_2)t, y_2 - S_y(x_2, y_2)t) \approx (0.05 + 0.03t, 0.90 - 0.26t)$

$S(0.05 + 0.03t, 0.90 - 0.26t) = \sqrt{(0.05 + 0.03t)^2 + (0.90 - 0.26t)^2} + \sqrt{(2.05 + 0.03t)^2 + (-1.10 - 0.26t)^2}$

$$+ \sqrt{(-3.95 + 0.03t)^2 + (-1.10 - 0.26t)^2}$$

Using a computer algebra system, we find that the minimum occurs when $t \approx 1.78$. Thus $(x_3, y_3) \approx (0.10, 0.44)$.

$(x_4, y_4) = (x_3 - S_x(x_3, y_3)t, y_3 - S_y(x_3, y_3)t) \approx (0.10 - 0.09t, 0.44 - 0.01t)$

$S(0.10 - 0.09t, 0.45 - 0.01t) = \sqrt{(0.10 - 0.09t)^2 + (0.45 - 0.01t)^2} + \sqrt{(2.10 - 0.09t)^2 + (-1.55 - 0.01t)^2}$

$$+ \sqrt{(-3.90 - 0.09t)^2 + (-1.55 - 0.01t)^2}$$

Using a computer algebra system, we find that the minimum occurs when $t \approx 0.44$. Thus, $(x_4, y_4) \approx (0.06, 0.44)$.

Note: The minimum occurs at $(x, y) = (0.0555, 0.3992)$.

(f) $-\nabla S(x, y)$ points in the direction that S *decreases* most rapidly. You would use $\nabla S(x, y)$ for maximization problems.

25. Write the equation to be maximized or minimized as a function of two variables. Set the partial derivatives equal to zero (or undefined) to obtain the critical points. Use the Second Partials Test to test for relative extrema using the critical points. Check the boundary points, too.

27. (a)

x	y	xy	x^2
-2	0	0	4
0	1	0	0
2	3	6	4
$\sum x_i = 0$	$\sum y_i = 4$	$\sum x_i y_i = 6$	$\sum x_i^2 = 8$

$$a = \frac{3(6) - 0(4)}{3(8) - 0^2} = \frac{3}{4}, \ b = \frac{1}{3}\left[4 - \frac{3}{4}(0)\right] = \frac{4}{3},$$

$$y = \frac{3}{4}x + \frac{4}{3}$$

(b) $S = \left(-\frac{3}{2} + \frac{4}{3} - 0\right)^2 + \left(\frac{4}{3} - 1\right)^2 + \left(\frac{3}{2} + \frac{4}{3} - 3\right)^2$

$$= \frac{1}{6}$$

29. (a)

x	y	xy	x^2
0	4	0	0
1	3	3	1
1	1	1	1
2	0	0	4
$\sum x_i = 4$	$\sum y_i = 8$	$\sum x_i y_i = 4$	$\sum x_i^2 = 6$

$$a = \frac{4(4) - 4(8)}{4(6) - 4^2} = -2, \ b = \frac{1}{4}[8 + 2(4)] = 4,$$

$$y = -2x + 4$$

(b) $S = (4 - 4)^2 + (2 - 3)^2 + (2 - 1)^2 + (0 - 0)^2 = 2$

31. $(0, 0), (1, 1), (3, 4), (4, 2), (5, 5)$

$$\sum x_i = 13, \qquad \sum y_i = 12,$$
$$\sum x_i y_i = 46, \qquad \sum x_i^2 = 51$$

$$a = \frac{5(46) - 13(12)}{5(51) - (13)^2} = \frac{74}{86} = \frac{37}{43}$$

$$b = \frac{1}{5}\left[12 - \frac{37}{43}(13)\right] = \frac{7}{43}$$

$$y = \frac{37}{43}x + \frac{7}{43}$$

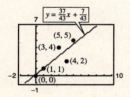

33. $(0, 6), (4, 3), (5, 0), (8, -4), (10, -5)$

$$\sum x_i = 27, \qquad \sum y_i = 0,$$
$$\sum x_i y_i = -70, \qquad \sum x_i^2 = 205$$

$$a = \frac{5(-70) - (27)(0)}{5(205) - (27)^2} = \frac{-350}{296} = -\frac{175}{148}$$

$$b = \frac{1}{5}\left[0 - \left(-\frac{175}{148}\right)(27)\right] = \frac{945}{148}$$

$$y = -\frac{175}{148}x + \frac{945}{148}$$

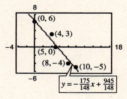

35. (a) $y = 1.7236x + 79.7334$

(b)

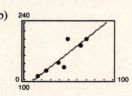

(c) For each one-year increase in age, the pressure changes by 1.7236 (slope of line).

37. $(1.0, 32), (1.5, 41), (2.0, 48), (2.5, 53)$

$$\sum x_i = 7, \sum y_i = 174, \sum x_i y_i = 322, \sum x_i^2 = 13.5$$

$$a = 14, b = 19, y = 14x + 19$$

When $x = 1.6$, $y = 41.4$ bushels per acre.

39. $S(a, b, c) = \sum_{i=1}^{n} (y_i - ax_i^2 - bx_i - c)^2$

$\dfrac{\partial S}{\partial a} = \sum_{i=1}^{n} -2x_i^2(y_i - ax_i^2 - bx_i - c) = 0$

$\dfrac{\partial S}{\partial b} = \sum_{i=1}^{n} -2x_i(y_i - ax_i^2 - bx_i - c) = 0$

$\dfrac{\partial S}{\partial c} = -2\sum_{i=1}^{n} (y_i - ax_i^2 - bx_i - c) = 0$

$a\sum_{i=1}^{n} x_i^4 + b\sum_{i=1}^{n} x_i^3 + c\sum_{i=1}^{n} x_i^2 = \sum_{i=1}^{n} x_i^2 y_i$

$a\sum_{i=1}^{n} x_i^3 + b\sum_{i=1}^{n} x_i^2 + c\sum_{i=1}^{n} x_i = \sum_{i=1}^{n} x_i y_i$

$a\sum_{i=1}^{n} x_i^2 + b\sum_{i=1}^{n} x_i + cn = \sum_{i=1}^{n} y_i$

41. $(-2, 0), (-1, 0), (0, 1), (1, 2), (2, 5)$

$\sum x_i = 0$

$\sum y_i = 8$

$\sum x_i^2 = 10$

$\sum x_i^3 = 0$

$\sum x_i^4 = 34$

$\sum x_i y_i = 12$

$\sum x_i^2 y_i = 22$

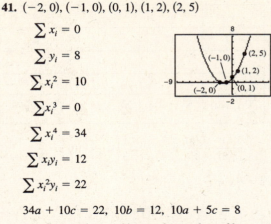

$34a + 10c = 22, \ 10b = 12, \ 10a + 5c = 8$

$a = \frac{3}{7}, \ b = \frac{6}{5}, \ c = \frac{26}{35}, \ y = \frac{3}{7}x^2 + \frac{6}{5}x + \frac{26}{35}$

43. $(0, 0), (2, 2), (3, 6), (4, 12)$

$\sum x_i = 9$

$\sum y_i = 20$

$\sum x_i^2 = 29$

$\sum x_i^3 = 99$

$\sum x_i^4 = 353$

$\sum x_i y_i = 70$

$\sum x_i^2 y_i = 254$

$353a + 99b + 29c = 254$

$99a + 29b + 9c = 70$

$29a + 9b + 4c = 20$

$a = 1, \ b = -1, \ c = 0, \ y = x^2 - x$

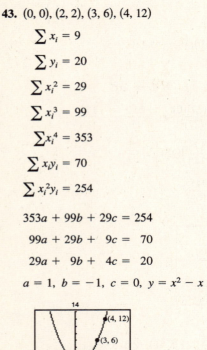

45. $(0, 0), (2, 15), (4, 30), (6, 50), (8, 65), (10, 70)$

$\sum x_i = 30,$

$\sum y_i = 230,$

$\sum x_i^2 = 220,$

$\sum x_i^3 = 1{,}800,$

$\sum x_i^4 = 15{,}664,$

$\sum x_i y_i = 1{,}670,$

$\sum x_i^2 y_i = 13{,}500$

$15{,}664a + 1{,}800b + 220c = 13{,}500$

$1{,}800a + 220b + 30c = 1{,}670$

$220a + 30b + 6c = 230$

$y = -\frac{25}{112}x^2 + \frac{541}{56}x - \frac{25}{14} \approx -0.22x^2 + 9.66x - 1.79$

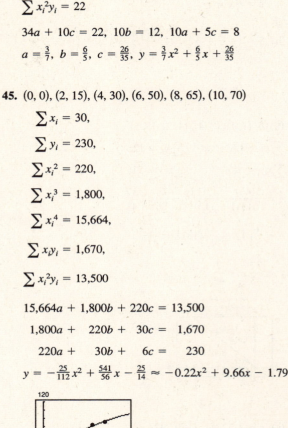

47. (a) $\ln P = -0.1499h + 9.3018$

(b) $\ln P = -0.1499h + 9.3018$

$P = e^{-0.1499h + 9.3018} = 10{,}957.7e^{-0.1499h}$

(c)

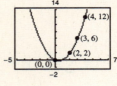

(d) Same answers.

Section 12.10 Lagrange Multipliers

1. Maximize $f(x, y) = xy$.

Constraint: $x + y = 10$

$\nabla f = \lambda \nabla g$

$y\mathbf{i} + x\mathbf{j} = \lambda(\mathbf{i} + \mathbf{j})$

$\left.\begin{array}{l} y = \lambda \\ x = \lambda \end{array}\right\} x = y$

$x + y = 10 = \implies x = y = 5$

$f(5, 5) = 25$

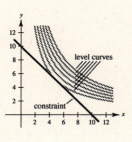

3. Minimize $f(x, y) = x^2 + y^2$.

Constraint: $x + y = 4$

$\nabla f = \lambda \nabla g$

$2x\mathbf{i} + 2y\mathbf{j} = \lambda\mathbf{i} + \lambda\mathbf{j}$

$\left.\begin{array}{l} 2x = \lambda \\ 2y = \lambda \end{array}\right\} x = y$

$x + y = 4 \implies x = y = 2$

$f(2, 2) = 8$

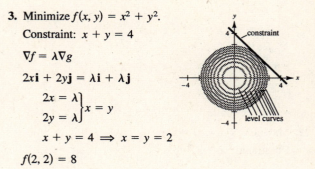

5. Minimize $f(x, y) = x^2 - y^2$.

Constraint: $x - 2y = -6$

$\nabla f = \lambda \nabla g$

$2x\mathbf{i} - 2y\mathbf{j} = \lambda\mathbf{i} - 2\lambda\mathbf{j}$

$2x = \lambda \quad \implies x = \dfrac{\lambda}{2}$

$-2y = -2\lambda \implies y = \lambda$

$x - 2y = -6 \quad \implies -\dfrac{3}{2}\lambda = -6$

$\lambda = 4, \ x = 2, \ y = 4$

$f(2, 4) = -12$

7. Maximize $f(x, y) = 2x + 2xy + y$.

Constraint: $2x + y = 100$

$\nabla f = \lambda \nabla g$

$(2 + 2y)\mathbf{i} + (2x + 1)\mathbf{j} = 2\lambda\mathbf{i} + \lambda\mathbf{j}$

$\left.\begin{array}{l} 2 + 2y = 2\lambda \implies y = \lambda - 1 \\ 2x + 1 = \lambda \quad \implies x = \dfrac{\lambda - 1}{2} \end{array}\right\} y = 2x$

$2x + y = 100 \implies 4x = 100$

$x = 25, \ y = 50$

$f(25, 50) = 2600$

9. Note: $f(x, y) = \sqrt{6 - x^2 - y^2}$ is maximum when $g(x, y)$ is maximum.

Maximize $g(x, y) = 6 - x^2 - y^2$.

Constraint: $x + y = 2$

$\left.\begin{array}{l} -2x = \lambda \\ -2y = \lambda \end{array}\right\} x = y$

$x + y = 2 \implies x = y = 1$

$f(1, 1) = \sqrt{g(1, 1)} = 2$

11. Maximize $f(x, y) = e^{xy}$.

Constraint: $x^2 + y^2 = 8$

$\left.\begin{array}{l} ye^{xy} = 2x\lambda \\ xe^{xy} = 2y\lambda \end{array}\right\} x = y$

$x^2 + y^2 = 8 \implies 2x^2 = 8$

$x = y = 2$

$f(2, 2) = e^4$

13. Maximize or minimize $f(x, y) = x^2 + 3xy + y^2$.

Constraint: $x^2 + y^2 \le 1$

Case 1: On the circle $x^2 + y^2 = 1$

$\left.\begin{array}{l} 2x + 3y = 2x\lambda \\ 3x + 2y = 2y\lambda \end{array}\right\} x^2 = y^2$

$x^2 + y^2 = 1 \implies x = \pm\dfrac{\sqrt{2}}{2}, y = \pm\dfrac{\sqrt{2}}{2}$

Maxima: $f\left(\pm\dfrac{\sqrt{2}}{2}, \pm\dfrac{\sqrt{2}}{2}\right) = \dfrac{5}{2}$

Minima: $f\left(\pm\dfrac{\sqrt{2}}{2}, \mp\dfrac{\sqrt{2}}{2}\right) = -\dfrac{1}{2}$

Case 2: Inside the circle

$\left.\begin{array}{l} f_x = 2x + 3y = 0 \\ f_y = 3x + 2y = 0 \end{array}\right\} x = y = 0$

$f_{xx} = 2, \ f_{yy} = 2, \ f_{xy} = 3, \ f_{xx}f_{yy} - (f_{xy})^2 \le 0$

Saddle point: $f(0, 0) = 0$

By combining these two cases, we have a maximum of $\frac{5}{2}$ at

$\left(\pm\dfrac{\sqrt{2}}{2}, \pm\dfrac{\sqrt{2}}{2}\right)$

and a minimum of $-\frac{1}{2}$ at

$\left(\pm\dfrac{\sqrt{2}}{2}, \mp\dfrac{\sqrt{2}}{2}\right)$.

15. Minimize $f(x, y, z) = x^2 + y^2 + z^2$.

Constraint: $x + y + z = 6$

$$\left.\begin{array}{l} 2x = \lambda \\ 2y = \lambda \\ 2z = \lambda \end{array}\right\} x = y = z$$

$x + y + z = 6 \implies x = y = z = 2$

$f(2, 2, 2) = 12$

17. Minimize $f(x, y, z) = x^2 + y^2 + z^2$.

Constraint: $x + y + z = 1$

$$\left.\begin{array}{l} 2x = \lambda \\ 2y = \lambda \\ 2z = \lambda \end{array}\right\} x = y = z$$

$x + y + z = 1 \implies x = y = z = \frac{1}{3}$

$f\left(\frac{1}{3}, \frac{1}{3}, \frac{1}{3}\right) = \frac{1}{3}$

19. Maximize $f(x, y, z) = xyz$.

Constraints: $x + y + z = 32$

$$x - y + z = 0$$

$\nabla f = \lambda \nabla g + \mu \nabla h$

$yz\mathbf{i} + xz\mathbf{j} + xy\mathbf{k} = \lambda(\mathbf{i} + \mathbf{j} + \mathbf{k}) + \mu(\mathbf{i} - \mathbf{j} + \mathbf{k})$

$$\left.\begin{array}{l} yz = \lambda + \mu \\ xz = \lambda - \mu \\ xy = \lambda + \mu \end{array}\right\} yz = xy \implies x = z$$

$$\left.\begin{array}{l} x + y + z = 32 \\ x - y + z = 0 \end{array}\right\} 2x + 2z = 32 \implies x = z = 8$$

$$y = 16$$

$f(8, 16, 8) = 1024$

21. Maximize $f(x, y, z) = xy + yz$.

Constraints: $x + 2y = 6$

$$x - 3z = 0$$

$\nabla f = \lambda \nabla g + \mu \nabla h$

$y\mathbf{i} + (x + z)\mathbf{j} + y\mathbf{k} = \lambda(\mathbf{i} + 2\mathbf{j}) + \mu(\mathbf{i} - 3\mathbf{k})$

$$\left.\begin{array}{l} y = \lambda + \mu \\ x + z = 2\lambda \\ y = -3\mu \end{array}\right\} y = \frac{3}{4}\lambda \implies x + z = \frac{8}{3}y$$

$x + 2y = 6 \implies y = 3 - \frac{x}{2}$

$x - 3z = 0 \implies z = \frac{x}{3}$

$$x + \frac{x}{3} = \frac{8}{3}\left(3 - \frac{x}{2}\right)$$

$$x = 3, y = \frac{3}{2}, z = 1$$

$f\left(3, \frac{3}{2}, 1\right) = 6$

23. Minimize the square of the distance $f(x, y) = x^2 + y^2$ subject to the constraint $2x + 3y = -1$.

$$\left.\begin{array}{l} 2x = 2\lambda \\ 2y = 3\lambda \end{array}\right\} y = \frac{3x}{2}$$

$2x + 3y = -1 \implies x = -\frac{2}{13}, y = -\frac{3}{13}$

The point on the line is $\left(-\frac{2}{13}, -\frac{3}{13}\right)$ and the desired distance is

$$d = \sqrt{\left(-\frac{2}{13}\right)^2 + \left(-\frac{3}{13}\right)^2} = \frac{\sqrt{13}}{13}.$$

25. Minimize the square of the distance

$$f(x, y, z) = (x - 2)^2 + (y - 1)^2 + (z - 1)^2$$

subject to the constraint $x + y + z = 1$.

$$\left.\begin{array}{l} 2(x - 2) = \lambda \\ 2(y - 1) = \lambda \\ 2(z - 1) = \lambda \end{array}\right\} y = z \text{ and } y = x - 1$$

$x + y + z = 1 \implies x + 2(x - 1) = 1$

$$x = 1, y = z = 0$$

The point on the plane is $(1, 0, 0)$ and the desired distance is

$$d = \sqrt{(1 - 2)^2 + (0 - 1)^2 + (0 - 1)^2} = \sqrt{3}.$$

27. Maximize $f(x, y, z) = z$ subject to the constraints
$x^2 + y^2 + z^2 = 36$ and $2x + y - z = 2$.

$$\left. \begin{array}{l} 0 = 2x\lambda + 2\mu \\ 0 = 2y\lambda + \mu \\ 1 = 2z\lambda - \mu \end{array} \right\} x = 2y$$

$x^2 + y^2 + z^2 = 36$

$2x + y - z = 2 \Rightarrow z = 2x + y - 2 = 5y - 2$

$(2y)^2 + y^2 + (5y - 2)^2 = 36$

$30y^2 - 20y - 32 = 0$

$15y^2 - 10y - 16 = 0$

$$y = \frac{5 \pm \sqrt{265}}{15}$$

Choosing the positive value for y we have the point

$$\left(\frac{10 + 2\sqrt{265}}{15}, \frac{5 + \sqrt{265}}{15}, \frac{-1 + \sqrt{265}}{3} \right).$$

29. Optimization problems that have restrictions or contstraints on the values that can be used to produce the optimal solution are called contrained optimization problems.

31. Maximize $V(x, y, z) = xyz$ subject to the constraint
$x + 2y + 2z = 108$.

$$\left. \begin{array}{l} yz = \lambda \\ xz = 2\lambda \\ xy = 2\lambda \end{array} \right\} y = z \text{ and } x = 2y$$

$x + 2y + 2z = 108 \Rightarrow 6y = 108, y = 18$

$x = 36, y = z = 18$

Volume is maximum when the dimensions are
$36 \times 18 \times 18$ inches

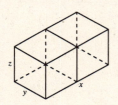

33. Minimize $C(x, y, z) = 5xy + 3(2xz + 2yz + xy)$ subject to the constraint $xyz = 480$.

$$\left. \begin{array}{l} 8y + 6z = yz\lambda \\ 8x + 6z = xz\lambda \\ 6x + 6y = xy\lambda \end{array} \right\} x = y, 4y = 3z$$

$xyz = 480 \Rightarrow \frac{4}{3}y^3 = 480$

$x = y = \sqrt[3]{360}, z = \frac{4}{3}\sqrt[3]{360}$

Dimensions: $\sqrt[3]{360} \times \sqrt[3]{360} \times \frac{4}{3}\sqrt[3]{360}$ feet

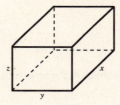

35. Maximize $V(x, y, z) = (2x)(2y)(2z) = 8xyz$ subject to the constraint $\dfrac{x^2}{a^2} + \dfrac{y^2}{b^2} + \dfrac{z^2}{c^2} = 1$.

$$\left. \begin{array}{l} 8yz = \dfrac{2x}{a^2}\lambda \\ 8xz = \dfrac{2y}{b^2}\lambda \\ 8xy = \dfrac{2z}{c^2}\lambda \end{array} \right\} \dfrac{x^2}{a^2} = \dfrac{y^2}{b^2} = \dfrac{z^2}{c^2}$$

$\dfrac{x^2}{a^2} + \dfrac{y^2}{b^2} + \dfrac{z^2}{c^2} = 1 \Rightarrow \dfrac{3x^2}{a^2} = 1, \dfrac{3y^2}{b^2} = 1, \dfrac{3z^2}{c^2} = 1$

$$x = \frac{a}{\sqrt{3}}, y = \frac{b}{\sqrt{3}}, z = \frac{c}{\sqrt{3}}$$

Therefore, the dimensions of the box are $\dfrac{2\sqrt{3}a}{3} \times \dfrac{2\sqrt{3}b}{3} \times \dfrac{2\sqrt{3}c}{3}$.

37. Using the formula Time $= \dfrac{\text{Distance}}{\text{Rate}}$, minimize $T(x, y) = \dfrac{\sqrt{d_1^2 + x^2}}{v_1} + \dfrac{\sqrt{d_2^2 + y^2}}{v_2}$ subject to the constraint $x + y = a$.

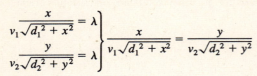

$$\left.\begin{array}{c} \dfrac{x}{v_1\sqrt{d_1^2 + x^2}} = \lambda \\[2mm] \dfrac{y}{v_2\sqrt{d_2^2 + y^2}} = \lambda \end{array}\right\} \quad \dfrac{x}{v_1\sqrt{d_1^2 + x^2}} = \dfrac{y}{v_2\sqrt{d_2^2 + y^2}}$$

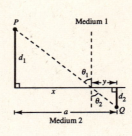

$x + y = a$

Since $\sin \theta_1 = \dfrac{x}{\sqrt{d_1^2 + x^2}}$ and $\sin \theta_2 = \dfrac{y}{\sqrt{d_2^2 + y^2}}$, we have

$$\dfrac{x/\sqrt{d_1^2 + x^2}}{v_1} = \dfrac{y/\sqrt{d_2^2 + y^2}}{v_2} \quad \text{or} \quad \dfrac{\sin \theta_1}{v_1} = \dfrac{\sin \theta_2}{v_2}.$$

39. Maximize $P(p, q, r) = 2pq + 2pr + 2qr$.

Constraint: $p + q + r = 1$

$\nabla P = \lambda \nabla g$

$$\left.\begin{array}{c} 2q + 2r = \lambda \\ 2p + 2r = \lambda \\ 2p + 2q = \lambda \end{array}\right\} \Rightarrow 3\lambda = 4(p + q + r) = 4(1)$$
$$\Rightarrow \lambda = \tfrac{4}{3}$$

$p + q + r = 1$

$$\left.\begin{array}{c} q + r = \tfrac{2}{3} \\ p + q + r = 1 \end{array}\right\} \Rightarrow p = \tfrac{1}{3}, q = \tfrac{1}{3}, r = \tfrac{1}{3}$$

$P\left(\tfrac{1}{3}, \tfrac{1}{3}, \tfrac{1}{3}\right) = 2\left(\tfrac{1}{3}\right)\left(\tfrac{1}{3}\right) + 2\left(\tfrac{1}{3}\right)\left(\tfrac{1}{3}\right) + 2\left(\tfrac{1}{3}\right)\left(\tfrac{1}{3}\right) = \tfrac{2}{3}$.

41. Maximize $P(x, y) = 100x^{0.25}y^{0.75}$

subject to the constraint $48x + 36y = 100{,}000$.

$$25x^{-0.75}y^{0.75} = 48\lambda \implies \left(\dfrac{y}{x}\right)^{0.75} = \dfrac{48\lambda}{25}$$

$$75x^{0.25}y^{-0.25} = 36\lambda \implies \left(\dfrac{x}{y}\right)^{0.25} = \dfrac{36\lambda}{75}$$

$$\left(\dfrac{y}{x}\right)^{0.75}\left(\dfrac{y}{x}\right)^{0.25} = \left(\dfrac{48\lambda}{25}\right)\left(\dfrac{75}{36\lambda}\right)$$

$$\dfrac{y}{x} = 4$$

$$y = 4x$$

$48x + 36y = 100{,}000 \implies 192x = 100{,}000$

$$x = \dfrac{3125}{6}, y = \dfrac{6250}{3}$$

Therefore, $P\left(\dfrac{3125}{6}, \dfrac{6250}{3}\right) \approx 147{,}314$.

43. Minimize $C(x, y) = 48x + 36y$ subject to the constraint $100x^{0.25}y^{0.75} = 20{,}000$.

$$48 = 25x^{-0.75}y^{0.75}\lambda \implies \left(\dfrac{y}{x}\right)^{0.75} = \dfrac{48}{25\lambda}$$

$$36 = 75x^{0.25}y^{-0.25}\lambda \implies \left(\dfrac{x}{y}\right)^{0.25} = \dfrac{36}{75\lambda}$$

$$\left(\dfrac{y}{x}\right)^{0.75}\left(\dfrac{y}{x}\right)^{0.25} = \left(\dfrac{48}{25\lambda}\right)\left(\dfrac{75\lambda}{36}\right)$$

$$\dfrac{y}{x} = 4 \implies y = 4x$$

$100x^{0.25}y^{0.75} = 20{,}000 \implies x^{0.25}(4x)^{0.75} = 200$

$$x = \dfrac{200}{4^{0.75}} = \dfrac{200}{2\sqrt{2}} = 50\sqrt{2}$$

$$y = 4x = 200\sqrt{2}$$

Therefore, $C\left(50\sqrt{2}, 200\sqrt{2}\right) \approx \$13{,}576.45$.

45. (a) Maximize $g(\alpha, \beta, \gamma) = \cos \alpha \cos \beta \cos \gamma$ subject to the constraint $\alpha + \beta + \gamma = \pi$.

$$\left.\begin{array}{r} -\sin \alpha \cos \beta \cos \gamma = \lambda \\ -\cos \alpha \sin \beta \cos \gamma = \lambda \\ -\cos \alpha \cos \beta \sin \gamma = \lambda \end{array}\right\} \tan \alpha = \tan \beta = \tan \gamma \Rightarrow \alpha = \beta = \gamma$$

$$\alpha + \beta + \gamma = \pi \Rightarrow \alpha = \beta = \gamma = \frac{\pi}{3}$$

$$g\left(\frac{\pi}{3}, \frac{\pi}{3}, \frac{\pi}{3}\right) = \frac{1}{8}$$

(b) $\alpha + \beta + \gamma = \pi \Rightarrow \gamma = \pi - (\alpha + \beta)$

$$g(\alpha + \beta) = \cos \alpha \cos \beta \cos (\pi - (\alpha + \beta))$$

$$= \cos \alpha \cos \beta [\cos \pi \cos(\alpha + \beta) + \sin \pi \sin(\alpha + \beta)]$$

$$= -\cos \alpha \cos \beta \cos(\alpha + \beta)$$

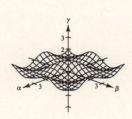

Review Exercises for Chapter 12

1. No, it is not the graph of a function.

3. $f(x, y) = e^{x^2 + y^2}$

The level curves are of the form

$$c = e^{x^2 + y^2}$$

$$\ln c = x^2 + y^2.$$

The level curves are circles centered at the origin.

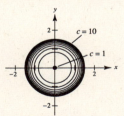

Generated by Mathematica

5. $f(x, y) = x^2 - y^2$

The level curves are of the form

$$c = x^2 - y^2$$

$$1 = \frac{x^2}{c} - \frac{y^2}{c}.$$

The level curves are hyperbolas.

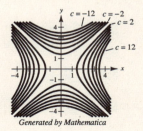

Generated by Mathematica

7. $f(x, y) = e^{-(x^2 + y^2)}$

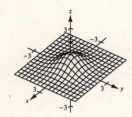

9. $f(x, y, z) = x^2 - y + z^2 = 1$

$$y = x^2 + z^2 - 1$$

Elliptic paraboloid

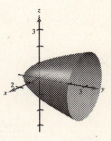

11. $\displaystyle\lim_{(x, y) \to (1, 1)} \frac{xy}{x^2 + y^2} = \frac{1}{2}$

Continuous except at $(0, 0)$.

13. $\displaystyle\lim_{(x, y) \to (0, 0)} \frac{-4x^2 y}{x^4 + y^2}$

For $y = x^2$, $\dfrac{-4x^2 y}{x^4 + y^2} = \dfrac{-4x^4}{x^4 + x^4} = -2$, for $x \neq 0$

For $y = 0$, $\dfrac{-4x^2 y}{x^4 + y^2} = 0$, for $x \neq 0$

Thus, the limit does not exist. Continuous except at $(0, 0)$.

15. $f(x, y) = e^x \cos y$

$f_x = e^x \cos y$

$f_y = -e^x \sin y$

17. $z = xe^y + ye^x$

$\dfrac{\partial z}{\partial x} = e^y + ye^x$

$\dfrac{\partial z}{\partial y} = xe^y + e^x$

19. $g(x, y) = \dfrac{xy}{x^2 + y^2}$

$g_x = \dfrac{y(x^2 + y^2) - xy(2x)}{(x^2 + y^2)^2} = \dfrac{y(y^2 - x^2)}{(x^2 + y^2)^2}$

$g_y = \dfrac{x(x^2 - y^2)}{(x^2 + y^2)^2}$

21. $f(x, y, z) = z \arctan \dfrac{y}{x}$

$f_x = \dfrac{z}{1 + (y^2/x^2)}\left(-\dfrac{y}{x^2}\right) = \dfrac{-yz}{x^2 + y^2}$

$f_y = \dfrac{z}{1 + (y^2/x^2)}\left(\dfrac{1}{x}\right) = \dfrac{xz}{x^2 + y^2}$

$f_z = \arctan \dfrac{y}{x}$

23. $u(x, t) = ce^{-n^2 t} \sin(nx)$

$\dfrac{\partial u}{\partial x} = cne^{-n^2 t} \cos(nx)$

$\dfrac{\partial u}{\partial t} = -cn^2 e^{-n^2 t} \sin(nx)$

25.

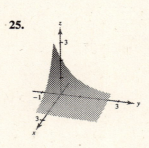

27. $f(x, y) = 3x^2 - xy + 2y^3$

$f_x = 6x - y$

$f_y = -x + 6y^2$

$f_{xx} = 6$

$f_{yy} = 12y$

$f_{xy} = -1$

$f_{yx} = -1$

29. $h(x, y) = x \sin y + y \cos x$

$h_x = \sin y - y \sin x$

$h_y = x \cos y + \cos x$

$h_{xx} = -y \cos x$

$h_{yy} = -x \sin y$

$h_{xy} = \cos y - \sin x$

$h_{yx} = \cos y - \sin x$

31. $z = x^2 - y^2$

$\dfrac{\partial z}{\partial x} = 2x$

$\dfrac{\partial^2 z}{\partial x^2} = 2$

$\dfrac{\partial z}{\partial y} = -2y$

$\dfrac{\partial^2 z}{\partial y^2} = -2$

Therefore, $\dfrac{\partial^2 z}{\partial x^2} + \dfrac{\partial^2 z}{\partial y^2} = 0$.

33. $z = \dfrac{y}{x^2 + y^2}$

$\dfrac{\partial z}{\partial x} = \dfrac{-2xy}{(x^2 + y^2)^2}$

$\dfrac{\partial^2 z}{\partial x^2} = -2y\left[\dfrac{-4x^2}{(x^2 + y^2)^3} + \dfrac{1}{(x^2 + y^2)^2}\right] = 2y\dfrac{3x^2 - y^2}{(x^2 + y^2)^3}$

$\dfrac{\partial z}{\partial y} = \dfrac{(x^2 + y^2) - 2y}{(x^2 + y^2)^2} = \dfrac{x^2 - y^2}{(x^2 + y^2)^2}$

$\dfrac{\partial^2 z}{\partial y^2} = \dfrac{(x^2 + y^2)^2(-2y) - 2(x^2 - y^2)(x^2 + y^2)(2y)}{(x^2 + y^2)^4}$

$= -2y\dfrac{3x^2 - y^2}{(x^2 + y^2)^3}$

Therefore, $\dfrac{\partial^2 z}{\partial x^2} + \dfrac{\partial^2 z}{\partial y^2} = 0$.

35. $z = x \sin \dfrac{y}{x}$

$dz = \dfrac{\partial z}{\partial x} dx + \dfrac{\partial z}{\partial y} dy = \left(\sin \dfrac{y}{x} - \dfrac{y}{x} \cos \dfrac{y}{x}\right) dx + \left(\cos \dfrac{y}{x}\right) dy$

37. $z^2 = x^2 + y^2$

$2z\,dx = 2x\,dx + 2y\,dy$

$dz = \dfrac{x}{z}dx + \dfrac{y}{z}dy = \dfrac{5}{13}\left(\dfrac{1}{2}\right) + \dfrac{12}{13}\left(\dfrac{1}{2}\right) = \dfrac{17}{26} \approx 0.654$ cm

Percentage error: $\dfrac{dz}{z} = \dfrac{17/26}{13} \approx 0.0503 \approx 5\%$

39. $V = \frac{1}{3}\pi r^2 h$

$dV = \frac{2}{3}\pi rh\,dr + \frac{1}{3}\pi r^2\,dh = \frac{2}{3}\pi(2)(5)\left(\pm\frac{1}{8}\right) + \frac{1}{3}\pi(2)^2\left(\pm\frac{1}{8}\right)$

$\qquad = \pm\frac{5}{6}\pi \pm \frac{1}{6}\pi = \pm\pi$ in.^{3}

41. $w = \ln(x^2 + y^2),\ x = 2t + 3,\ y = 4 - t$

Chain Rule: $\dfrac{dw}{dt} = \dfrac{\partial w}{\partial x}\dfrac{dx}{dt} + \dfrac{\partial w}{\partial y}\dfrac{dy}{dt}$

$\qquad = \dfrac{2x}{x^2 + y^2}(2) + \dfrac{2y}{x^2 + y^2}(-1)$

$\qquad = \dfrac{2(2t + 3)2}{(2t + 3)^2 + (4 - t)^2} - \dfrac{2(4 - t)}{(2t + 3)^2 + (4 - t)^2}$

$\qquad = \dfrac{10t + 4}{5t^2 + 4t + 25}$

Substitution: $w = \ln(x^2 + y^2) = \ln[(2t + 3)^2 + (4 - t)^2]$

$\dfrac{dw}{dt} = \dfrac{2(2t + 3)(2) - 2(4 - t)}{(2t + 3)^2 + (4 - t)^2} = \dfrac{10t + 4}{5t^2 + 4t + 25}$

43. $u = x^2 + y^2 + z^2,\ x = r\cos t,\ y = r\sin t,\ z = t$

Chain Rule: $\dfrac{\partial u}{\partial r} = \dfrac{\partial u}{\partial x}\dfrac{\partial x}{\partial r} + \dfrac{\partial u}{\partial y}\dfrac{\partial y}{\partial r} + \dfrac{\partial u}{\partial z}\dfrac{\partial z}{\partial r}$

$\qquad = 2x\cos t + 2y\sin t + 2z(0)$

$\qquad = 2(r\cos^2 t + r\sin^2 t) = 2r$

$\dfrac{\partial u}{\partial t} = \dfrac{\partial u}{\partial x}\dfrac{\partial x}{\partial t} + \dfrac{\partial u}{\partial y}\dfrac{\partial y}{\partial t} + \dfrac{\partial u}{\partial z}\dfrac{\partial z}{\partial t}$

$\qquad = 2x(-r\sin t) + 2y(r\cos t) + 2z$

$\qquad = 2(-r^2\sin t\cos t + r^2\sin t\cos t) + 2t$

$\qquad = 2t$

Substitution: $u(r, t) = r^2\cos^2 t + r^2\sin^2 t + t^2 = r^2 + t^2$

$\qquad \dfrac{\partial u}{\partial r} = 2r$

$\qquad \dfrac{\partial u}{\partial t} = 2t$

45. $x^2 y - 2yz - xz - z^2 = 0$

$2xy - 2y\dfrac{\partial z}{\partial x} - x\dfrac{\partial z}{\partial x} - z - 2z\dfrac{\partial z}{\partial x} = 0$

$\qquad\qquad \dfrac{\partial z}{\partial x} = \dfrac{-2xy + z}{-2y - x - 2z} = \dfrac{2xy - z}{x + 2y + 2z}$

$x^2 - 2y\dfrac{\partial z}{\partial y} - 2z - x\dfrac{\partial z}{\partial y} - 2z\dfrac{\partial z}{\partial y} = 0$

$\qquad\qquad \dfrac{\partial z}{\partial y} = \dfrac{-x^2 + 2z}{-2y - x - 2z} = \dfrac{x^2 - 2z}{x + 2y + 2z}$

47. $f(x, y) = x^2y$

$$\nabla f = 2xy\mathbf{i} + x^2\mathbf{j}$$

$$\nabla f(2, 1) = 4\mathbf{i} + 4\mathbf{j}$$

$$\mathbf{u} = \frac{1}{\sqrt{2}}\mathbf{v} = \frac{\sqrt{2}}{2}\mathbf{i} - \frac{\sqrt{2}}{2}\mathbf{j}$$

$$D_{\mathbf{u}}f(2, 1) = \nabla f(2, 1) \cdot \mathbf{u} = 2\sqrt{2} - 2\sqrt{2} = 0$$

49. $w = y^2 + xz$

$$\nabla w = z\mathbf{i} + 2y\mathbf{j} + x\mathbf{k}$$

$$\nabla w(1, 2, 2) = 2\mathbf{i} + 4\mathbf{j} + \mathbf{k}$$

$$\mathbf{u} = \frac{1}{3}\mathbf{v} = \frac{2}{3}\mathbf{i} - \frac{1}{3}\mathbf{j} + \frac{2}{3}\mathbf{k}$$

$$D_{\mathbf{u}}w(1, 2, 2) = \nabla w(1, 2, 2) \cdot \mathbf{u} = \frac{4}{3} - \frac{4}{3} + \frac{2}{3} = \frac{2}{3}$$

51. $z = \dfrac{y}{x^2 + y^2}$

$$\nabla z = -\frac{2xy}{(x^2 + y^2)^2}\mathbf{i} + \frac{x^2 - y^2}{(x^2 + y^2)^2}\mathbf{j}$$

$$\nabla z(1, 1) = -\frac{1}{2}\mathbf{i} = \left\langle -\frac{1}{2}, 0 \right\rangle$$

$$\|\nabla z(1, 1)\| = \frac{1}{2}$$

53. $z = e^{-x}\cos y$

$$\nabla z = -e^{-x}\cos y\mathbf{i} - e^{-x}\sin y\mathbf{j}$$

$$\nabla z\left(0, \frac{\pi}{4}\right) = -\frac{\sqrt{2}}{2}\mathbf{i} - \frac{\sqrt{2}}{2}\mathbf{j} = \left\langle -\frac{\sqrt{2}}{2}, -\frac{\sqrt{2}}{2} \right\rangle$$

$$\left\| \nabla z\left(0, \frac{\pi}{4}\right) \right\| = 1$$

55. $9x^2 - 4y^2 = 65$

$$f(x, y) = 9x^2 - 4y^2$$

$$\nabla f(x, y) = 18x\mathbf{i} - 8y\mathbf{j}$$

$$\nabla f(3, 2) = 54\mathbf{i} - 16\mathbf{j}$$

Unit normal: $\dfrac{54\mathbf{i} - 16\mathbf{j}}{\|54\mathbf{i} - 16\mathbf{j}\|} = \dfrac{1}{\sqrt{793}}(27\mathbf{i} - 8\mathbf{j})$

57. $F(x, y, z) = x^2y - z = 0$

$$\nabla F = 2xy\mathbf{i} + x^2\mathbf{j} - \mathbf{k}$$

$$\nabla F(2, 1, 4) = 4\mathbf{i} + 4\mathbf{j} - \mathbf{k}$$

Therefore, the equation of the tangent plane is

$$4(x - 2) + 4(y - 1) - (z - 4) = 0 \quad \text{or}$$

$$4x + 4y - z = 8,$$

and the equation of the normal line is

$$x = 4t + 2, \, y = 4t + 1, \, z = -t + 4.$$

59. $F(x, y, z) = x^2 + y^2 - 4x + 6y + z + 9 = 0$

$$\nabla F = (2x - 4)\mathbf{i} + (2y + 6)\mathbf{j} + \mathbf{k}$$

$$\nabla F(2, -3, 4) = \mathbf{k}$$

Therefore, the equation of the tangent plane is

$$z - 4 = 0 \quad \text{or} \quad z = 4,$$

and the equation of the normal line is

$$x = 2, \, y = -3, \, z = 4 + t.$$

61. $F(x, y, z) = x^2 - y^2 - z = 0$

$$G(x, y, z) = 3 - z = 0$$

$$\nabla F = 2x\mathbf{i} - 2y\mathbf{j} - \mathbf{k}$$

$$\nabla G = -\mathbf{k}$$

$$\nabla F(2, 1, 3) = 4\mathbf{i} - 2\mathbf{j} - \mathbf{k}$$

$$\nabla F \times \nabla G = \begin{vmatrix} \mathbf{i} & \mathbf{j} & \mathbf{k} \\ 4 & -2 & -1 \\ 0 & 0 & -1 \end{vmatrix} = 2(\mathbf{i} + 2\mathbf{j})$$

Therefore, the equation of the tangent line is

$$\frac{x - 2}{1} = \frac{y - 1}{2}, \, z = 3.$$

63. $f(x, y, z) = x^2 + y^2 + z^2 - 14$

$$\nabla f(x, y, z) = 2x\mathbf{i} + 2y\mathbf{j} + 2z\mathbf{k}$$

$$\nabla f(2, 1, 3) = 4\mathbf{i} + 2\mathbf{j} + 6\mathbf{k} \text{ Normal vector to plane.}$$

$$\cos \theta = \frac{|\mathbf{n} \cdot \mathbf{k}|}{\|\mathbf{n}\|} = \frac{6}{\sqrt{56}} = \frac{3\sqrt{14}}{14}$$

$$\theta = 36.7°$$

65. $f(x, y) = x^3 - 3xy + y^2$

$\qquad f_x = 3x^2 - 3y = 3(x^2 - y) = 0$

$\qquad f_y = -3x + 2y = 0$

$\qquad f_{xx} = 6x$

$\qquad f_{yy} = 2$

$\qquad f_{xy} = -3$

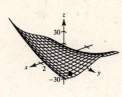

From $f_x = 0$, we have $y = x^2$. Substituting this into $f_y = 0$, we have $-3x + 2x^2 = x(2x - 3) = 0$. Thus, $x = 0$ or $\frac{3}{2}$.

At the critical point $(0, 0)$, $f_{xx}f_{yy} - (f_{xy})^2 < 0$. Therefore, $(0, 0, 0)$ is a saddle point.

At the critical point $\left(\frac{3}{2}, \frac{9}{4}\right)$, $f_{xx}f_{yy} - (f_{xy})^2 > 0$ and $f_{xx} > 0$. Therefore, $\left(\frac{3}{2}, \frac{9}{4}, -\frac{27}{16}\right)$ is a relative minimum.

67. $f(x, y) = xy + \dfrac{1}{x} + \dfrac{1}{y}$

$\qquad f_x = y - \dfrac{1}{x^2} = 0, \ x^2 y = 1$

$\qquad f_y = x - \dfrac{1}{y^2} = 0, \ xy^2 = 1$

Thus, $x^2 y = xy^2$ or $x = y$ and substitution yields the critical point $(1, 1)$.

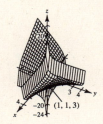

$\qquad f_{xx} = \dfrac{2}{x^3}$

$\qquad f_{xy} = 1$

$\qquad f_{yy} = \dfrac{2}{y^3}$

At the critical point $(1, 1)$, $f_{xx} = 2 > 0$ and $f_{xx}f_{yy} - (f_{xy})^2 = 3 > 0$. Thus, $(1, 1, 3)$ is a relative minimum.

69. The level curves are hyperbolas. There is a critical point at $(0, 0)$, but there are no relative extrema. The gradient is normal to the level curve at any given point at (x_0, y_0).

71. $P(x_1, x_2) = R - C_1 - C_2$

$\qquad\qquad = [225 - 0.4(x_1 + x_2)](x_1 + x_2) - (0.05x_1^2 + 15x_1 + 5400) - (0.03x_2^2 + 15x_2 + 6100)$

$\qquad\qquad = -0.45x_1^2 - 0.43x_2^2 - 0.8x_1 x_2 + 210x_1 + 210x_2 - 11,500$

$\qquad P_{x_1} = -0.9x_1 - 0.8x_2 + 210 = 0$

$\qquad\qquad\qquad 0.9x_1 + 0.8x_2 = 210$

$\qquad P_{x_2} = -0.86x_2 - 0.8x_1 + 210 = 0$

$\qquad\qquad\qquad 0.8x_1 + 0.86x_2 = 210$

Solving this system yields $x_1 \approx 94$ and $x_2 \approx 157$.

$\qquad\qquad P_{x_1 x_1} = -0.9$

$\qquad\qquad P_{x_1 x_2} = -0.8$

$\qquad\qquad P_{x_2 x_2} = -0.86$

$\qquad\qquad P_{x_1 x_1} < 0$

$\qquad P_{x_1 x_1} P_{x_2 x_2} - (P_{x_1 x_2})^2 > 0$

Therefore, profit is maximum when $x_1 \approx 94$ and $x_2 \approx 157$.

73. Maximize $f(x, y) = 4x + xy + 2y$ subject to the constraint $20x + 4y = 2000$.

$$\left. \begin{array}{r} 4 + y = 20\lambda \\ x + 2 = 4\lambda \end{array} \right\} 5x - y = -6$$

$$20x + 4y = 2000 \implies \quad 5x + y = \ 500$$

$$\underline{ \quad 5x - y = \ -6}$$

$$ \quad 10x = \ 494$$

$$ \quad x = 49.4$$

$$ \quad y = 253$$

$f(49.4, 253) = 13{,}201.8$

75. (a) $y = 2.29t + 2.34$

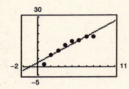

(b)

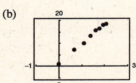

Yes, the data appears more linear.

(c) $y = 8.37 \ln t + 1.54$

(d)

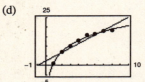

The logarithmic model is a better fit.

77. Optimize $f(x, y, z) = xy + yz + xz$ subject to the constraint $x + y + z = 1$.

$$\left. \begin{array}{l} y + z = \lambda \\ x + z = \lambda \\ x + y = \lambda \end{array} \right\} x = y = z$$

$$x + y + z = 1 \implies x = y = z = \tfrac{1}{3}$$

Maximum: $f\left(\tfrac{1}{3}, \tfrac{1}{3}, \tfrac{1}{3}\right) = \tfrac{1}{3}$

79. $PQ = \sqrt{x^2 + 4}, QR = \sqrt{y^2 + 1}, RS = z; x + y + z = 10$

$C = 3\sqrt{x^2 + 4} + 2\sqrt{y^2 + 1} + z$

Constraint: $x + y + z = 10$

$\nabla C = \lambda \nabla g$

$$\frac{3x}{\sqrt{x^2 + 4}}\mathbf{i} + \frac{2y}{\sqrt{y^2 + 1}}\mathbf{j} + \mathbf{k} = \lambda[\mathbf{i} + \mathbf{j} + \mathbf{k}]$$

$3x = \lambda\sqrt{x^2 + 4}$

$2y = \lambda\sqrt{y^2 + 1}$

$\ 1 = \lambda$

$9x^2 = x^2 + 4 \implies x^2 = \dfrac{1}{2}$

$4y^2 = y^2 + 1 \implies y^2 = \dfrac{1}{3}$

Hence, $x = \dfrac{\sqrt{2}}{2}, y = \dfrac{\sqrt{3}}{3}, z = 10 - \dfrac{\sqrt{2}}{2} - \dfrac{\sqrt{3}}{3} \approx 8.716$ m.

Problem Solving for Chapter 12

1. (a) The three sides have lengths 5, 6, and 5.

Thus, $s = \frac{16}{2} = 8$ and $A = \sqrt{8(3)(2)(3)} = 12$

(b) Let $f(a, b, c) = (\text{area})^2 = s(s - a)(s - b)(s - c)$, subject to the constraint $a + b + c = \text{constant (perimeter)}$.

Using Lagrange multipliers,

$-s(s - b)(s - c) = \lambda$

$-s(s - a)(s - c) = \lambda$

$-s(s - a)(s - b) = \lambda$

From the first 2 equations $s - b = s - a \implies a = b$.

Similarly, $b = c$ and hence $a = b = c$ which is an equilateral triangle.

(c) Let $f(a, b, c) = a + b + c$, subject to $(\text{Area})^2 = s(s - a)(s - b)(s - c)$ constant.

Using Lagrange multipliers,

$1 = -\lambda s(s - b)(s - c)$

$1 = -\lambda s(s - a)(s - c)$

$1 = -\lambda s(s - a)(s - b)$

Hence, $s - a = s - b \implies a = b$ and $a = b = c$.

3. (a) $F(x, y, z) = xyz - 1 = 0$

$F_x = yz, F_y = xz, F_z = xy$

Tangent plane:

$y_0 z_0(x - x_0) + x_0 z_0(y - y_0) + x_0 y_0(z - z_0) = 0$

$y_0 z_0 x + x_0 z_0 y + x_0 y_0 z = 3x_0 y_0 z_0 = 3$

(b) $V = \frac{1}{3}(\text{base})(\text{height})$

$= \frac{1}{3}\left(\frac{1}{2}\frac{3}{y_0 z_0}\frac{3}{x_0 z_0}\right)\left(\frac{3}{x_0 y_0}\right) = \frac{9}{2}$

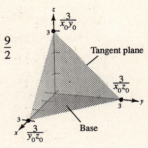

5. We cannot use Theorem 12.9 since f is not a differentiable function of x and y. Hence, we use the definition of directional derivatives.

$D_{\mathbf{u}} f(x, y) = \lim_{t \to 0} \frac{f(x + t\cos\theta, y + t\sin\theta) - f(x, y)}{t}$

$D_{\mathbf{u}} f(0, 0) = \lim_{t \to 0} \frac{f\left[0 + \left(\frac{t}{\sqrt{2}}\right), 0 + \left(\frac{t}{\sqrt{2}}\right)\right] - f(0, 0)}{t}$

$= \lim_{t \to 0} \frac{1}{t}\left[\frac{4\left(\frac{t}{\sqrt{2}}\right)\left(\frac{t}{\sqrt{2}}\right)}{\left(\frac{t^2}{2}\right) + \left(\frac{t^2}{2}\right)}\right] = \lim_{t \to 0} \frac{1}{t}\left[\frac{2t^2}{t^2}\right] = \lim_{t \to 0} \frac{2}{t}$ which does not exist.

If $f(0, 0) = 2$, then

$D_{\mathbf{u}} f(0, 0) = \lim_{t \to 0} \frac{f\left(0 + \frac{t}{\sqrt{2}}, 0 + \frac{t}{\sqrt{2}}\right) - 2}{t} = \lim_{t \to 0} \frac{1}{t}\left[\frac{2t^2}{t^2} - 2\right] = 0$

which implies that the directional derivative exists.

7. $H = k(5xy + 6xz + 6yz)$

$z = \dfrac{1000}{xy} \implies H = k\left(5xy + \dfrac{6000}{y} + \dfrac{6000}{x}\right).$

$H_x = 5y - \dfrac{6000}{x^2} = 0 \implies 5yx^2 = 6000$

By symmetry, $x = y \implies x^3 = y^3 = 1200.$

Thus, $x = y = 2\sqrt[3]{150}$ and $z = \dfrac{5}{3}\sqrt[3]{150}.$

9. (a) $\dfrac{\partial f}{\partial x} = Cax^{a-1}y^{1-a}, \dfrac{\partial f}{\partial y} = C(1-a)x^a y^{-a}$

$x\dfrac{\partial f}{\partial x} + y\dfrac{\partial f}{\partial y} = Cax^a y^{1-a} + C(1-a)x^a y^{1-a}$

$= [Ca + C(1-a)]x^a y^{1-a}$

$= Cx^a y^{1-a} = f$

(b) $f(tx, ty) = C(tx)^a(ty)^{1-a} = Ct^a x^a t^{1-a}y^{1-a}$

$= Cx^a y^{1-a}(t) = tf(x, y)$

11. (a) $x = 64(\cos 45°)t = 32\sqrt{2}t$

$y = 64(\sin 45°)t - 16t^2 = 32\sqrt{2}t - 16t^2$

(b) $\tan \alpha = \dfrac{y}{x + 50}$

$\alpha = \arctan\left(\dfrac{y}{x+50}\right) = \arctan\left(\dfrac{32\sqrt{2}t - 16t^2}{32\sqrt{2}t + 50}\right)$

(c) $\dfrac{d\alpha}{dt} = \dfrac{1}{1 + \left(\dfrac{32\sqrt{2}t - 16t^2}{32\sqrt{2}t + 50}\right)^2} \cdot \dfrac{-64\left(8\sqrt{2}t^2 + 25t - 25\sqrt{2}\right)}{(32\sqrt{2}t + 50)^2} = \dfrac{-16\left(8\sqrt{2}t^2 + 25t - 25\sqrt{2}\right)}{64t^4 - 256\sqrt{2}t^3 + 1024t^2 + 800\sqrt{2}t + 625}$

(d)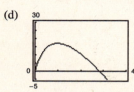

No. The rate of change of α is greatest when the projectile is closest to the camera.

(e) $\dfrac{d\alpha}{dt} = 0$ when

$8\sqrt{2}t^2 + 25t - 25\sqrt{2} = 0$

$t = \dfrac{-25 + \sqrt{25^2 - 4(8\sqrt{2})(-25\sqrt{2})}}{2(8\sqrt{2})} \approx 0.98$ second.

No, the projectile is at its maximum height when $dy/dt = 32\sqrt{2} - 32t = 0$ or $t = \sqrt{2} \approx 1.41$ seconds.

13. (a) There is a minimum at $(0, 0, 0)$, maxima at $(0, \pm1, 2/e)$ and saddle point at $(\pm1, 0, 1/e)$: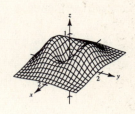

$f_x = (x^2 + 2y^2)e^{-(x^2+y^2)}(-2x) + (2x)e^{-(x^2+y^2)}$

$= e^{-(x^2+y^2)}[(x^2 + 2y^2)(-2x) + 2x]$

$= e^{-(x^2+y^2)}[-2x^3 + 4xy^2 + 2x] = 0 \implies x^3 + 2xy^2 - x = 0$

$f_y = (x^2 + 2y^2)e^{-(x^2+y^2)}(-2y) + (4y)e^{-(x^2+y^2)}$

$= e^{-(x^2+y^2)}[(x^2 + 2y^2)(-2y) + 4y]$

$= e^{-(x^2+y^2)}[-4y^3 - 2x^2y + 4y] = 0 \implies 2y^3 + x^2y - 2y = 0$

Solving the two equations $x^3 + 2xy^2 - x = 0$ and $2y^3 + x^2y - 2y = 0$, you obtain the following critical points: $(0, \pm1)$, $(\pm1, 0)$, $(0, 0)$. Using the second derivative test, you obtain the results above.

—CONTINUED—

13. —CONTINUED—

(b) As in part (a), you obtain

$$f_x = e^{-(x^2+y^2)}[2x(x^2 - 1 - 2y^2)]$$
$$f_y = e^{-(x^2+y^2)}[2y(2 + x^2 - 2y^2)]$$

The critical numbers are $(0, 0)$, $(0, \pm 1)$, $(\pm 1, 0)$. These yield

$(\pm 1, 0, -1/e)$ minima
$(0, \pm 1, 2/e)$ maxima
$(0, 0, 0)$ saddle

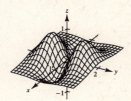

(c) In general, for $\alpha > 0$ you obtain

$(0, 0, 0)$ minimum

$(0, \pm 1, \beta/e)$ maxima

$(\pm 1, 0, \alpha/e)$ saddle

For $\alpha < 0$, you obtain

$(\pm 1, 0, \alpha/e)$ minima

$(0, \pm 1, \beta/e)$ maxima

$(0, 0, 0)$ saddle

15. (a)
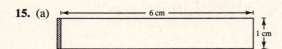

(c) The height has more effect since the shaded region in (b) is larger than the shaded region in (a).

(b)
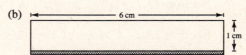

(d) $A = hl \implies dA = l\,dh + h\,dl$

If $dl = 0.01$ and $dh = 0$, then $dA = 1(0.01) = 0.01$.
If $dh = 0.01$ and $dl = 0$, then $dA = 6(0.01) = 0.06$.

17. Essay

19. $u(x, t) = \dfrac{1}{2}[f(x - ct) + f(x + ct)]$

Let $r = x - ct$ and $s = x + ct$. Then $u(r, s) = \dfrac{1}{2}[f(r) + f(s)]$.

$$\frac{\partial u}{\partial t} = \frac{\partial u}{\partial r}\frac{\partial r}{\partial t} + \frac{\partial u}{\partial s}\frac{\partial s}{\partial t} = \frac{1}{2}\frac{df}{dr}(-c) + \frac{1}{2}\frac{df}{ds}(c)$$

$$\frac{\partial^2 u}{\partial t^2} = \frac{1}{2}\frac{d^2f}{dr^2}(-c)^2 + \frac{1}{2}\frac{d^2f}{ds^2}(c)^2 = \frac{c^2}{2}\left[\frac{d^2f}{dr^2} + \frac{d^2f}{ds^2}\right]$$

$$\frac{\partial u}{\partial x} = \frac{\partial u}{\partial r}\frac{\partial r}{\partial x} + \frac{\partial u}{\partial s}\frac{\partial s}{\partial x} = \frac{1}{2}\frac{df}{dr}(1) + \frac{1}{2}\frac{df}{ds}(1)$$

$$\frac{\partial^2 u}{\partial x^2} = \frac{1}{2}\frac{d^2f}{dr^2}(1)^2 + \frac{1}{2}\frac{d^2f}{ds^2}(1)^2 = \frac{1}{2}\left[\frac{d^2f}{dr^2} + \frac{d^2f}{ds^2}\right]$$

Thus, $\dfrac{\partial^2 u}{\partial t^2} = c^2\dfrac{\partial^2 u}{\partial x^2}$.

C H A P T E R 13
Multiple Integration

CHAPTER 13
Multiple Integration

Section 13.1 Iterated Integrals and Area in the Plane

Solutions to Odd-Numbered Exercises

1. $\displaystyle\int_0^x (2x - y)\, dy = \left[2xy - \frac{1}{2}y^2\right]_0^x = \frac{3}{2}x^2$

3. $\displaystyle\int_1^{2y} \frac{y}{x}\, dx = \left[y \ln x\right]_1^{2y} = y \ln 2y - 0 = y \ln 2y,\ (y > 0)$

5. $\displaystyle\int_0^{\sqrt{4-x^2}} x^2 y\, dy = \left[\frac{1}{2}x^2 y^2\right]_0^{\sqrt{4-x^2}} = \frac{4x^2 - x^4}{2}$

7. $\displaystyle\int_{e^y}^{y} \frac{y \ln x}{x}\, dx = \left[\frac{1}{2}y \ln^2 x\right]_{e^y}^{y} = \frac{1}{2}y[\ln^2 y - \ln^2 e^y] = \frac{y}{2}[(\ln y)^2 - y^2],\ (y > 0)$

9. $\displaystyle\int_0^{x^3} ye^{-y/x}\, dy = \left[-xye^{-y/x}\right]_0^{x^3} + x\int_0^{x^3} e^{-y/x}\, dy = -x^4 e^{-x^2} - \left[x^2 e^{-y/x}\right]_0^{x^3} = x^2(1 - e^{-x^2} - x^2 e^{-x^2})$

$u = y,\, du = dy,\, dv = e^{-y/x}\, dy,\, v = -xe^{-y/x}$

11. $\displaystyle\int_0^1 \int_0^2 (x + y)\, dy\, dx = \int_0^1 \left[xy + \frac{1}{2}y^2\right]_0^2 dx = \int_0^1 (2x + 2)\, dx = \left[x^2 + 2x\right]_0^1 = 3$

13. $\displaystyle\int_0^1 \int_0^x \sqrt{1 - x^2}\, dy\, dx = \int_0^1 \left[y\sqrt{1 - x^2}\right]_0^x dx = \int_0^1 x\sqrt{1 - x^2}\, dx = \left[-\frac{1}{2}\left(\frac{2}{3}\right)(1 - x^2)^{3/2}\right]_0^1 = \frac{1}{3}$

15. $\displaystyle\int_1^2 \int_0^4 (x^2 - 2y^2 + 1)\, dx\, dy = \int_1^2 \left[\frac{1}{3}x^3 - 2xy^2 + x\right]_0^4 dy$

$\displaystyle = \int_1^2 \left(\frac{64}{3} - 8y^2 + 4\right) dy = \frac{4}{3}\int_1^2 (19 - 6y^2)\, dy = \left[\frac{4}{3}(19y - 2y^3)\right]_1^2 = \frac{20}{3}$

17. $\displaystyle\int_0^1 \int_0^{\sqrt{1-y^2}} (x + y)\, dx\, dy = \int_0^1 \left[\frac{1}{2}x^2 + xy\right]_0^{\sqrt{1-y^2}} dy$

$\displaystyle = \int_0^1 \left[\frac{1}{2}(1 - y^2) + y\sqrt{1 - y^2}\right] dy = \left[\frac{1}{2}y - \frac{1}{6}y^3 - \frac{1}{2}\left(\frac{2}{3}\right)(1 - y^2)^{3/2}\right]_0^1 = \frac{2}{3}$

19. $\displaystyle\int_0^2 \int_0^{\sqrt{4-y^2}} \frac{2}{\sqrt{4 - y^2}}\, dx\, dy = \int_0^2 \left[\frac{2x}{\sqrt{4 - y^2}}\right]_0^{\sqrt{4-y^2}} dy = \int_0^2 2\, dy = \left[2y\right]_0^2 = 4$

21. $\displaystyle\int_0^{\pi/2} \int_0^{\sin\theta} \theta r\, dr\, d\theta = \int_0^{\pi/2} \left[\theta\frac{r^2}{2}\right]_0^{\sin\theta} d\theta = \int_0^{\pi/2} \frac{1}{2}\theta \sin^2\theta\, d\theta$

$\displaystyle = \frac{1}{4}\int_0^{\pi/2} (\theta - \theta\cos 2\theta)\, d\theta = \frac{1}{4}\left[\frac{\theta^2}{2} - \left(\frac{1}{4}\cos 2\theta + \frac{\theta}{2}\sin 2\theta\right)\right]_0^{\pi/2} = \frac{\pi^2}{32} + \frac{1}{8}$

23. $\displaystyle\int_1^\infty \int_0^{1/x} y \, dy \, dx = \int_1^\infty \left[\frac{y^2}{2}\right]_0^{1/x} dx = \frac{1}{2}\int_1^\infty \frac{1}{x^2}\, dx = \left[-\frac{1}{2x}\right]_1^\infty = 0 + \frac{1}{2} = \frac{1}{2}$

25. $\displaystyle\int_1^\infty \int_1^\infty \frac{1}{xy}\, dx \, dy = \int_1^\infty \left[\frac{1}{y}\ln x\right]_1^\infty dy = \int_1^\infty \left[\frac{1}{y}(\infty) - \frac{1}{y}(0)\right] dy$

Diverges

27. $\displaystyle A = \int_0^8 \int_0^3 dy \, dx = \int_0^8 \left[y\right]_0^3 dx = \int_0^8 3 \, dx = \left[3x\right]_0^8 = 24$

$\displaystyle A = \int_0^3 \int_0^8 dx \, dy = \int_0^3 \left[x\right]_0^8 dy = \int_0^3 8 \, dy = \left[8y\right]_0^3 = 24$

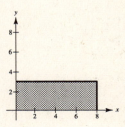

29. $\displaystyle A = \int_0^2 \int_0^{4-x^2} dy \, dx = \int_0^2 \left[y\right]_0^{4-x^2} dx$

$\displaystyle = \int_0^2 (4 - x^2)\, dx$

$\displaystyle = \left[4x - \frac{x^3}{3}\right]_0^2 = \frac{16}{3}$

$\displaystyle A = \int_0^4 \int_0^{\sqrt{4-y}} dx \, dy$

$\displaystyle = \int_0^4 \left[x\right]_0^{\sqrt{4-y}} dy = \int_0^4 \sqrt{4-y}\, dy = -\int_0^4 (4-y)^{1/2}(-1)\, dy = \left[-\frac{2}{3}(4-y)^{3/2}\right]_0^4 = \frac{2}{3}(8) = \frac{16}{3}$

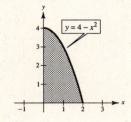

31. $\displaystyle A = \int_{-2}^1 \int_{x+2}^{4-x^2} dy \, dx$

$\displaystyle = \int_{-2}^1 \left[y\right]_{x+2}^{4-x^2} dx$

$\displaystyle = \int_{-2}^1 (4 - x^2 - x - 2)\, dx$

$\displaystyle = \int_{-2}^1 (2 - x - x^2)\, dx$

$\displaystyle = \left[2x - \frac{1}{2}x^2 - \frac{1}{3}x^3\right]_{-2}^1 = \frac{9}{2}$

$\displaystyle A = \int_0^3 \int_{-\sqrt{4-y}}^{y-2} dx \, dy + 2\int_3^4 \int_0^{\sqrt{4-y}} dx \, dy$

$\displaystyle = \int_0^3 \left[x\right]_{-\sqrt{4-y}}^{y-2} dy + 2\int_3^4 \left[x\right]_0^{\sqrt{4-y}} dy$

$\displaystyle = \int_0^3 \left(y - 2 + \sqrt{4-y}\right) dy + 2\int_3^4 \sqrt{4-y}\, dy$

$\displaystyle = \left[\frac{1}{2}y^2 - 2y - \frac{2}{3}(4-y)^{3/2}\right]_0^3 - \left[\frac{4}{3}(4-y)^{3/2}\right]_3^4 = \frac{9}{2}$

33. $\displaystyle\int_0^4 \int_0^{(2-\sqrt{x})^2} dy \, dx = \int_0^4 \left[y\right]_0^{(2-\sqrt{x})^2} dx$

$\displaystyle = \int_0^4 (4 - 4\sqrt{x} + x)\, dx$

$\displaystyle = \left[4x - \frac{8}{3}x\sqrt{x} + \frac{x^2}{2}\right]_0^4 = \frac{8}{3}$

$\displaystyle\int_0^4 \int_0^{(2-\sqrt{y})^2} dx \, dy = \frac{8}{3}$

Integration steps are similar to those above.

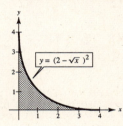

35. $A = \int_0^3 \int_0^{2x/3} dy\, dx + \int_3^5 \int_0^{5-x} dy\, dx$

$= \int_0^3 \left[y \right]_0^{2x/3} dx + \int_3^5 \left[y \right]_0^{5-x} dx$

$= \int_0^3 \frac{2x}{3}\, dx + \int_3^5 (5 - x)\, dx$

$= \left[\frac{1}{3}x^2 \right]_0^3 + \left[5x - \frac{1}{2}x^2 \right]_3^5 = 5$

$A = \int_0^2 \int_{3y/2}^{5-y} dx\, dy$

$= \int_0^2 \left[x \right]_{3y/2}^{5-y} dy$

$= \int_0^2 \left(5 - y - \frac{3y}{2} \right) dy$

$= \int_0^2 \left(5 - \frac{5y}{2} \right) dy = \left[5y - \frac{5}{4}y^2 \right]_0^2 = 5$

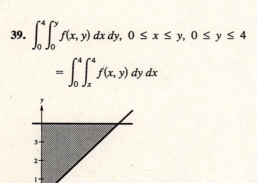

37. $\frac{A}{4} = \int_0^a \int_0^{(b/a)\sqrt{a^2-x^2}} dy\, dx = \int_0^a \left[y \right]_0^{(b/a)\sqrt{a^2-x^2}} dx$

$= \frac{b}{a} \int_0^a \sqrt{a^2 - x^2}\, dx = ab \int_0^{\pi/2} \cos^2 \theta\, d\theta$

$(x = a \sin \theta, dx = a \cos \theta\, d\theta)$

$= \frac{ab}{2} \int_0^{\pi/2} (1 + \cos 2\theta)\, d\theta = \left[\frac{ab}{2} \left(\theta + \frac{1}{2} \sin 2\theta \right) \right]_0^{\pi/2}$

$= \frac{\pi ab}{4}$

Therefore, $A = \pi ab$.

$\frac{A}{4} = \int_0^b \int_0^{(a/b)\sqrt{b^2-y^2}} dx\, dy = \frac{\pi ab}{4}$

Therefore, $A = \pi ab$. Integration steps are similar to those above.

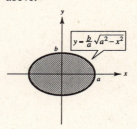

39. $\int_0^4 \int_0^y f(x, y)\, dx\, dy,\ 0 \le x \le y,\ 0 \le y \le 4$

$= \int_0^4 \int_x^4 f(x, y)\, dy\, dx$

41. $\int_{-2}^2 \int_0^{\sqrt{4-x^2}} f(x, y)\, dy\, dx,\ 0 \le y \le \sqrt{4 - x^2},\ -2 \le x \le 2$

$= \int_0^2 \int_{-\sqrt{4-y^2}}^{\sqrt{4-y^2}} dx\, dy$

43. $\int_1^{10} \int_0^{\ln y} f(x, y)\, dx\, dy,\ 0 \le x \le \ln y,\ 1 \le y \le 10$

$= \int_0^{\ln 10} \int_{e^x}^{10} f(x, y)\, dy\, dx$

45. $\int_{-1}^1 \int_{x^2}^1 f(x, y)\, dy\, dx,\ x^2 \le y \le 1,\ -1 \le x \le 1$

$= \int_0^1 \int_{-\sqrt{y}}^{\sqrt{y}} f(x, y)\, dx\, dy$

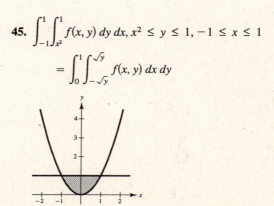

47. $\int_0^1 \int_0^2 dy\, dx = \int_0^2 \int_0^1 dx\, dy = 2$

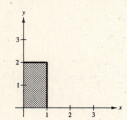

49. $\int_0^1 \int_{-\sqrt{1-y^2}}^{\sqrt{1-y^2}} dx\, dy = \int_{-1}^1 \int_0^{\sqrt{1-x^2}} dy\, dx = \frac{\pi}{2}$

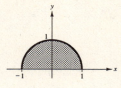

51. $\int_0^2 \int_0^x dy\, dx + \int_2^4 \int_0^{4-x} dy\, dx = \int_0^2 \int_y^{4-y} dx\, dy = 4$

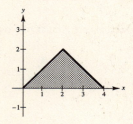

53. $\int_0^2 \int_{x/2}^1 dy\, dx = \int_0^1 \int_0^{2y} dx\, dy = 1$

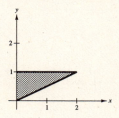

55. $\int_0^1 \int_{y^2}^{\sqrt[3]{y}} dx\, dy = \int_0^1 \int_{x^3}^{\sqrt{x}} dy\, dx = \frac{5}{12}$

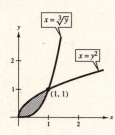

57. The first integral arises using vertical representative rectangles. The second two integrals arise using horizontal representative rectangles.

$$\int_0^5 \int_x^{\sqrt{50-x^2}} x^2 y^2\, dy\, dx = \int_0^5 \left[\frac{1}{3} x^2 (50-x^2)^{3/2} - \frac{1}{3} x^5 \right] dx$$

$$= \frac{15625}{24} \pi$$

$$\int_0^5 \int_0^y x^2 y^2\, dx\, dy + \int_5^{5\sqrt{2}} \int_0^{\sqrt{50-y^2}} x^2 y^2\, dx\, dy = \int_0^5 \frac{1}{3} y^5\, dy + \int_5^{5\sqrt{2}} \frac{1}{3} (50-y^2)^{3/2} y^2\, dy = \frac{15625}{18} + \left(\frac{15625}{18} \pi - \frac{15625}{18} \right)$$

$$= \frac{15625}{24} \pi$$

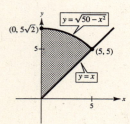

59. $\int_0^2 \int_x^2 x\sqrt{1 + y^3}\, dy\, dx = \int_0^2 \int_0^y x\sqrt{1 + y^3}\, dx\, dy = \int_0^2 \left[\sqrt{1 + y^3} \cdot \frac{x^2}{2}\right]_0^y dy$

$$= \frac{1}{2}\int_0^2 \sqrt{1 + y^3}\, y^2\, dy = \left[\frac{1}{2} \cdot \frac{1}{3} \cdot \frac{2}{3}(1 + y^3)^{3/2}\right]_0^2 = \frac{1}{9}(27) - \frac{1}{9}(1) = \frac{26}{9}$$

61. $\int_0^1 \int_y^1 \sin(x^2)\, dx\, dy = \int_0^1 \int_0^x \sin(x^2)\, dy\, dx = \int_0^1 \left[y \sin(x^2)\right]_0^x dx$

$$= \int_0^1 x \sin(x^2)\, dx = \left[-\frac{1}{2}\cos(x^2)\right]_0^1 = -\frac{1}{2}\cos 1 + \frac{1}{2}(1) = \frac{1}{2}(1 - \cos 1) \approx 0.2298$$

63. $\int_0^2 \int_{x^2}^{2x} (x^3 + 3y^2)\, dy\, dx = \frac{1664}{105} \approx 15.848$ **65.** $\int_0^4 \int_0^y \frac{2}{(x + 1)(y + 1)}\, dx\, dy = (\ln 5)^2 \approx 2.590$

67. (a) $x = y^3 \iff y = x^{1/3}$

$\quad\quad x = 4\sqrt{2y} \iff x^2 = 32y \iff y = \dfrac{x^2}{32}$

(b) $\int_0^8 \int_{x^2/32}^{x^{1/3}} (x^2 y - xy^2)\, dy\, dx$

(c) Both integrals equal $67520/693 \approx 97.43$

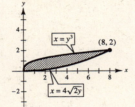

69. $\int_0^2 \int_0^{4-x^2} e^{xy}\, dy\, dx \approx 20.5648$ **71.** $\int_0^{2\pi} \int_0^{1+\cos\theta} 6r^2 \cos\theta\, dr\, d\theta = \dfrac{15\pi}{2}$

73. An iterated integral is a double integral of a function of two variables. First integrate with respect to one variable while holding the other variable constant. Then integrate with respect to the second variable.

75. The region is a rectangle.   **77.** True

Section 13.2   Double Integrals and Volume

For Exercise 1–3, $\Delta x_i = \Delta y_i = 1$ and the midpoints of the squares are

$\left(\dfrac{1}{2}, \dfrac{1}{2}\right)$, $\left(\dfrac{3}{2}, \dfrac{1}{2}\right)$, $\left(\dfrac{5}{2}, \dfrac{1}{2}\right)$, $\left(\dfrac{7}{2}, \dfrac{1}{2}\right)$, $\left(\dfrac{1}{2}, \dfrac{3}{2}\right)$, $\left(\dfrac{3}{2}, \dfrac{3}{2}\right)$, $\left(\dfrac{5}{2}, \dfrac{3}{2}\right)$, $\left(\dfrac{7}{2}, \dfrac{3}{2}\right)$.

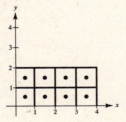

1. $f(x, y) = x + y$

$\quad \sum_{i=1}^8 f(x_i, y_i)\, \Delta x_i \Delta y_i = 1 + 2 + 3 + 4 + 2 + 3 + 4 + 5 = 24$

$\quad \int_0^4 \int_0^2 (x + y)\, dy\, dx = \int_0^4 \left[xy + \frac{y^2}{2}\right]_0^2 dx = \int_0^4 (2x + 2)\, dx = \left[x^2 + 2x\right]_0^4 = 24$

3. $f(x, y) = x^2 + y^2$

$$\sum_{i=1}^{8} f(x_i, y_i) \, \Delta x_i \, \Delta y_i = \frac{2}{4} + \frac{10}{4} + \frac{26}{4} + \frac{50}{4} + \frac{10}{4} + \frac{18}{4} + \frac{34}{4} + \frac{58}{4} = 52$$

$$\int_0^4 \int_0^2 (x^2 + y^2) \, dy \, dx = \int_0^4 \left[x^2 y + \frac{y^3}{3} \right]_0^2 dx = \int_0^4 \left(2x^2 + \frac{8}{3} \right) dx = \left[\frac{2x^3}{3} + \frac{8x}{3} \right]_0^4 = \frac{160}{3}$$

5. $\int_0^4 \int_0^4 f(x, y) dy \, dx \approx (32 + 31 + 28 + 23) + (31 + 30 + 27 + 22) + (28 + 27 + 24 + 19) + (23 + 22 + 19 + 14)$

$\qquad = 400$

Using the corner of the ith square furthest from the origin, you obtain 272.

7. $\int_0^2 \int_0^1 (1 + 2x + 2y) \, dy \, dx = \int_0^2 \left[y + 2xy + y^2 \right]_0^1 dx$

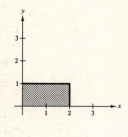

$\qquad\qquad = \int_0^2 (2 + 2x) \, dx$

$\qquad\qquad = \left[2x + x^2 \right]_0^2$

$\qquad\qquad = 8$

9. $\int_0^6 \int_{y/2}^3 (x + y) \, dx \, dy = \int_0^6 \left[\frac{1}{2} x^2 + xy \right]_{y/2}^3 dy$

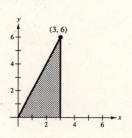

$\qquad\qquad = \int_0^6 \left(\frac{9}{2} + 3y - \frac{5}{8} y^2 \right) dy$

$\qquad\qquad = \left[\frac{9}{2} y + \frac{3}{2} y^2 - \frac{5}{24} y^3 \right]_0^6$

$\qquad\qquad = 36$

11. $\int_{-a}^a \int_{-\sqrt{a^2-x^2}}^{\sqrt{a^2-x^2}} (x + y) \, dy \, dx = \int_{-a}^a \left[xy + \frac{1}{2} y^2 \right]_{-\sqrt{a^2-x^2}}^{\sqrt{a^2-x^2}} dx$

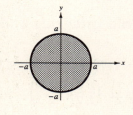

$\qquad\qquad = \int_{-a}^a 2x \sqrt{a^2 - x^2} \, dx$

$\qquad\qquad = \left[-\frac{2}{3} (a^2 - x^2)^{3/2} \right]_{-a}^a = 0$

13. $\int_0^5 \int_0^3 xy \, dx \, dy = \int_0^3 \int_0^5 xy \, dy \, dx$

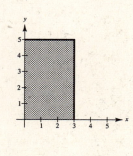

$\qquad\qquad = \int_0^3 \left[\frac{1}{2} xy^2 \right]_0^5 dx$

$\qquad\qquad = \frac{25}{2} \int_0^3 x \, dx$

$\qquad\qquad = \left[\frac{25}{4} x^2 \right]_0^3 = \frac{225}{4}$

15. $\displaystyle\int_0^2\int_{y/2}^y \frac{y}{x^2+y^2}\,dx\,dy + \int_2^4\int_{y/2}^2 \frac{y}{x^2+y^2}\,dx\,dy = \int_0^2\int_x^{2x} \frac{y}{x^2+y^2}\,dy\,dx$

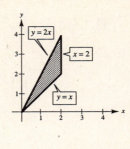

$\displaystyle = \frac{1}{2}\int_0^2\left[\ln(x^2+y^2)\right]_x^{2x}dx$

$\displaystyle = \frac{1}{2}\int_0^2(\ln 5x^2 - \ln 2x^2)\,dx$

$\displaystyle = \frac{1}{2}\ln\frac{5}{2}\int_0^2 dx$

$\displaystyle = \left[\frac{1}{2}\left(\ln\frac{5}{2}\right)x\right]_0^2 = \ln\frac{5}{2}$

17. $\displaystyle\int_3^4\int_{4-y}^{\sqrt{4-y}} -2y\ln x\,dx\,dy = \int_0^1\int_{4-x}^{4-x^2} -2y\ln x\,dy\,dx$

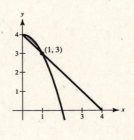

$\displaystyle = -\int_0^1\left[\ln x\cdot y^2\right]_{4-x}^{4-x^2}dx$

$\displaystyle = -\int_0^1\left[\ln x[(4-x^2)^2 - (4-x)^2]\right]dx$

$\displaystyle = \frac{26}{25}$

19. $\displaystyle\int_0^4\int_0^{3x/4} x\,dy\,dx + \int_4^5\int_0^{\sqrt{25-x^2}} x\,dy\,dx = \int_0^3\int_{4y/3}^{\sqrt{25-y^2}} x\,dx\,dy$

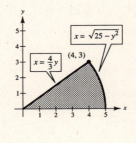

$\displaystyle = \int_0^3\left[\frac{1}{2}x^2\right]_{4y/3}^{\sqrt{25-y^2}}dy$

$\displaystyle = \frac{25}{18}\int_0^3 (9-y^2)\,dy$

$\displaystyle = \left[\frac{25}{18}\left(9y - \frac{1}{3}y^3\right)\right]_0^3 = 25$

21. $\displaystyle\int_0^4\int_0^2 \frac{y}{2}\,dy\,dx = \int_0^4\left[\frac{y^2}{4}\right]_0^2 dx$

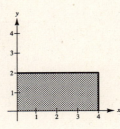

$\displaystyle = \int_0^4 dx = 4$

23. $\displaystyle\int_0^2\int_0^y (4-x-y)\,dx\,dy = \int_0^2\left[4x - \frac{x^2}{2} - xy\right]_0^y dy$

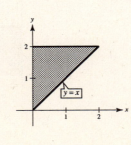

$\displaystyle = \int_0^2\left(4y - \frac{y^2}{2} - y^2\right)dy$

$\displaystyle = \left[2y^2 - \frac{y^3}{6} - \frac{y^3}{3}\right]_0^2$

$\displaystyle = 8 - \frac{8}{6} - \frac{8}{3} = 4$

25. $\int_0^6 \int_0^{(-2/3)x+4} \left(\dfrac{12 - 2x - 3y}{4} \right) dy\, dx = \int_0^6 \left[\dfrac{1}{4}\left(12y - 2xy - \dfrac{3}{2}y^2 \right) \right]_0^{(-2/3)x+4} dx$

$$= \int_0^6 \left(\dfrac{1}{6}x^2 - 2x + 6 \right) dx$$

$$= \left[\dfrac{1}{18}x^3 - x^2 + 6x \right]_0^6$$

$$= 12$$

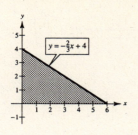

27. $\int_0^1 \int_0^y (1 - xy)\, dx\, dy = \int_0^1 \left[x - \dfrac{x^2 y}{2} \right]_0^y dy$

$$= \int_0^1 \left(y - \dfrac{y^3}{2} \right) dy$$

$$= \left[\dfrac{y^2}{2} - \dfrac{y^4}{8} \right]_0^1$$

$$= \dfrac{3}{8}$$

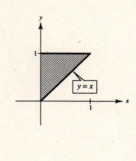

29. $\int_0^\infty \int_0^\infty \dfrac{1}{(x+1)^2(y+1)^2}\, dy\, dx = \int_0^\infty \left[-\dfrac{1}{(x+1)^2(y+1)} \right]_0^\infty dx = \int_0^\infty \dfrac{1}{(x+1)^2}\, dx = \left[-\dfrac{1}{(x+1)} \right]_0^\infty = 1$

31. $4\int_0^2 \int_0^{\sqrt{4-x^2}} (4 - x^2 - y^2)\, dy\, dx = 8\pi$

33. $V = \int_0^1 \int_0^x xy\, dy\, dx$

$$= \int_0^1 \left[\dfrac{1}{2}xy^2 \right]_0^x dx = \dfrac{1}{2}\int_0^1 x^3\, dx$$

$$= \left[\dfrac{1}{8}x^4 \right]_0^1 = \dfrac{1}{8}$$

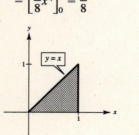

35. $V = \int_0^2 \int_0^4 x^2\, dy\, dx$

$$= \int_0^2 \left[x^2 y \right]_0^4 dx = \int_0^2 4x^2\, dx$$

$$= \left[\dfrac{4x^3}{3} \right]_0^2 = \dfrac{32}{3}$$

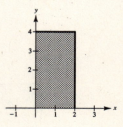

37. Divide the solid into two equal parts.

$$V = 2\int_0^1 \int_0^x \sqrt{1 - x^2}\, dy\, dx$$

$$= 2\int_0^1 \left[y\sqrt{1 - x^2} \right]_0^x dx$$

$$= 2\int_0^1 x\sqrt{1 - x^2}\, dx$$

$$= \left[-\dfrac{2}{3}(1 - x^2)^{3/2} \right]_0^1 = \dfrac{2}{3}$$

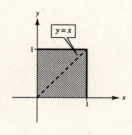

39. $V = \int_0^2 \int_0^{\sqrt{4-x^2}} (x + y) \, dy \, dx$

$= \int_0^2 \left[xy + \frac{1}{2}y^2 \right]_0^{\sqrt{4-x^2}} dx$

$= \int_0^2 \left(x\sqrt{4 - x^2} + 2 - \frac{1}{2}x^2 \right) dx$

$= \left[-\frac{1}{3}(4 - x^2)^{3/2} + 2x - \frac{1}{6}x^3 \right]_0^2 = \frac{16}{3}$

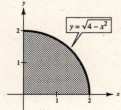

41. $V = 4 \int_0^2 \int_0^{\sqrt{4-x^2}} (x^2 + y^2) \, dy \, dx$

$= 4 \int_0^2 \left[x^2\sqrt{4 - x^2} + \frac{1}{3}(4 - x^2)^{3/2} \right] dx, \quad x = 2 \sin \theta$

$= 4 \int_0^{\pi/2} \left(16 \cos^2 \theta - \frac{32}{3} \cos^4 \theta \right) d\theta$

$= 4 \left[16\left(\frac{\pi}{4}\right) - \frac{32}{3}\left(\frac{3\pi}{16}\right) \right]$

$= 8\pi$

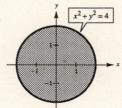

43. $V = 4 \int_0^2 \int_0^{\sqrt{4-x^2}} (4 - x^2 - y^2) \, dy \, dx = 8\pi$

45. $V = \int_0^2 \int_0^{-0.5x+1} \frac{2}{1 + x^2 + y^2} \, dy \, dx \approx 1.2315$

47. f is a continuous function such that $0 \le f(x, y) \le 1$ over a region R of area 1. Let $f(m, n) = $ the minimum value of f over R and $f(M, N) = $ the maximum value of f over R. Then

$$f(m, n) \iint_R dA \le \iint_R f(x, y) \, dA \le f(M, N) \iint_R dA.$$

Since $\iint_R dA = 1$ and $0 \le f(m, n) \le f(M, N) \le 1$, we have $0 \le f(m, n)(1) \le \iint_R f(x, y) \, dA \le f(M, N)(1) \le 1$.

Therefore, $0 \le \iint_R f(x, y) \, dA \le 1$.

49. $\int_0^1 \int_{y/2}^{1/2} e^{-x^2} \, dx \, dy = \int_0^{1/2} \int_0^{2x} e^{-x^2} \, dy \, dx$

$= \int_0^{1/2} 2xe^{-x^2} \, dx$

$= \left[-e^{-x^2} \right]_0^{1/2}$

$= -e^{-1/4} + 1$

$= 1 - e^{-1/4} \approx 0.221$

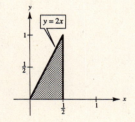

51. $\int_0^1 \int_0^{\arccos y} \sin x\sqrt{1 + \sin^2 x} \, dx \, dy$

$= \int_0^{\pi/2} \int_0^{\cos x} \sin x\sqrt{1 + \sin^2 x} \, dy \, dx$

$= \int_0^{\pi/2} (1 + \sin^2 x)^{1/2} \sin x \cos x \, dx$

$= \left[\frac{1}{2} \cdot \frac{2}{3}(1 + \sin^2 x)^{3/2} \right]_0^{\pi/2} = \frac{1}{3}\left[2\sqrt{2} - 1 \right]$

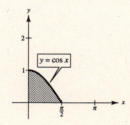

53. Average $= \dfrac{1}{8}\displaystyle\int_0^4\int_0^2 x\,dy\,dx = \dfrac{1}{8}\int_0^4 2x\,dx = \left[\dfrac{x^2}{8}\right]_0^4 = 2$

55. Average $= \dfrac{1}{4}\displaystyle\int_0^2\int_0^2 (x^2 + y^2)\,dx\,dy$

$$= \dfrac{1}{4}\int_0^2\left[\dfrac{x^3}{3} + xy^2\right]_0^2 dy = \dfrac{1}{4}\int_0^2\left(\dfrac{8}{3} + 2y^2\right) dy$$

$$= \left[\dfrac{1}{4}\left(\dfrac{8}{3}y + \dfrac{2}{3}y^3\right)\right]_0^2 = \dfrac{8}{3}$$

57. See the definition on page 946.

59. The value of $\displaystyle\int_R\int f(x, y)\,dA$ would be kB.

61. Average $= \dfrac{1}{1250}\displaystyle\int_{300}^{325}\int_{200}^{250} 100x^{0.6}y^{0.4}\,dx\,dy$

$$= \dfrac{1}{1250}\int_{300}^{325}\left[(100y^{0.4})\dfrac{x^{1.6}}{1.6}\right]_{200}^{250} dy = \dfrac{128{,}844.1}{1250}\int_{300}^{325} y^{0.4}\,dy = 103.0753\left[\dfrac{y^{1.4}}{1.4}\right]_{300}^{325} \approx 25{,}645.24$$

63. $f(x, y) \geq 0$ for all (x, y) and

$$\int_{-\infty}^{\infty}\int_{-\infty}^{\infty} f(x, y)\,dA = \int_0^5\int_0^2 \dfrac{1}{10}\,dy\,dx = \int_0^5 \dfrac{1}{5}\,dx = 1$$

$$P(0 \leq x \leq 2, 1 \leq y \leq 2) = \int_0^2\int_1^2 \dfrac{1}{10}\,dy\,dx = \int_0^2 \dfrac{1}{10}\,dx = \dfrac{1}{5}.$$

65. $f(x, y) \geq 0$ for all (x, y) and

$$\int_{-\infty}^{\infty}\int_{-\infty}^{\infty} f(x, y)\,dA = \int_0^3\int_3^6 \dfrac{1}{27}(9 - x - y)\,dy\,dx$$

$$= \int_0^3 \dfrac{1}{27}\left[9y - xy - \dfrac{y^2}{2}\right]_3^6 dx = \int_0^3\left(\dfrac{1}{2} - \dfrac{1}{9}x\right) dx = \left[\dfrac{x}{2} - \dfrac{x^2}{18}\right]_0^3 = 1$$

$$P(0 \leq x \leq 1, 4 \leq y \leq 6) = \int_0^1\int_4^6 \dfrac{1}{27}(9 - x - y)\,dy\,dx = \int_0^1 \dfrac{2}{27}(4 - x)\,dx = \dfrac{7}{27}.$$

67. Divide the base into six squares, and assume the height at the center of each square is the height of the entire square.
   Thus,

$$V \approx (4 + 3 + 6 + 7 + 3 + 2)(100) = 2500m^3.$$

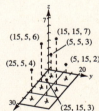

69. $\displaystyle\int_0^1\int_0^2 \sin\sqrt{x + y}\,dy\,dx$ $m = 4, n = 8$

   (a) 1.78435

   (b) 1.7879

71. $\displaystyle\int_4^6\int_0^2 y\cos\sqrt{x}\,dx\,dy$ $m = 4, n = 8$

   (a) 11.0571

   (b) 11.0414

73. $V \approx 125$

Matches d.

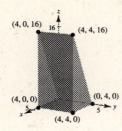

75. False

$$V = 8\int_0^1\int_0^{\sqrt{1-y^2}}\sqrt{1-x^2-y^2}\,dx\,dy$$

77. Average $= \displaystyle\int_0^1 f(x)\,dx = \int_0^1\int_1^x e^{t^2}\,dt\,dx = -\int_0^1\int_x^1 e^{t^2}\,dt\,dx$

$$= -\int_0^1\int_0^t e^{t^2}\,dx\,dt = -\int_0^1 te^{t^2}\,dt$$

$$= \left[-\frac{1}{2}e^{t^2}\right]_0^1 = -\frac{1}{2}(e-1) = \frac{1}{2}(1-e)$$

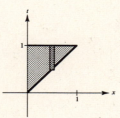

Section 13.3   Change of Variables: Polar Coordinates

1. Rectangular coordinates

3. Polar coordinates

5. $R = \{(r,\theta):\ 0\le r\le 8, 0\le\theta\le\pi\}$

7. $R = \{(r,\theta):\ 0\le r\le 3+3\sin\theta, 0\le\theta\le 2\pi\}$ Cardioid

9. $\displaystyle\int_0^{2\pi}\int_0^6 3r^2\sin\theta\,dr\,d\theta = \int_0^{2\pi}\left[r^3\sin\theta\right]_0^6 d\theta$

$$= \int_0^{2\pi} 216\sin\theta\,d\theta$$

$$= \left[-216\cos\theta\right]_0^{2\pi} = 0$$

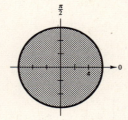

11. $\displaystyle\int_0^{\pi/2}\int_2^3 \sqrt{9-r^2}\,r\,dr\,d\theta = \int_0^{\pi/2}\left[-\frac{1}{3}(9-r^2)^{3/2}\right]_2^3 d\theta$

$$= \left[\frac{5\sqrt{5}}{3}\theta\right]_0^{\pi/2}$$

$$= \frac{5\sqrt{5}\pi}{6}$$

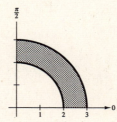

13. $\displaystyle\int_0^{\pi/2}\int_0^{1+\sin\theta}\theta r\,dr\,d\theta = \int_0^{\pi/2}\left[\frac{\theta r^2}{2}\right]_0^{1+\sin\theta} d\theta$

$$= \int_0^{\pi/2}\frac{1}{2}\theta(1+\sin\theta)^2\,d\theta$$

$$= \left[\frac{1}{8}\theta^2 + \sin\theta - \theta\cos\theta + \frac{1}{2}\theta\left(-\frac{1}{2}\cos\theta\cdot\sin\theta + \frac{1}{2}\theta\right) + \frac{1}{8}\sin^2\theta\right]_0^{\pi/2}$$

$$= \frac{3}{32}\pi^2 + \frac{9}{8}$$

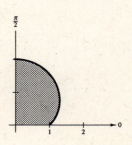

15. $\displaystyle\int_0^a \int_0^{\sqrt{a^2-y^2}} y\, dx\, dy = \int_0^{\pi/2} \int_0^a r^2 \sin\theta\, dr\, d\theta = \frac{a^3}{3} \int_0^{\pi/2} \sin\theta\, d\theta = \left[\frac{a^3}{3}(-\cos\theta)\right]_0^{\pi/2} = \frac{a^3}{3}$

17. $\displaystyle\int_0^3 \int_0^{\sqrt{9-x^2}} (x^2+y^2)^{3/2}\, dy\, dx = \int_0^{\pi/2} \int_0^3 r^4\, dr\, d\theta = \frac{243}{5} \int_0^{\pi/2} d\theta = \frac{243\pi}{10}$

19. $\displaystyle\int_0^2 \int_0^{\sqrt{2x-x^2}} xy\, dy\, dx = \int_0^{\pi/2} \int_0^{2\cos\theta} r^3 \cos\theta \sin\theta\, dr\, d\theta = 4\int_0^{\pi/2} \cos^5\theta \sin d\theta = \left[-\frac{4\cos^6\theta}{6}\right]_0^{\pi/2} = \frac{2}{3}$

21. $\displaystyle\int_0^2 \int_0^x \sqrt{x^2+y^2}\, dy\, dx + \int_2^{2\sqrt{2}} \int_0^{\sqrt{8-x^2}} \sqrt{x^2+y^2}\, dy\, dx = \int_0^{\pi/4} \int_0^{2\sqrt{2}} r^2\, dr\, d\theta$

$\displaystyle = \int_0^{\pi/4} \frac{16\sqrt{2}}{3}\, d\theta$

$\displaystyle = \frac{4\sqrt{2}\pi}{3}$

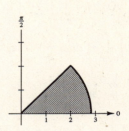

23. $\displaystyle\int_0^2 \int_0^{\sqrt{4-x^2}} (x+y)\, dy\, dx = \int_0^{\pi/2} \int_0^2 (r\cos\theta + r\sin\theta)r\, dr\, d\theta = \int_0^{\pi/2} \int_0^2 (\cos\theta + \sin\theta)r^2\, dr\, d\theta$

$\displaystyle = \frac{8}{3} \int_0^{\pi/2} (\cos\theta + \sin\theta)\, d\theta = \left[\frac{8}{3}(\sin\theta - \cos\theta)\right]_0^{\pi/2} = \frac{16}{3}$

25. $\displaystyle\int_0^{1/\sqrt{2}} \int_{\sqrt{1-y^2}}^{\sqrt{4-y^2}} \arctan\frac{y}{x}\, dx\, dy + \int_{1/\sqrt{2}}^{\sqrt{2}} \int_y^{\sqrt{4-y^2}} \arctan\frac{y}{x}\, dx\, dy$

$\displaystyle = \int_0^{\pi/4} \int_1^2 \theta r\, dr\, d\theta$

$\displaystyle = \int_0^{\pi/4} \frac{3}{2}\theta\, d\theta = \left[\frac{3\theta^2}{4}\right]_0^{\pi/4} = \frac{3\pi^2}{64}$

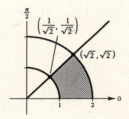

27. $\displaystyle V = \int_0^{\pi/2} \int_0^1 (r\cos\theta)(r\sin\theta)r\, dr\, d\theta$

$\displaystyle = \frac{1}{2}\int_0^{\pi/2} \int_0^1 r^3 \sin 2\theta\, dr\, d\theta = \frac{1}{8}\int_0^{\pi/2} \sin 2\theta\, d\theta = \left[-\frac{1}{16}\cos 2\theta\right]_0^{\pi/2} = \frac{1}{8}$

29. $\displaystyle V = \int_0^{2\pi} \int_0^5 r^2\, dr\, d\theta = \int_0^{2\pi} \frac{125}{3}\, d\theta = \frac{250\pi}{3}$

31. $\displaystyle V = 2\int_0^{\pi/2} \int_0^{4\cos\theta} \sqrt{16-r^2}\, r\, dr\, d\theta = 2\int_0^{\pi/2} \left[-\frac{1}{3}\left(\sqrt{16-r^2}\right)^3\right]_0^{4\cos\theta} d\theta = -\frac{2}{3}\int_0^{\pi/2} (64\sin^3\theta - 64)\, d\theta$

$\displaystyle = \frac{128}{3}\int_0^{\pi/2} [1 - \sin\theta(1 - \cos^2\theta)]\, d\theta = \frac{128}{3}\left[\theta + \cos\theta - \frac{\cos^3\theta}{3}\right]_0^{\pi/2} = \frac{64}{9}(3\pi - 4)$

33. $V = \int_0^{2\pi} \int_a^4 \sqrt{16 - r^2}\, r\, dr\, d\theta = \int_0^{2\pi} \left[-\frac{1}{3}\left(\sqrt{16 - r^2}\right)^3 \right]_a^4 d\theta = \frac{1}{3}\left(\sqrt{16 - a^2}\right)^3 (2\pi)$

One-half the volume of the hemisphere is $(64\pi)/3$.

$$\frac{2\pi}{3}(16 - a^2)^{3/2} = \frac{64\pi}{3}$$

$$(16 - a^2)^{3/2} = 32$$

$$16 - a^2 = 32^{2/3}$$

$$a^2 = 16 - 32^{2/3} = 16 - 8\sqrt[3]{2}$$

$$a = \sqrt{4\left(4 - 2\sqrt[3]{2}\right)} = 2\sqrt{4 - 2\sqrt[3]{2}} \approx 2.4332$$

35. Total Volume $= V = \int_0^{2\pi} \int_0^4 25 e^{-r^2/4}\, r\, dr\, d\theta$

$$= \int_0^{2\pi} \left[-50 e^{-r^2/4} \right]_0^4 d\theta$$

$$= \int_0^{2\pi} -50(e^{-4} - 1)\, d\theta$$

$$= (1 - e^{-4})\, 100\pi \approx 308.40524$$

Let c be the radius of the hole that is removed.

$$\frac{1}{10} V = \int_0^{2\pi} \int_0^c 25 e^{-r^2/4}\, r\, dr\, d\theta = \int_0^{2\pi} \left[-50 e^{-r^2/4} \right]_0^c d\theta$$

$$= \int_0^{2\pi} -50(e^{-c^2/4} - 1)\, d\theta \Longrightarrow 30.84052 = 100\pi(1 - e^{-c^2/4})$$

$$\Longrightarrow e^{-c^2/4} = 0.90183$$

$$-\frac{c^2}{4} = -0.10333$$

$$c^2 = 0.41331$$

$$c = 0.6429$$

$$\Longrightarrow \text{diameter} = 2c = 1.2858$$

37. $A = \int_0^{\pi} \int_0^{6\cos\theta} r\, dr\, d\theta = \int_0^{\pi} 18\cos^2\theta\, d\theta = 9 \int_0^{\pi} (1 + \cos 2\theta)\, d\theta = \left[9\left(\theta + \frac{1}{2}\sin 2\theta\right) \right]_0^{\pi} = 9\pi$

39. $\int_0^{2\pi} \int_0^{1+\cos\theta} r\, dr\, d\theta = \frac{1}{2} \int_0^{2\pi} (1 + 2\cos\theta + \cos^2\theta)\, d\theta$

$$= \frac{1}{2} \int_0^{2\pi} \left(1 + 2\cos\theta + \frac{1 + \cos 2\theta}{2} \right) d\theta = \frac{1}{2} \left[\theta + 2\sin\theta + \frac{1}{2}\left(\theta + \frac{1}{2}\sin 2\theta\right) \right]_0^{2\pi} = \frac{3\pi}{2}$$

41. $3 \int_0^{\pi/3} \int_0^{2\sin 3\theta} r\, dr\, d\theta = \frac{3}{2} \int_0^{\pi/3} 4\sin^2 3\theta\, d\theta = 3 \int_0^{\pi/3} (1 - \cos 6\theta)\, d\theta = 3\left[\theta - \frac{1}{6}\sin 6\theta \right]_0^{\pi/3} = \pi$

43. Let R be a region bounded by the graphs of $r = g_1(\theta)$ and $r = g_2(\theta)$, and the lines $\theta = a$ and $\theta = b$.

When using polar coordinates to evaluate a double integral over R, R can be partitioned into small polar sectors.

45. r-simple regions have fixed bounds for θ.

θ-simple regions have fixed bounds for r.

47. You would need to insert a factor of r because of the $r\,dr\,d\theta$ nature of polar coordinate integrals. The plane regions would be sectors of circles.

49. $\displaystyle\int_{\pi/4}^{\pi/2}\int_0^5 r\sqrt{1+r^3}\,\sin\sqrt{\theta}\,dr\,d\theta \approx 56.051$

$\left[\text{\textbf{Note:} This integral equals }\left(\int_{\pi/4}^{\pi/2}\sin\sqrt{\theta}\,d\theta\right)\left(\int_0^5 r\sqrt{1+r^3}\,dr\right)\right]$

51. Volume = base × height

$\approx 8\pi \times 12 \approx 300$

Answer (c)

53. False

Let $f(r,\theta) = r - 1$ where R is the circular sector $0 \le r \le 6$ and $0 \le \theta \le \pi$. Then,

$$\int_R\int (r-1)\,dA > 0 \quad\text{but}\quad r - 1 \not> 0 \text{ for all } r.$$

55. (a) $\displaystyle I^2 = \int_{-\infty}^{\infty}\int_{-\infty}^{\infty} e^{-(x^2+y^2)/2}\,dA = 4\int_0^{\pi/2}\int_0^{\infty} e^{-r^2/2}r\,dr\,d\theta = 4\int_0^{\pi/2}\left[-e^{-r^2/2}\right]_0^{\infty}d\theta = 4\int_0^{\pi/2}d\theta = 2\pi$

(b) Therefore, $I = \sqrt{2\pi}$.

57. $\displaystyle\int_{-7}^{7}\int_{-\sqrt{49-x^2}}^{\sqrt{49-x^2}} 4000e^{-0.01(x^2+y^2)}\,dy\,dx = \int_0^{2\pi}\int_0^7 4000e^{-0.01r^2}r\,dr\,d\theta = \int_0^{2\pi}\left[-200{,}000e^{-0.01r^2}\right]_0^7 d\theta$

$$= 2\pi(-200{,}000)(e^{-0.49}-1) = 400{,}000\pi(1-e^{-0.49}) \approx 486{,}788$$

59. (a) $\displaystyle\int_2^4\int_{y/\sqrt{3}}^{y} f\,dx\,dy$

(b) $\displaystyle\int_{2/\sqrt{3}}^{2}\int_2^{\sqrt{3}x} f\,dy\,dx + \int_2^{4/\sqrt{3}}\int_x^{\sqrt{3}x} f\,dy\,dx + \int_{4/\sqrt{3}}^{4}\int_x^{4} f\,dy\,dx$

(c) $\displaystyle\int_{\pi/4}^{\pi/3}\int_{2\csc\theta}^{4\csc\theta} fr\,dr\,d\theta$

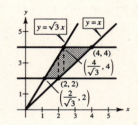

61. $\displaystyle A = \frac{\Delta\theta r_2^2}{2} - \frac{\Delta\theta r_1^2}{2} = \Delta\theta\left(\frac{r_1+r_2}{2}\right)(r_2-r_1) = r\Delta r\Delta\theta$

Section 13.4 Center of Mass and Moments of Inertia

1. $\displaystyle m = \int_0^4\int_0^3 xy\,dy\,dx = \int_0^4\left[\frac{xy^2}{2}\right]_0^3 dx = \int_0^4\frac{9}{2}x\,dx = \left[\frac{9x^2}{4}\right]_0^4 = 36$

3. $\displaystyle m = \int_0^{\pi/2}\int_0^2 (r\cos\theta)(r\sin\theta)r\,dr\,d\theta = \int_0^{\pi/2}\int_0^2 \cos\theta\sin\theta\cdot r^3\,dr\,d\theta$

$$= \int_0^{\pi/2} 4\cos\theta\sin\theta\,d\theta$$

$$= \left[4\frac{\sin^2\theta}{2}\right]_0^{\pi/2} = 2$$

5. (a) $m = \int_0^a \int_0^b k\,dy\,dx = kab$

$M_x = \int_0^a \int_0^b ky\,dy\,dx = \dfrac{kab^2}{2}$

$M_y = \int_0^a \int_0^b kx\,dy\,dx = \dfrac{ka^2b}{2}$

$\bar{x} = \dfrac{M_y}{m} = \dfrac{ka^2b/2}{kab} = \dfrac{a}{2}$

$\bar{y} = \dfrac{M_x}{m} = \dfrac{kab^2/2}{kab} = \dfrac{b}{2}$

$(\bar{x}, \bar{y}) = \left(\dfrac{a}{2}, \dfrac{b}{2}\right)$

(b) $m = \int_0^a \int_0^b ky\,dy\,dx = \dfrac{kab^2}{2}$

$M_x = \int_0^a \int_0^b ky^2\,dy\,dx = \dfrac{kab^3}{3}$

$M_y = \int_0^a \int_0^b kxy\,dy\,dx = \dfrac{ka^2b^2}{4}$

$\bar{x} = \dfrac{M_y}{m} = \dfrac{ka^2b^2/4}{kab^2/2} = \dfrac{a}{2}$

$\bar{y} = \dfrac{M_x}{m} = \dfrac{kab^3/3}{kab^2/2} = \dfrac{2}{3}b$

$(\bar{x}, \bar{y}) = \left(\dfrac{a}{2}, \dfrac{2}{3}b\right)$

(c) $m = \int_0^a \int_0^b kx\,dy\,dx = k\int_0^a xb\,dx = \dfrac{1}{2}ka^2b$

$M_x = \int_0^a \int_0^b kxy\,dy\,dx = \dfrac{ka^2b^2}{4}$

$M_y = \int_0^a \int_0^b kx^2\,dy\,dx = \dfrac{ka^3b}{3}$

$\bar{x} = \dfrac{M_y}{m} = \dfrac{ka^3b/3}{ka^2b/2} = \dfrac{2}{3}a$

$\bar{y} = \dfrac{M_x}{m} = \dfrac{ka^2b^2/4}{ka^2b/2} = \dfrac{b}{2}$

$(\bar{x}, \bar{y}) = \left(\dfrac{2}{3}a, \dfrac{b}{2}\right)$

7. (a) $m = \dfrac{k}{2}bh$

$\bar{x} = \dfrac{b}{2}$ by symmetry

$M_x = \int_0^{b/2} \int_0^{2hx/b} ky\,dy\,dx + \int_{b/2}^b \int_0^{-2h(x-b)/b} ky\,dy\,dx$

$ = \dfrac{kbh^2}{12} + \dfrac{kbh^2}{12} = \dfrac{kbh^2}{6}$

$\bar{y} = \dfrac{M_x}{m} = \dfrac{kbh^2/6}{kbh/2} = \dfrac{h}{3}$

$(\bar{x}, \bar{y}) = \left(\dfrac{b}{2}, \dfrac{h}{3}\right)$

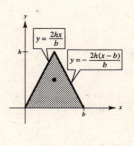

—**CONTINUED**—

7. —CONTINUED—

(b) $m = \int_0^{b/2} \int_0^{2hx/b} ky \, dy \, dx + \int_{b/2}^b \int_0^{-2h(x-b)/b} ky \, dy \, dx = \dfrac{kbh^2}{6}$

$M_x = \int_0^{b/2} \int_0^{2hx/b} ky^2 \, dy \, dx + \int_{b/2}^b \int_0^{-2h(x-b)/b} ky^2 \, dy \, dx = \dfrac{kbh^3}{12}$

$M_y = \int_0^{b/2} \int_0^{2hx/b} kxy \, dy \, dx + \int_{b/2}^b \int_0^{-2h(x-b)/b} kxy \, dy \, dx = \dfrac{kb^2h^2}{12}$

$\bar{x} = \dfrac{M_y}{m} = \dfrac{kb^2h^2/12}{kbh^2/6} = \dfrac{b}{2}$

$\bar{y} = \dfrac{M_x}{m} = \dfrac{kbh^3/12}{kbh^2/6} = \dfrac{h}{2}$

(c) $m = \int_0^{b/2} \int_0^{2hx/b} kx \, dy \, dx + \int_{b/2}^b \int_0^{-2h(x-b)/b} kx \, dy \, dx$

$= \dfrac{1}{12}kb^2h + \dfrac{1}{6}kb^2h = \dfrac{1}{4}kb^2h$

$M_x = \int_0^{b/2} \int_0^{2hx/b} kxy \, dy \, dx + \int_{b/2}^b \int_0^{-2h(x-b)/b} kxy \, dy \, dx$

$= \dfrac{1}{32}kh^2b^2 + \dfrac{5}{96}kh^2b^2 = \dfrac{1}{12}kh^2b^2$

$M_y = \int_0^{b/2} \int_0^{2hx/b} kx^2 \, dy \, dx + \int_{b/2}^b \int_0^{-2h(x-b)/b} kx^2 \, dy \, dx$

$= \dfrac{1}{32}kb^3h + \dfrac{11}{96}kb^3h = \dfrac{7}{48}kb^3h$

$\bar{x} = \dfrac{M_y}{m} = \dfrac{7kb^3h/48}{kb^2h/4} = \dfrac{7}{12}b$

$\bar{y} = \dfrac{M_x}{m} = \dfrac{kh^2b^2/12}{kb^2h/4} = \dfrac{h}{3}$

9. (a) The x-coordinate changes by 5: $(\bar{x}, \bar{y}) = \left(\dfrac{a}{2} + 5, \dfrac{b}{2}\right)$

(b) The x-coordinate changes by 5: $(\bar{x}, \bar{y}) = \left(\dfrac{a}{2} + 5, \dfrac{2b}{3}\right)$

(c) $m = \int_5^{a+5} \int_0^b kx \, dy \, dx = \dfrac{1}{2}k(a+5)^2 b - \dfrac{25}{2}kb$

$M_x = \int_5^{a+5} \int_0^b kxy \, dy \, dx = \dfrac{1}{4}k(a+5)^2 b^2 - \dfrac{25}{4}kb^2$

$M_y = \int_5^{a+5} \int_0^b kx^2 \, dy \, dx = \dfrac{1}{3}k(a+5)^3 b - \dfrac{125}{3}kb$

$\bar{x} = \dfrac{M_y}{m} = \dfrac{2(a^2 + 15a + 75)}{3(a+10)}$

$\bar{y} = \dfrac{M_x}{m} = \dfrac{b}{2}$

11. (a) $\bar{x} = 0$ by symmetry

$m = \dfrac{\pi a^2 k}{2}$

$M_x = \int_{-a}^a \int_0^{\sqrt{a^2 - x^2}} yk \, dy \, dx = \dfrac{2a^3 k}{3}$

$\bar{y} = \dfrac{M_x}{m} = \dfrac{2a^3 k}{3} \cdot \dfrac{2}{\pi a^2 k} = \dfrac{4a}{3\pi}$

(b) $m = \int_{-a}^a \int_0^{\sqrt{a^2 - x^2}} k(a - y) y \, dy \, dx = \dfrac{a^4 k}{24}(16 - 3\pi)$

$M_x = \int_{-a}^a \int_0^{\sqrt{a^2 - x^2}} k(a - y) y^2 \, dy \, dx = \dfrac{a^5 k}{120}(15\pi - 32)$

$M_y = \int_{-a}^a \int_0^{\sqrt{a^2 - x^2}} kx(a - y) y \, dy \, dx = 0$

$\bar{x} = \dfrac{M_y}{m} = 0$

$\bar{y} = \dfrac{M_x}{m} = \dfrac{a}{5}\left[\dfrac{15\pi - 32}{16 - 3\pi}\right]$

13. $m = \int_0^4 \int_0^{\sqrt{x}} kxy \, dy \, dx = \dfrac{32k}{3}$

$M_x = \int_0^4 \int_0^{\sqrt{x}} kxy^2 \, dy \, dx = \dfrac{256k}{21}$

$M_y = \int_0^4 \int_0^{\sqrt{x}} kx^2 y \, dy \, dx = 32k$

$\bar{x} = \dfrac{M_y}{m} = \dfrac{32k}{1} \cdot \dfrac{3}{32k} = 3$

$\bar{y} = \dfrac{M_x}{m} = \dfrac{256k}{21} \cdot \dfrac{3}{32k} = \dfrac{8}{7}$

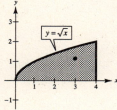

15. $\bar{x} = 0$ by symmetry

$m = \int_{-1}^1 \int_0^{1/(1+x^2)} k \, dy \, dx = \dfrac{k\pi}{2}$

$M_x = \int_{-1}^1 \int_0^{1/(1+x^2)} ky \, dy \, dx = \dfrac{k}{8}(2 + \pi)$

$\bar{y} = \dfrac{M_x}{m} = \dfrac{k}{8}(2 + \pi) \cdot \dfrac{2}{k\pi} = \dfrac{2 + \pi}{4\pi}$

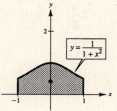

17. $\bar{y} = 0$ by symmetry

$m = \int_{-4}^4 \int_0^{16-y^2} kx \, dx \, dy = \dfrac{8192k}{15}$

$M_y = \int_{-4}^4 \int_0^{16-y^2} kx^2 \, dx \, dy = \dfrac{524{,}288k}{105}$

$\bar{x} = \dfrac{M_y}{m} = \dfrac{524{,}288k}{105} \cdot \dfrac{15}{8192k} = \dfrac{64}{7}$

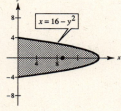

19. $\bar{x} = \dfrac{L}{2}$ by symmetry

$m = \int_0^L \int_0^{\sin \pi x/L} ky \, dy \, dx = \dfrac{kL}{4}$

$M_x = \int_0^L \int_0^{\sin \pi x/L} ky^2 \, dy \, dx = \dfrac{4kL}{9\pi}$

$\bar{y} = \dfrac{M_x}{m} = \dfrac{4kL}{9\pi} \cdot \dfrac{4}{kL} = \dfrac{16}{9\pi}$

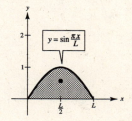

21. $m = \dfrac{\pi a^2 k}{8}$

$M_x = \iint_R ky \, dA = \int_0^{\pi/4} \int_0^a kr^2 \sin \theta \, dr \, d\theta = \dfrac{ka^3(2 - \sqrt{2})}{6}$

$M_y = \iint_R kx \, dA = \int_0^{\pi/4} \int_0^a kr^2 \cos \theta \, dr \, d\theta = \dfrac{ka^3 \sqrt{2}}{6}$

$\bar{x} = \dfrac{M_y}{m} = \dfrac{ka^3 \sqrt{2}}{6} \cdot \dfrac{8}{\pi a^2 k} = \dfrac{4a\sqrt{2}}{3\pi}$

$\bar{y} = \dfrac{M_x}{m} = \dfrac{ka^3(2 - \sqrt{2})}{6} \cdot \dfrac{8}{\pi a^2 k} = \dfrac{4a(2 - \sqrt{2})}{3\pi}$

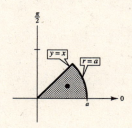

23. $m = \int_0^2 \int_0^{e^{-x}} ky \, dy \, dx = \frac{k}{4}(1 - e^{-4})$

$M_x = \int_0^2 \int_0^{e^{-x}} ky^2 \, dy \, dx = \frac{k}{9}(1 - e^{-6})$

$M_y = \int_0^2 \int_0^{e^{-x}} kxy \, dy \, dx = \frac{k(1 - 5e^{-4})}{8}$

$\bar{x} = \frac{M_y}{m} = \frac{k(e^4 - 5)}{8e^4} \cdot \frac{4e^4}{k(e^4 - 1)} = \frac{e^4 - 5}{2(e^4 - 1)} \approx 0.46$

$\bar{y} = \frac{M_x}{m} = \frac{k(e^6 - 1)}{9e^6} \cdot \frac{4e^4}{k(e^4 - 1)} = \frac{4}{9}\left[\frac{e^6 - 1}{e^6 - e^2}\right] \approx 0.45$

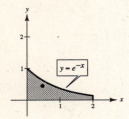

25. $\bar{y} = 0$ by symmetry

$m = \iint_R k \, dA = \int_{-\pi/6}^{\pi/6} \int_0^{2\cos 3\theta} kr \, dr \, d\theta = \frac{k\pi}{3}$

$M_y = \iint_R kx \, dA$

$\quad = \int_{-\pi/6}^{\pi/6} \int_0^{2\cos 3\theta} kr^2 \cos\theta \, dr \, d\theta = \frac{27\sqrt{3}}{40}k \approx 1.17k$

$\bar{x} = \frac{M_y}{m} = \frac{81\sqrt{3}}{40\pi} \approx 1.12$

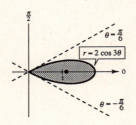

27. $m = bh$

$I_x = \int_0^b \int_0^h y^2 \, dy \, dx = \frac{bh^3}{3}$

$I_y = \int_0^b \int_0^h x^2 \, dy \, dx = \frac{b^3h}{3}$

$\bar{\bar{x}} = \sqrt{\frac{I_y}{m}} = \sqrt{\frac{b^3h}{3} \cdot \frac{1}{bh}} = \sqrt{\frac{b^2}{3}} = \frac{b}{\sqrt{3}} = \frac{\sqrt{3}}{3}b$

$\bar{\bar{y}} = \sqrt{\frac{I_x}{m}} = \sqrt{\frac{bh^3}{3} \cdot \frac{1}{bh}} = \sqrt{\frac{h^2}{3}} = \frac{h}{\sqrt{3}} = \frac{\sqrt{3}}{3}h$

29. $m = \pi a^2$

$I_x = \iint_R y^2 \, dA = \int_0^{2\pi} \int_0^a r^3 \sin^2\theta \, dr \, d\theta = \frac{a^4\pi}{4}$

$I_y = \iint_R x^2 \, dA = \int_0^{2\pi} \int_0^a r^3 \cos^2\theta \, dr \, d\theta = \frac{a^4\pi}{4}$

$I_0 = I_x + I_y = \frac{a^4\pi}{4} + \frac{a^4\pi}{4} = \frac{a^4\pi}{2}$

$\bar{\bar{x}} = \bar{\bar{y}} = \sqrt{\frac{I_x}{m}} = \sqrt{\frac{a^4\pi}{4} \cdot \frac{1}{\pi a^2}} = \frac{a}{2}$

31. $m = \frac{\pi a^2}{4}$

$I_x = \iint_R y^2 \, dA = \int_0^{\pi/2} \int_0^a r^3 \sin^2\theta \, dr \, d\theta = \frac{\pi a^4}{16}$

$I_y = \iint_R x^2 \, dA = \int_0^{\pi/2} \int_0^a r^3 \cos^2\theta \, dr \, d\theta = \frac{\pi a^4}{16}$

$I_0 = I_x + I_y = \frac{\pi a^4}{16} + \frac{\pi a^4}{16} = \frac{\pi a^4}{8}$

$\bar{\bar{x}} = \bar{\bar{y}} = \sqrt{\frac{I_x}{m}} = \sqrt{\frac{\pi a^4}{16} \cdot \frac{4}{\pi a^2}} = \frac{a}{2}$

33. $\rho = ky$

$m = k\int_0^a \int_0^b y \, dy \, dx = \frac{kab^2}{2}$

$I_x = k\int_0^a \int_0^b y^3 \, dy \, dx = \frac{kab^4}{4}$

$I_y = k\int_0^a \int_0^b x^2 y \, dy \, dx = \frac{ka^3b^2}{6}$

$I_0 = I_x + I_y = \frac{3kab^4 + 2kb^2a^3}{12}$

$\bar{x} = \sqrt{\frac{I_y}{m}} = \sqrt{\frac{ka^3b^2/6}{kab^2/2}} = \sqrt{\frac{a^2}{3}} = \frac{a}{\sqrt{3}} = \frac{\sqrt{3}}{3}a$

$\bar{y} = \sqrt{\frac{I_x}{m}} = \sqrt{\frac{kab^4/4}{kab^2/2}} = \sqrt{\frac{b^2}{2}} = \frac{b}{\sqrt{2}} = \frac{\sqrt{2}}{2}b$

35. $\rho = kx$

$$m = k\int_0^2 \int_0^{4-x^2} x\,dy\,dx = 4k$$

$$I_x = k\int_0^2 \int_0^{4-x^2} xy^2\,dy\,dx = \frac{32k}{3}$$

$$I_y = k\int_0^2 \int_0^{4-x^2} x^3\,dy\,dx = \frac{16k}{3}$$

$$I_0 = I_x + I_y = 16k$$

$$\bar{\bar{x}} = \sqrt{\frac{I_y}{m}} = \sqrt{\frac{16k/3}{4k}} = \sqrt{\frac{4}{3}} = \frac{2}{\sqrt{3}} = \frac{2\sqrt{3}}{3}$$

$$\bar{\bar{y}} = \sqrt{\frac{I_x}{m}} = \sqrt{\frac{32k/3}{4k}} = \sqrt{\frac{8}{3}} = \frac{4}{\sqrt{6}} = \frac{2\sqrt{6}}{3}$$

37. $\rho = kxy$

$$m = \int_0^4 \int_0^{\sqrt{x}} kxy\,dy\,dx = \frac{32k}{3}$$

$$I_x = \int_0^4 \int_0^{\sqrt{x}} kxy^3\,dy\,dx = 16k$$

$$I_y = \int_0^4 \int_0^{\sqrt{x}} kx^3 y\,dy\,dx = \frac{512k}{5}$$

$$I_0 = I_x + I_y = \frac{592k}{5}$$

$$\bar{\bar{x}} = \sqrt{\frac{I_y}{m}} = \sqrt{\frac{512k}{5} \cdot \frac{3}{32k}} = \sqrt{\frac{48}{5}} = \frac{4\sqrt{15}}{5}$$

$$\bar{\bar{y}} = \sqrt{\frac{I_x}{m}} = \sqrt{\frac{16k}{1} \cdot \frac{3}{32k}} = \sqrt{\frac{3}{2}} = \frac{\sqrt{6}}{2}$$

39. $\rho = kx$

$$m = \int_0^1 \int_{x^2}^{\sqrt{x}} kx\,dy\,dx = \frac{3k}{20}$$

$$I_x = \int_0^1 \int_{x^2}^{\sqrt{x}} kxy^2\,dy\,dx = \frac{3k}{56}$$

$$I_y = \int_0^1 \int_{x^2}^{\sqrt{x}} kx^3\,dy\,dx = \frac{k}{18}$$

$$I_0 = I_x + I_y = \frac{55k}{504}$$

$$\bar{\bar{x}} = \sqrt{\frac{I_y}{m}} = \sqrt{\frac{k}{18} \cdot \frac{20}{3k}} = \frac{\sqrt{30}}{9}$$

$$\bar{\bar{y}} = \sqrt{\frac{I_x}{m}} = \sqrt{\frac{3k}{56} \cdot \frac{20}{3k}} = \frac{\sqrt{70}}{14}$$

41. $I = 2k\int_{-b}^{b} \int_0^{\sqrt{b^2-x^2}} (x-a)^2\,dy\,dx = 2k\int_{-b}^{b} (x-a)^2 \sqrt{b^2-x^2}\,dx$

$$= 2k\left[\int_{-b}^{b} x^2\sqrt{b^2-x^2}\,dx - 2a\int_{-b}^{b} x\sqrt{b^2-x^2}\,dx + a^2\int_{-b}^{b} \sqrt{b^2-x^2}\,dx\right]$$

$$= 2k\left[\frac{\pi b^4}{8} + 0 + \frac{\pi a^2 b^2}{2}\right] = \frac{k\pi b^2}{4}(b^2 + 4a^2)$$

43. $I = \int_0^4 \int_0^{\sqrt{x}} kx(x-6)^2\,dy\,dx = \int_0^4 kx\sqrt{x}(x^2 - 12x + 36)\,dx = k\left[\frac{2}{9}x^{9/2} - \frac{24}{7}x^{7/2} + \frac{72}{5}x^{5/2}\right]_0^4 = \frac{42{,}752k}{315}$

45. $I = \int_0^a \int_0^{\sqrt{a^2-x^2}} k(a-y)(y-a)^2 \, dy \, dx = \int_0^a \int_0^{\sqrt{a^2-x^2}} k(a-y)^3 \, dy \, dx = \int_0^a \left[-\frac{k}{4}(a-y)^4 \right]_0^{\sqrt{a^2-x^2}} dx$

$\quad = -\frac{k}{4} \int_0^a \left[a^4 - 4a^3 y + 6a^2 y^2 - 4ay^3 + y^4 \right]_0^{\sqrt{a^2-x^2}} dx$

$\quad = -\frac{k}{4} \int_0^a \left[a^4 - 4a^3 \sqrt{a^2-x^2} + 6a^2(a^2-x^2) - 4a(a^2-x^2)\sqrt{a^2-x^2} + (a^4 - 2a^2 x^2 + x^4) - a^4 \right] dx$

$\quad = -\frac{k}{4} \int_0^a \left[7a^4 - 8a^2 x^2 + x^4 - 8a^3 \sqrt{a^2-x^2} + 4ax^2 \sqrt{a^2-x^2} \right] dx$

$\quad = -\frac{k}{4} \left[7a^4 x - \frac{8a^2}{3}x^3 + \frac{x^5}{5} - 4a^3 \left(x\sqrt{a^2-x^2} + a^2 \arcsin \frac{x}{a} \right) + \frac{a}{2}\left(x(2x^2 - a^2)\sqrt{a^2-x^2} + a^4 \arcsin \frac{x}{a} \right) \right]_0^a$

$\quad = -\frac{k}{4} \left(7a^5 - \frac{8}{3}a^5 + \frac{1}{5}a^5 - 2a^5 \pi + \frac{1}{4}a^5 \pi \right) = a^5 k \left(\frac{7\pi}{16} - \frac{17}{15} \right)$

47. $\rho(x, y) = ky.$ $\bar{y}$ will increase

49. $\rho(x, y) = kxy.$

Both $\bar{x}$ and $\bar{y}$ will increase

51. Let $\rho(x, y)$ be a continuous density function on the planar lamina R.

The movements of mass with respect to the x- and y-axes are

$$M_x = \int_R \int y \, \rho(x, y) dA \text{ and } M_y = \int_R \int x \, \rho(x, y) \, dA.$$

If m is the mass of the lamina, then the center of mass is

$$(\bar{x}, \bar{y}) = \left(\frac{M_y}{m}, \frac{M_x}{m} \right).$$

53. See the definition on page 968

55. $\bar{y} = \frac{L}{2}, A = bL, h = \frac{L}{2}$

$I_{\bar{y}} = \int_0^b \int_0^L \left(y - \frac{L}{2} \right)^2 dy \, dx$

$\quad = \int_0^b \left[\frac{[y - (L/2)]^3}{3} \right]_0^L dx = \frac{L^3 b}{12}$

$y_a = \bar{y} - \frac{I_{\bar{y}}}{hA} = \frac{L}{2} - \frac{L^3 b/12}{(L/2)(bL)} = \frac{L}{3}$

57. $\bar{y} = \frac{2L}{3}, A = \frac{bL}{2}, h = \frac{L}{3}$

$I_{\bar{y}} = 2 \int_0^{b/2} \int_{2Lx/b}^L \left(y - \frac{2L}{3} \right)^2 dy \, dx$

$\quad = \frac{2}{3} \int_0^{b/2} \left[\left(y - \frac{2L}{3} \right)^3 \right]_{2Lx/b}^L dx$

$\quad = \frac{2}{3} \int_0^{b/2} \left[\frac{L}{27} - \left(\frac{2Lx}{b} - \frac{2L}{3} \right)^3 \right] dx$

$\quad = \frac{2}{3} \left[\frac{L^3 x}{27} - \frac{b}{8L} \left(\frac{2Lx}{b} - \frac{2L}{3} \right)^4 \right]_0^{b/2} = \frac{L^3 b}{36}$

$y_a = \frac{2L}{3} - \frac{L^3 b/36}{L^2 b/6} = \frac{L}{2}$

Section 13.5   Surface Area

1. $f(x, y) = 2x + 2y$

R = triangle with vertices $(0, 0)$, $(2, 0)$, $(0, 2)$

$f_x = 2, f_y = 2$

$\sqrt{1 + (f_x)^2 + (f_y)^2} = 3$

$S = \int_0^2 \int_0^{2-x} 3 \, dy \, dx = 3 \int_0^2 (2 - x) \, dx$

$= \left[3\left(2x - \frac{x^2}{2} \right) \right]_0^2 = 6$

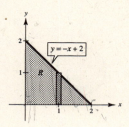

3. $f(x, y) = 8 + 2x + 2y$

$R = \{(x, y): x^2 + y^2 \leq 4\}$

$f_x = 2, f_y = 2$

$\sqrt{1 + (f_x)^2 + (f_y)^2} = 3$

$S = \int_{-2}^2 \int_{-\sqrt{4-x^2}}^{\sqrt{4-x^2}} 3 \, dy \, dx = \int_0^{2\pi} \int_0^2 3r \, dr \, d\theta = 12\pi$

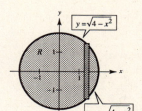

5. $f(x, y) = 9 - x^2$

R = square with vertices, $(0, 0)$, $(3, 0)$, $(0, 3)$, $(3, 3)$

$f_x = -2x, \ f_y = 0$

$\sqrt{1 + (f_x)^2 + (f_y)^2} = \sqrt{1 + 4x^2}$

$S = \int_0^3 \int_0^3 \sqrt{1 + 4x^2} \, dy \, dx = \int_0^3 3\sqrt{1 + 4x^2} \, dx$

$= \left[\frac{3}{4}(2x\sqrt{1 + 4x^2} + \ln|2x + \sqrt{1 + 4x^2}|) \right]_0^3 = \frac{3}{4}(6\sqrt{37} + \ln|6 + \sqrt{37}|)$

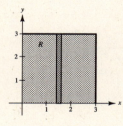

7. $f(x, y) = 2 + x^{3/2}$

R = rectangle with vertices $(0, 0)$, $(0, 4)$, $(3, 4)$, $(3, 0)$

$f_x = \frac{3}{2}x^{1/2}, \ f_y = 0$

$\sqrt{1 + (f_x)^2 + (f_y)^2} = \sqrt{1 + \left(\frac{9}{4} \right)x} = \frac{\sqrt{4 + 9x}}{2}$

$S = \int_0^3 \int_0^4 \frac{\sqrt{4 + 9x}}{2} \, dy \, dx = \int_0^3 4\left(\frac{\sqrt{4 + 9x}}{2} \right) dx$

$= \left[\frac{4}{27}(4 + 9x)^{3/2} \right]_0^3 = \frac{4}{27}(31\sqrt{31} - 8)$

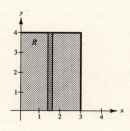

9. $f(x, y) = \ln|\sec x|$

$R = \left\{ (x, y): 0 \leq x \leq \frac{\pi}{4}, \ 0 \leq y \leq \tan x \right\}$

$f_x = \tan x, \ f_y = 0$

$\sqrt{1 + (f_x)^2 + (f_y)^2} = \sqrt{1 + \tan^2 x} = \sec x$

$S = \int_0^{\pi/4} \int_0^{\tan x} \sec x \, dy \, dx = \int_0^{\pi/4} \sec x \tan x \, dx = \left[\sec x \right]_0^{\pi/4} = \sqrt{2} - 1$

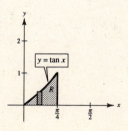

11. $f(x, y) = \sqrt{x^2 + y^2}$

$R = \{(x, y): 0 \leq f(x, y) \leq 1\}$

$0 \leq \sqrt{x^2 + y^2} \leq 1, \; x^2 + y^2 \leq 1$

$f_x = \dfrac{x}{\sqrt{x^2 + y^2}}, \; f_y = \dfrac{y}{\sqrt{x^2 + y^2}}$

$\sqrt{1 + (f_x)^2 + (f_y)^2} = \sqrt{1 + \dfrac{x^2}{x^2 + y^2} + \dfrac{y^2}{x^2 + y^2}} = \sqrt{2}$

$S = \displaystyle\int_{-1}^{1}\int_{-\sqrt{1-x^2}}^{\sqrt{1-x^2}} \sqrt{2} \, dy \, dx = \int_{0}^{2\pi}\int_{0}^{1} \sqrt{2} \, r \, dr \, d\theta = \sqrt{2}\pi$

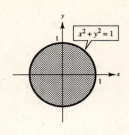

13. $f(x, y) = \sqrt{a^2 - x^2 - y^2}$

$R = \{(x, y): x^2 + y^2 \leq b^2, \, b < a\}$

$f_x = \dfrac{-x}{\sqrt{a^2 - x^2 - y^2}}, \; f_y = \dfrac{-y}{\sqrt{a^2 - x^2 - y^2}}$

$\sqrt{1 + (f_x)^2 + (f_y)^2} = \sqrt{1 + \dfrac{x^2}{a^2 - x^2 - y^2} + \dfrac{y^2}{a^2 - x^2 - y^2}} = \dfrac{a}{\sqrt{a^2 - x^2 - y^2}}$

$S = \displaystyle\int_{-b}^{b}\int_{-\sqrt{b^2-x^2}}^{\sqrt{b^2-x^2}} \dfrac{a}{\sqrt{a^2 - x^2 - y^2}} \, dy \, dx = \int_{0}^{2\pi}\int_{0}^{b} \dfrac{a}{\sqrt{a^2 - r^2}} \, r \, dr \, d\theta = 2\pi a\left(a - \sqrt{a^2 - b^2}\right)$

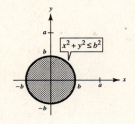

15. $z = 24 - 3x - 2y$

$\sqrt{1 + (f_x)^2 + (f_y)^2} = \sqrt{14}$

$S = \displaystyle\int_{0}^{8}\int_{0}^{-(3/2)x + 12} \sqrt{14} \, dy \, dx = 48\sqrt{14}$

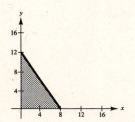

17. $z = \sqrt{25 - x^2 - y^2}$

$\sqrt{1 + (f_x)^2 + (f_y)^2} = \sqrt{1 + \dfrac{x^2}{25 - x^2 - y^2} + \dfrac{y^2}{25 - x^2 - y^2}} = \dfrac{5}{\sqrt{25 - x^2 - y^2}}$

$S = 2\displaystyle\int_{-3}^{3}\int_{-\sqrt{9-x^2}}^{\sqrt{9-x^2}} \dfrac{5}{\sqrt{25 - (x^2 + y^2)}} \, dy \, dx$

$= 2\displaystyle\int_{0}^{2\pi}\int_{0}^{3} \dfrac{5}{\sqrt{25 - r^2}} \, r \, dr \, d\theta = 20\pi$

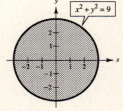

19. $f(x, y) = 2y + x^2$

$R = $ triangle with vertices $(0, 0), (1, 0), (1, 1)$

$\sqrt{1 + (f_x)^2 + (f_y)^2} = \sqrt{5 + 4x^2}$

$S = \displaystyle\int_{0}^{1}\int_{0}^{x} \sqrt{5 + 4x^2} \, dy \, dx = \dfrac{1}{12}\left(27 - 5\sqrt{5}\right)$

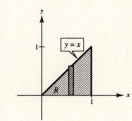

21. $f(x, y) = 4 - x^2 - y^2$

$R = \{(x, y): 0 \leq f(x, y)\}$

$0 \leq 4 - x^2 - y^2, \; x^2 + y^2 \leq 4$

$f_x = -2x, \; f_y = -2y$

$\sqrt{1 + (f_x)^2 + (f_y)^2} = \sqrt{1 + 4x^2 + 4y^2}$

$S = \displaystyle\int_{-2}^{2} \int_{-\sqrt{4-x^2}}^{\sqrt{4-x^2}} \sqrt{1 + 4x^2 + 4y^2} \, dy \, dx$

$= \displaystyle\int_{0}^{2\pi} \int_{0}^{2} \sqrt{1 + 4r^2} \, r \, dr \, d\theta = \dfrac{(17\sqrt{17} - 1)\pi}{6}$

23. $f(x, y) = 4 - x^2 - y^2$

$R = \{(x, y): 0 \leq x \leq 1, \; 0 \leq y \leq 1\}$

$f_x = -2x, \; f_y = -2y$

$\sqrt{1 + (f_x)^2 + (f_y)^2} = \sqrt{1 + 4x^2 + 4y^2}$

$S = \displaystyle\int_{0}^{1} \int_{0}^{1} \sqrt{(1 + 4x^2) + 4y^2} \, dy \, dx \approx 1.8616$

25. Surface area $> (4) \cdot (6) = 24$.

Matches (e)

27. $f(x, y) = e^x$

$R = \{(x, y): 0 \leq x \leq 1, \; 0 \leq y \leq 1\}$

$f_x = e^x, \; f_y = 0$

$\sqrt{1 + (f_x)^2 + (f_y)^2} = \sqrt{1 + e^{2x}}$

$S = \displaystyle\int_{0}^{1} \int_{0}^{1} \sqrt{1 + e^{2x}} \, dy \, dx$

$= \displaystyle\int_{0}^{1} \sqrt{1 + e^{2x}} \approx 2.0035$

29. $f(x, y) = x^3 - 3xy + y^3$

$R = $ square with vertices $(1, 1), (-1, 1), (-1, -1), (1, -1)$

$f_x = 3x^2 - 3y = 3(x^2 - y), f_y = -3x + 3y^2 = 3(y^2 - x)$

$S = \displaystyle\int_{-1}^{1} \int_{-1}^{1} \sqrt{1 + 9(x^2 - y)^2 + 9(y^2 - x)^2} \, dy \, dx$

31. $f(x, y) = e^{-x} \sin y$

$f_x = -e^{-x} \sin y, f_y = e^{-x} \cos y$

$\sqrt{1 + f_x^2 + f_y^2} = \sqrt{1 + e^{-2x} \sin^2 y + e^{-2x} \cos^2 y}$

$= \sqrt{1 + e^{-2x}}$

$S = \displaystyle\int_{-2}^{2} \int_{-\sqrt{4-x^2}}^{\sqrt{4-x^2}} \sqrt{1 + e^{-2x}} \, dy \, dx$

33. $f(x, y) = e^{xy}$

$R = \{(x, y): 0 \leq x \leq 4, \; 0 \leq y \leq 10\}$

$f_x = ye^{xy}, \; f_y = xe^{xy}$

$\sqrt{1 + (f_x)^2 + (f_y)^2} = \sqrt{1 + y^2 e^{2xy} + x^2 e^{2xy}} = \sqrt{1 + e^{2xy}(x^2 + y^2)}$

$S = \displaystyle\int_{0}^{4} \int_{0}^{10} \sqrt{1 + e^{2xy}(x^2 + y^2)} \, dy \, dx$

35. See the definition on page 972.

37. $f(x, y) = \sqrt{1 - x^2}; f_x = \dfrac{-x}{\sqrt{1^2 - x^2}}, f_y = 0$

$$S = \iint_R \sqrt{1 + f_x^2 + f_y^2} \, dA$$

$$= 16 \int_0^1 \int_0^x \frac{1}{\sqrt{1 - x^2}} \, dy \, dx$$

$$= 16 \int_0^1 \frac{x}{\sqrt{1 - x^2}} \, dx = \left[-16(1 - x^2)^{1/2} \right]_0^1 = 16$$

39. (a) $V = \displaystyle\int_0^{50} \int_0^{\sqrt{50^2 - x^2}} \left(20 + \frac{xy}{100} - \frac{x + y}{5} \right) dy \, dx$

$$= \int_0^{50} \left[20\sqrt{50^2 - x^2} + \frac{x}{200}(50^2 - x^2) - \frac{x}{5}\sqrt{50^2 - x^2} - \frac{50^2 - x^2}{10} \right] dy$$

$$= \left[10\left(x\sqrt{50 - x^2} + 50^2 \arcsin \frac{x}{50} \right) + \frac{25}{4}x^2 - \frac{x^4}{800} + \frac{1}{15}(50^2 - x^2)^{3/2} - 250x + \frac{x^3}{30} \right]_0^{50}$$

$$\approx 30{,}415.74 \text{ ft}^3$$

(b) $z = 20 + \dfrac{xy}{100}$

$$\sqrt{1 + (f_x)^2 + (f_y)^2} = \sqrt{1 + \frac{y^2}{100^2} + \frac{x^2}{100^2}} = \frac{\sqrt{100^2 + x^2 + y^2}}{100}$$

$$S = \frac{1}{100} \int_0^{50} \int_0^{\sqrt{50^2 - x^2}} \sqrt{100^2 + x^2 + y^2} \, dy \, dx$$

$$= \frac{1}{100} \int_0^{\pi/2} \int_0^{50} \sqrt{100^2 + r^2} \, r \, dr \, d\theta \approx 2081.53 \text{ ft}^2$$

41. (a) $V = \displaystyle\iint_R f(x, y)$

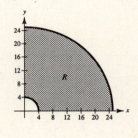

$$= 8 \iint_R \sqrt{625 - x^2 - y^2} \, dA \qquad \text{where } R \text{ is the region in the first quadrant}$$

$$= 8 \int_0^{\pi/2} \int_4^{25} \sqrt{625 - r^2} \, r \, dr \, d\theta$$

$$= -4 \int_0^{\pi/2} \left[\frac{2}{3}(625 - r^2)^{3/2} \right]_4^{25} d\theta$$

$$= -\frac{8}{3}\left[0 - 609\sqrt{609} \right] \cdot \frac{\pi}{2}$$

$$= 812\pi\sqrt{609} \text{ cm}^3$$

(b) $A = \displaystyle\iint_R \sqrt{1 + (f_x)^2 + (f_y)^2} \, dA = 8 \iint_R \sqrt{1 + \frac{x^2}{625 - x^2 - y^2} + \frac{y^2}{625 - x^2 - y^2}} \, dA$

$$= 8 \iint_R \frac{25}{\sqrt{625 - x^2 - y^2}} \, dA = 8 \int_0^{\pi/2} \int_4^{25} \frac{25}{\sqrt{625 - r^2}} \, r \, dr \, d\theta$$

$$= \lim_{b \to 25^-} \left[-200\sqrt{625 - r^2} \right]_4^b \cdot \frac{\pi}{2} = 100\pi\sqrt{609} \text{ cm}^2$$

Section 13.6   Triple Integrals and Applications

1. $\displaystyle\int_0^3\int_0^2\int_0^1 (x+y+z)\,dx\,dy\,dx = \int_0^3\int_0^2\left[\frac{1}{2}x^2+xy+xz\right]_0^1 dy\,dx$

$$= \int_0^3\int_0^2\left(\frac{1}{2}+y+z\right)dy\,dz = \int_0^3\left[\frac{1}{2}y+\frac{1}{2}y^2+yz\right]_0^2 dz = \left[3z+z^2\right]_0^3 = 18$$

3. $\displaystyle\int_0^1\int_0^x\int_0^{xy} x\,dz\,dy\,dx = \int_0^1\int_0^x\left[xz\right]_0^{xy} dy\,dx$

$$= \int_0^1\int_0^x x^2y\,dy\,dx = \int_0^1\left[\frac{x^2y^2}{2}\right]_0^x dx = \int_0^1\frac{x^4}{2}\,dx = \left[\frac{x^5}{10}\right]_0^1 = \frac{1}{10}$$

5. $\displaystyle\int_1^4\int_0^1\int_0^x 2ze^{-x^2}\,dy\,dx\,dz = \int_1^4\int_0^1\left[(2ze^{-x^2})y\right]_0^x dx\,dz = \int_1^4\int_0^1 2zxe^{-x^2}\,dx\,dz$

$$= \int_1^4\left[-ze^{-x^2}\right]_0^1 dz = \int_1^4 z(1-e^{-1})\,dz = \left[(1-e^{-1})\frac{z^2}{2}\right]_1^4 = \frac{15}{2}\left(1-\frac{1}{e}\right)$$

7. $\displaystyle\int_0^4\int_0^{\pi/2}\int_0^{1-x} x\cos y\,dz\,dy\,dx = \int_0^4\int_0^{\pi/2}\left[(x\cos y)z\right]_0^{1-x} dy\,dx = \int_0^4\int_0^{\pi/2} x(1-x)\cos y\,dy\,dx$

$$= \int_0^4\left[x(1-x)\sin y\right]_0^{\pi/2} dx = \int_0^4 x(1-x)\,dx = \left[\frac{x^2}{2}-\frac{x^3}{3}\right]_0^4 = 8-\frac{64}{3} = \frac{-40}{3}$$

9. $\displaystyle\int_0^2\int_{-\sqrt{4-x^2}}^{\sqrt{4-x^2}}\int_0^{x^2} x\,dz\,dy\,dx = \int_0^2\int_{-\sqrt{4-x^2}}^{\sqrt{4-x^2}} x^3\,dy\,dx = \frac{128}{15}$

11. $\displaystyle\int_0^2\int_0^{\sqrt{4-x^2}}\int_1^4 \frac{x^2\sin y}{z}\,dz\,dy\,dx = \int_0^2\int_0^{\sqrt{4-x^2}}\left[x^2\sin y\ln|z|\right]_1^4 dy\,dx$

$$= \int_0^2\left[x^2\ln4(-\cos y)\right]_0^{\sqrt{4-x^2}} dx = \int_0^2 x^2\ln 4\left[1-\cos\sqrt{4-x^2}\right]dx \approx 2.44167$$

13. $\displaystyle\int_0^4\int_0^{4-x}\int_0^{4-x-y} dz\,dy\,dx$ **15.** $\displaystyle\int_{-3}^3\int_{-\sqrt{9-x^2}}^{\sqrt{9-x^2}}\int_0^{9-x^2-y^2} dz\,dy\,dx$

17. $\displaystyle\int_{-2}^2\int_0^{4-y^2}\int_0^x dz\,dx\,dy = \int_{-2}^2\int_0^{4-y^2} x\,dx\,dy$

$$= \frac{1}{2}\int_{-2}^2 (4-y^2)^2\,dy = \int_0^2 (16-8y^2+y^4)\,dy = \left[16y-\frac{8}{3}y^3+\frac{1}{5}y^5\right]_0^2 = \frac{256}{15}$$

19. $\displaystyle 8\int_0^a\int_0^{\sqrt{a^2-x^2}}\int_0^{\sqrt{a^2-x^2-y^2}} dz\,dy\,dx = 8\int_0^a\int_0^{\sqrt{a^2-x^2}}\sqrt{a^2-x^2-y^2}\,dy\,dx$

$$= 4\int_0^a\left[y\sqrt{a^2-x^2-y^2}+(a^2-x^2)\arcsin\left(\frac{y}{\sqrt{a^2-x^2}}\right)\right]_0^{\sqrt{a^2-x^2}} dx$$

$$= 4\left(\frac{\pi}{2}\right)\int_0^a (a^2-x^2)\,dx = \left[2\pi\left(a^2x-\frac{1}{3}x^3\right)\right]_0^a = \frac{4}{3}\pi a^3$$

21. $\displaystyle\int_0^2\int_0^{4-x^2}\int_0^{4-x^2} dz\,dy\,dx = \int_0^2 (4-x^2)^2\,dx = \int_0^2 (16 - 8x^2 + x^4)\,dx = \left[16x - \frac{8}{3}x^3 + \frac{1}{5}x^5\right]_0^2 = \frac{256}{15}$

23. Plane: $3x + 6y + 4z = 12$

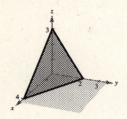

$$\int_0^3\int_0^{(12-4z)/3}\int_0^{(12-4z-3x)/6} dy\,dx\,dz$$

25. Top cylinder: $y^2 + z^2 = 1$

Side plane: $x = y$

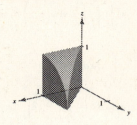

$$\int_0^1\int_0^x\int_0^{\sqrt{1-y^2}} dz\,dy\,dx$$

27. $Q = \{(x, y, z): 0 \le x \le 1, 0 \le y \le x, 0 \le z \le 3\}$

$\displaystyle\iiint_Q xyz\,dV = \int_0^3\int_0^1\int_y^1 xyz\,dx\,dy\,dz = \int_0^3\int_0^1\int_0^x xyz\,dy\,dx\,dz$

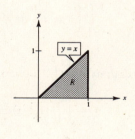

$\displaystyle\qquad\qquad = \int_0^1\int_0^3\int_y^1 xyz\,dx\,dz\,dy$

$\displaystyle\qquad\qquad = \int_0^1\int_0^3\int_0^x xyz\,dy\,dz\,dx$

$\displaystyle\qquad\qquad = \int_0^1\int_y^1\int_0^3 xyz\,dz\,dx\,dy$

$\displaystyle\qquad\qquad = \int_0^1\int_0^x\int_0^3 xyz\,dz\,dy\,dx\left(=\frac{9}{16}\right)$

29. $Q = \{(x, y, z): x^2 + y^2 \le 9, 0 \le z \le 4\}$

$\displaystyle\iiint_Q xyz\,dV = \int_0^4\int_{-3}^3\int_{-\sqrt{9-x^2}}^{\sqrt{9-x^2}} xyz\,dy\,dx\,dz$

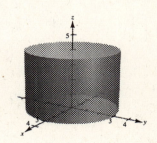

$\displaystyle\qquad\qquad = \int_0^4\int_{-3}^3\int_{-\sqrt{9-y^2}}^{\sqrt{9-y^2}} xyz\,dx\,dy\,dz$

$\displaystyle\qquad\qquad = \int_{-3}^3\int_0^4\int_{-\sqrt{9-y^2}}^{\sqrt{9-y^2}} xyz\,dx\,dz\,dy$

$\displaystyle\qquad\qquad = \int_{-3}^3\int_{-\sqrt{9-y^2}}^{\sqrt{9-y^2}}\int_0^4 xyz\,dz\,dx\,dy$

$\displaystyle\qquad\qquad = \int_{-3}^3\int_0^4\int_{-\sqrt{9-x^2}}^{\sqrt{9-x^2}} xyz\,dy\,dz\,dx$

$\displaystyle\qquad\qquad = \int_{-3}^3\int_{-\sqrt{9-x^2}}^{\sqrt{9-x^2}}\int_0^4 xyz\,dz\,dy\,dx\ (=0)$

31. $m = k \displaystyle\int_0^6 \int_0^{4-(2x/3)} \int_0^{2-(y/2)-(x/3)} dz\,dy\,dx$

$= 8k$

$M_{yz} = k \displaystyle\int_0^6 \int_0^{4-(2x/3)} \int_0^{2-(y/2)-(x/3)} x\,dz\,dy\,dx$

$= 12k$

$\bar{x} = \dfrac{M_{yz}}{m} = \dfrac{12k}{8k} = \dfrac{3}{2}$

33. $m = k \displaystyle\int_0^4 \int_0^4 \int_0^{4-x} x\,dz\,dy\,dx = k \int_0^4 \int_0^4 x(4-x)\,dy\,dx$

$= 4k \displaystyle\int_0^4 (4x - x^2)\,dx = \dfrac{128k}{3}$

$M_{xy} = k \displaystyle\int_0^4 \int_0^4 \int_0^{4-x} xz\,dz\,dy\,dx = k \int_0^4 \int_0^4 x\dfrac{(4-x)^2}{2}\,dy\,dx$

$= 2k \displaystyle\int_0^4 (16x - 8x^2 + x^3)\,dx = \dfrac{128k}{3}$

$\bar{z} = \dfrac{M_{xy}}{m} = 1$

35. $m = k \displaystyle\int_0^b \int_0^b \int_0^b xy\,dz\,dy\,dx = \dfrac{kb^5}{4}$

$M_{yz} = k \displaystyle\int_0^b \int_0^b \int_0^b x^2y\,dz\,dy\,dx = \dfrac{kb^6}{6}$

$M_{xz} = k \displaystyle\int_0^b \int_0^b \int_0^b xy^2\,dz\,dy\,dx = \dfrac{kb^6}{6}$

$M_{xy} = k \displaystyle\int_0^b \int_0^b \int_0^b xyz\,dz\,dy\,dx = \dfrac{kb^6}{8}$

$\bar{x} = \dfrac{M_{yz}}{m} = \dfrac{kb^6/6}{kb^5/4} = \dfrac{2b}{3}$

$\bar{y} = \dfrac{M_{xz}}{m} = \dfrac{kb^6/6}{kb^5/4} = \dfrac{2b}{3}$

$\bar{z} = \dfrac{M_{xy}}{m} = \dfrac{kb^6/8}{kb^5/4} = \dfrac{b}{2}$

37. $\bar{x}$ will be greater than 2, whereas $\bar{y}$ and $\bar{z}$ will be unchanged.

39. $\bar{y}$ will be greater than 0, whereas $\bar{x}$ and $\bar{z}$ will be unchanged.

41. $m = \dfrac{1}{3}k\pi r^2 h$

$\bar{x} = \bar{y} = 0$ by symmetry

$M_{xy} = 4k \displaystyle\int_0^r \int_0^{\sqrt{r^2-x^2}} \int_{h\sqrt{x^2+y^2}/r}^h z\,dz\,dy\,dx$

$= \dfrac{2kh^2}{r^2} \displaystyle\int_0^r \int_0^{\sqrt{r^2-x^2}} (r^2 - x^2 - y^2)\,dy\,dx$

$= \dfrac{4kh^2}{3r^2} \displaystyle\int_0^r (r^2 - x^2)^{3/2}\,dx$

$= \dfrac{k\pi r^2 h^2}{4}$

$\bar{z} = \dfrac{M_{xy}}{m} = \dfrac{k\pi r^2 h^2/4}{k\pi r^2 h/3} = \dfrac{3h}{4}$

43. $m = \dfrac{128k\pi}{3}$

$\bar{x} = \bar{y} = 0$ by symmetry

$z = \sqrt{4^2 - x^2 - y^2}$

$M_{xy} = 4k \displaystyle\int_0^4 \int_0^{\sqrt{4^2-x^2}} \int_0^{\sqrt{4^2-x^2-y^2}} z \, dz \, dy \, dx$

$\qquad = 2k \displaystyle\int_0^4 \int_0^{\sqrt{4^2-x^2}} (4^2 - x^2 - y^2) \, dy \, dx = 2k \int_0^4 \left[16y - x^2y - \frac{1}{3}y^3 \right]_0^{\sqrt{4^2-x^2}} dx = \frac{4k}{3} \int_0^4 (4^2 - x^2)^{3/2} \, dx$

$\qquad = \dfrac{1024k}{3} \displaystyle\int_0^{\pi/2} \cos^4 \theta \, d\theta \qquad (\text{let } x = 4 \sin \theta)$

$\qquad = 64\pi k \qquad \text{by Wallis's Formula}$

$\bar{z} = \dfrac{M_{xy}}{m} = \dfrac{64k\pi}{1} \cdot \dfrac{3}{128k\pi} = \dfrac{3}{2}$

45. $f(x, y) = \dfrac{5}{12}y$

$m = k \displaystyle\int_0^{20} \int_0^{-(3/5)x+12} \int_0^{(5/12)y} dz \, dy \, dx = 200k$

$M_{yz} = k \displaystyle\int_0^{20} \int_0^{-(3/5)x+12} \int_0^{(5/12)y} x \, dz \, dy \, dx = 1000k$

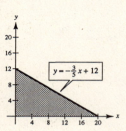

$M_{xz} = k \displaystyle\int_0^{20} \int_0^{-(3/5)x+12} \int_0^{(5/12)y} y \, dz \, dy \, dx = 1200k$

$M_{xy} = k \displaystyle\int_0^{20} \int_0^{-(3/5)x+12} \int_0^{(5/12)y} z \, dz \, dy \, dx = 250k$

$\bar{x} = \dfrac{M_{yz}}{m} = \dfrac{1000k}{200k} = 5$

$\bar{y} = \dfrac{M_{xz}}{m} = \dfrac{1200k}{200k} = 6$

$\bar{z} = \dfrac{M_{xy}}{m} = \dfrac{250k}{200k} = \dfrac{5}{4}$

47. (a) $I_x = k \displaystyle\int_0^a \int_0^a \int_0^a (y^2 + z^2) \, dx \, dy \, dz = ka \int_0^a \int_0^a (y^2 + z^2) \, dy \, dz$

$\qquad = ka \displaystyle\int_0^a \left[\frac{1}{3}y^3 + z^2y \right]_0^a dz = ka \int_0^a \left(\frac{1}{3}a^3 + az^2 \right) dz = \left[ka\left(\frac{1}{3}a^3z + \frac{1}{3}az^3 \right) \right]_0^a = \frac{2ka^5}{3}$

$\qquad I_x = I_y = I_z = \dfrac{2ka^5}{3}$ by symmetry

(b) $I_x = k \displaystyle\int_0^a \int_0^a \int_0^a (y^2 + z^2)xyz \, dx \, dy \, dz = \frac{ka^2}{2} \int_0^a \int_0^a (y^3z + yz^3) \, dy \, dz$

$\qquad = \dfrac{ka^2}{2} \displaystyle\int_0^a \left[\frac{y^4z}{4} + \frac{y^2z^3}{2} \right]_0^a dz = \frac{ka^4}{8} \int_0^a (a^2z + 2z^3) \, dz = \left[\frac{ka^4}{8}\left(\frac{a^2z^2}{2} + \frac{2z^4}{4} \right) \right]_0^a = \frac{ka^8}{8}$

$\qquad I_x = I_y = I_z = \dfrac{ka^8}{8}$ by symmetry

49. (a) $I_x = k \int_0^4 \int_0^4 \int_0^{4-x} (y^2 + z^2) \, dz \, dy \, dx = k \int_0^4 \int_0^4 \left[y^2(4-x) + \frac{1}{3}(4-x)^3 \right] dy \, dx$

$= k \int_0^4 \left[\frac{y^3}{3}(4-x) + \frac{y}{3}(4-x)^3 \right]_0^4 dx = k \int_0^4 \left[\frac{64}{3}(4-x) + \frac{4}{3}(4-x)^3 \right] dx$

$= k \left[-\frac{32}{3}(4-x)^2 - \frac{1}{3}(4-x)^4 \right]_0^4 = 256k$

$I_y = k \int_0^4 \int_0^4 \int_0^{4-x} (x^2 + z^2) \, dz \, dy \, dx = k \int_0^4 \int_0^4 \left[x^2(4-x) + \frac{1}{3}(4-x)^3 \right] dy \, dx$

$= 4k \int_0^4 \left[4x^2 - x^3 + \frac{1}{3}(4-x)^3 \right] dx = 4k \left[\frac{4}{3}x^3 - \frac{1}{4}x^4 - \frac{1}{12}(4-x)^4 \right]_0^4 = \frac{512k}{3}$

$I_z = k \int_0^4 \int_0^4 \int_0^{4-x} (x^2 + y^2) \, dz \, dy \, dx = k \int_0^4 \int_0^4 (x^2 + y^2)(4-x) \, dy \, dx$

$= k \int_0^4 \left[\left(x^2 y + \frac{y^3}{3} \right)(4-x) \right]_0^4 dx = k \int_0^4 \left(4x^2 + \frac{64}{3} \right)(4-x) \, dx = 256k$

(b) $I_x = k \int_0^4 \int_0^4 \int_0^{4-x} y(y^2 + z^2) \, dz \, dy \, dx = k \int_0^4 \int_0^4 \left[y^3(4-x) + \frac{1}{3}y(4-x)^3 \right] dy \, dx$

$= k \int_0^4 \left[\frac{y^4}{4}(4-x) + \frac{y^2}{6}(4-x)^3 \right]_0^4 dx = k \int_0^4 \left[64(4-x) + \frac{8}{3}(4-x)^3 \right] dx$

$= k \left[-32(4-x)^2 - \frac{2}{3}(4-x)^4 \right]_0^4 = \frac{2048k}{3}$

$I_y = k \int_0^4 \int_0^4 \int_0^{4-x} y(x^2 + z^2) \, dz \, dy \, dx = k \int_0^4 \int_0^4 \left[x^2 y(4-x) + \frac{1}{3}y(4-x)^3 \right] dy \, dx$

$= 8k \int_0^4 \left[4x^2 - x^3 + \frac{1}{3}(4-x)^3 \right] dx = 8k \left[\frac{4}{3}x^3 - \frac{1}{4}x^4 - \frac{1}{12}(4-x)^4 \right]_0^4 = \frac{1024k}{3}$

$I_z = k \int_0^4 \int_0^4 \int_0^{4-x} y(x^2 + y^2) \, dz \, dy \, dx = k \int_0^4 \int_0^4 (x^2 y + y^3)(4-x) \, dx$

$= k \int_0^4 \left[\left(\frac{x^2 y^2}{2} + \frac{y^4}{4} \right)(4-x) \right]_0^4 dx = k \int_0^4 (8x^2 + 64)(4-x) \, dx$

$= 8k \int_0^4 (32 - 8x + 4x^2 - x^3) \, dx = \left[8k \left(32x - 4x^2 + \frac{4}{3}x^3 - \frac{1}{4}x^4 \right) \right]_0^4 = \frac{2048k}{3}$

51. $I_{xy} = k \int_{-L/2}^{L/2} \int_{-a}^{a} \int_{-\sqrt{a^2-x^2}}^{\sqrt{a^2-x^2}} z^2 \, dz \, dx \, dy = k \int_{-L/2}^{L/2} \int_{-a}^{a} \frac{2}{3}(a^2 - x^2)\sqrt{a^2 - x^2} \, dx \, dy$

$= \frac{2}{3} \int_{-L/2}^{L/2} k \left[\frac{a^2}{2} \left(x\sqrt{a^2 - x^2} + a^2 \arcsin \frac{x}{a} \right) - \frac{1}{8} \left(x(2x^2 - a^2)\sqrt{x^2 - a^2} + a^4 \arcsin \frac{x}{a} \right) \right]_{-a}^{a} dy$

$= \frac{2k}{3} \int_{-L/2}^{L/2} 2 \left(\frac{a^4 \pi}{4} - \frac{a^4 \pi}{16} \right) dy = \frac{a^4 \pi L k}{4}$

Since $m = \pi a^2 L k$, $I_{xy} = ma^2/4$.

—CONTINUED—

51. —CONTINUED—

$$I_{xz} = k \int_{-L/2}^{L/2} \int_{-a}^{a} \int_{-\sqrt{a^2-x^2}}^{\sqrt{a^2-x^2}} y^2 \, dz \, dx \, dy = 2k \int_{-L/2}^{L/2} \int_{-a}^{a} y^2 \sqrt{a^2-x^2} \, dx \, dy$$

$$= 2k \int_{-L/2}^{L/2} \left[\frac{y^2}{2} \left(x\sqrt{a^2-x^2} + a^2 \arcsin \frac{x}{a} \right) \right]_{-a}^{a} dy = k\pi a^2 \int_{-L/2}^{L/2} y^2 \, dy = \frac{2k\pi a^2}{3} \left(\frac{L^3}{8} \right) = \frac{1}{12} mL^2$$

$$I_{yz} = k \int_{-L/2}^{L/2} \int_{-a}^{a} \int_{-\sqrt{a^2-x^2}}^{\sqrt{a^2-x^2}} x^2 \, dz \, dx \, dy = 2k \int_{-L/2}^{L/2} \int_{-a}^{a} x^2 \sqrt{a^2-x^2} \, dx \, dy$$

$$= 2k \int_{-L/2}^{L/2} \frac{1}{8} \left[x(2x^2 - a^2)\sqrt{a^2-x^2} + a^4 \arcsin \frac{x}{a} \right]_{-a}^{a} dy = \frac{ka^4\pi}{4} \int_{-L/2}^{L/2} dy = \frac{ka^4\pi L}{4} = \frac{ma^2}{4}$$

$$I_x = I_{xy} + I_{xz} = \frac{ma^2}{4} + \frac{mL^2}{12} = \frac{m}{12}(3a^2 + L^2)$$

$$I_y = I_{xy} + I_{yz} = \frac{ma^2}{4} + \frac{ma^2}{4} = \frac{ma^2}{2}$$

$$I_z = I_{xz} + I_{yz} = \frac{mL^2}{12} + \frac{ma^2}{4} = \frac{m}{12}(3a^2 + L^2)$$

53. $\displaystyle \int_{-1}^{1} \int_{-1}^{1} \int_{0}^{1-x} (x^2 + y^2)\sqrt{x^2 + y^2 + z^2} \, dz \, dy \, dx$

55. See the definition, page 978.

See Theorem 13.4, page 979.

57. (a) The annular solid on the right has the greater density.

(b) The annular solid on the right has the greater movement of inertia.

(c) The solid on the left will reach the bottom first. The solid on the right has a greater resistance to rotational motion.

Section 13.7   Triple Integrals in Cylindrical and Spherical Coordinates

1. $\displaystyle \int_{0}^{4} \int_{0}^{\pi/2} \int_{0}^{2} r \cos \theta \, dr \, d\theta \, dz = \int_{0}^{4} \int_{0}^{\pi/2} \left[\frac{r^2}{2} \cos \theta \right]_{0}^{2} d\theta \, dz$

$$= \int_{0}^{4} \int_{0}^{\pi/2} 2 \cos \theta \, d\theta \, dz = \int_{0}^{4} \left[2 \sin \theta \right]_{0}^{\pi/2} dz = \int_{0}^{4} 2 \, dz = 8$$

3. $\displaystyle \int_{0}^{\pi/2} \int_{0}^{2\cos^2 \theta} \int_{0}^{4-r^2} r \sin \theta \, dz \, dr \, d\theta = \int_{0}^{\pi/2} \int_{0}^{2\cos^2 \theta} r(4-r^2)\sin \theta \, dr \, d\theta = \int_{0}^{\pi/2} \left[\left(2r^2 - \frac{r^4}{4} \right) \sin \theta \right]_{0}^{2\cos^2 \theta} d\theta$

$$= \int_{0}^{\pi/2} [8 \cos^4 \theta - 4 \cos^8 \theta] \sin \theta \, d\theta = \left[-\frac{8 \cos^5 \theta}{5} + \frac{4 \cos^9 \theta}{9} \right]_{0}^{\pi/2} = \frac{52}{45}$$

5. $\displaystyle \int_{0}^{2\pi} \int_{0}^{\pi/4} \int_{0}^{\cos \phi} \rho^2 \sin \phi \, d\rho \, d\phi \, d\theta = \frac{1}{3} \int_{0}^{2\pi} \int_{0}^{\pi/4} \cos^3 \phi \sin \phi \, d\phi \, d\theta = -\frac{1}{12} \int_{0}^{2\pi} \left[\cos^4 \phi \right]_{0}^{\pi/4} d\theta = \frac{\pi}{8}$

7. $\displaystyle \int_{0}^{4} \int_{0}^{z} \int_{0}^{\pi/2} r e^r \, d\theta \, dr \, dz = \pi(e^4 + 3)$

9. $\displaystyle\int_0^{\pi/2}\int_0^3\int_0^{e^{-r^2}} r\,dz\,dr\,d\theta = \int_0^{\pi/2}\int_0^3 re^{-r^2}\,dr\,d\theta$

$$= \int_0^{\pi/2}\left[-\frac{1}{2}e^{-r^2}\right]_0^3 d\theta$$

$$= \int_0^{\pi/2}\frac{1}{2}(1-e^{-9})d\theta$$

$$= \frac{\pi}{4}(1-e^{-9})$$

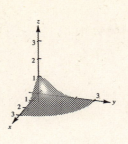

11. $\displaystyle\int_0^{2\pi}\int_{\pi/6}^{\pi/2}\int_0^4 \rho^2\sin\phi\,d\rho\,d\phi\,d\theta = \frac{64}{3}\int_0^{2\pi}\int_{\pi/6}^{\pi/2}\sin\phi\,d\phi\,d\theta$

$$= \frac{64}{3}\int_0^{2\pi}\left[-\cos\phi\right]_{\pi/6}^{\pi/2}d\theta$$

$$= \frac{32\sqrt{3}}{3}\int_0^{2\pi}d\theta$$

$$= \frac{64\sqrt{3}\,\pi}{3}$$

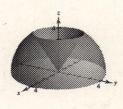

13. (a) $\displaystyle\int_0^{2\pi}\int_0^2\int_{r^2}^4 r^2\cos\theta\,dz\,dr\,d\theta = 0$

(b) $\displaystyle\int_0^{2\pi}\int_0^{\arctan(1/2)}\int_0^{4\sec\phi} \rho^3\sin^2\phi\cos\theta\,d\rho\,d\phi\,d\theta + \int_0^{2\pi}\int_{\arctan(1/2)}^{\pi/2}\int_0^{\cot\phi\csc\phi} \rho^3\sin^2\phi\cos\theta\,d\rho\,d\phi\,d\theta = 0$

15. (a) $\displaystyle\int_0^{2\pi}\int_0^a\int_a^{a+\sqrt{a^2-r^2}} r^2\cos\theta\,dz\,dr\,d\theta = 0$

(b) $\displaystyle\int_0^{\pi/4}\int_0^{2\pi}\int_{a\sec\phi}^{2a\cos\phi} \rho^3\sin^2\phi\cos\theta\,d\rho\,d\theta\,d\phi = 0$

17. $\displaystyle V = 4\int_0^{\pi/2}\int_0^{a\cos\theta}\int_0^{\sqrt{a^2-r^2}} r\,dz\,dr\,d\theta = 4\int_0^{\pi/2}\int_0^{a\cos\theta} r\sqrt{a^2-r^2}\,dr\,d\theta$

$$= \frac{4}{3}a^3\int_0^{\pi/2}(1-\sin^3\theta)\,d\theta = \frac{4}{3}a^3\left[\theta + \frac{1}{3}\cos\theta(\sin^2\theta+2)\right]_0^{\pi/2} = \frac{4}{3}a^3\left(\frac{\pi}{2}-\frac{2}{3}\right) = \frac{2a^3}{9}(3\pi-4)$$

19. $\displaystyle V = 2\int_0^{\pi}\int_0^{a\cos\theta}\int_0^{\sqrt{a^2-r^2}} r\,dz\,dr\,d\theta$

$$= 2\int_0^{\pi}\int_0^{a\cos\theta} r\sqrt{a^2-r^2}\,dr\,d\theta$$

$$= 2\int_0^{\pi}\left[-\frac{1}{3}(a^2-r^2)^{3/2}\right]_0^{a\cos\theta} d\theta$$

$$= \frac{2a^3}{3}\int_0^{\pi}(1-\sin^3\theta)\,d\theta$$

$$= \frac{2a^3}{3}\left[\theta+\cos\theta-\frac{\cos^3\theta}{3}\right]_0^{\pi}$$

$$= \frac{2a^3}{9}(3\pi-4)$$

21. $\displaystyle m = \int_0^{2\pi}\int_0^2\int_0^{9-r\cos\theta-2r\sin\theta}(kr)r\,dz\,dr\,d\theta$

$$= \int_0^{2\pi}\int_0^2 kr^2(9-r\cos\theta-2r\sin\theta)\,dr\,d\theta$$

$$= \int_0^{2\pi} k\left[3r^3-\frac{r^4}{4}\cos\theta-\frac{r^4}{2}\sin\theta\right]_0^2 d\theta$$

$$= \int_0^{2\pi} k[24-4\cos\theta-8\sin\theta]\,d\theta$$

$$= k\left[24\theta-4\sin\theta+8\cos\theta\right]_0^{2\pi}$$

$$= k[48\pi+8-8] = 48k\pi$$

23. $z = h - \dfrac{h}{r_0}\sqrt{x^2 + y^2} = \dfrac{h}{r_0}(r_0 - r)$

$V = 4 \displaystyle\int_0^{\pi/2} \int_0^{r_0} \int_0^{h(r_0-r)/r_0} r\, dz\, dr\, d\theta$

$= \dfrac{4h}{r_0} \displaystyle\int_0^{\pi/2} \int_0^{r_0} (r_0 r - r^2)\, dr\, d\theta$

$= \dfrac{4h}{r_0} \displaystyle\int_0^{\pi/2} \dfrac{r_0^3}{6}\, d\theta$

$= \dfrac{4h}{r_0} \left(\dfrac{r_0^3}{6}\right)\left(\dfrac{\pi}{2}\right) = \dfrac{1}{3}\pi r_0^2 h$

25. $\rho = k\sqrt{x^2 + y^2} = kr$

$\bar{x} = \bar{y} = 0$ by symmetry

$m = 4k \displaystyle\int_0^{\pi/2} \int_0^{r_0} \int_0^{h(r_0-r)/r_0} r^2\, dz\, dr\, d\theta$

$= \dfrac{1}{6}k\pi r_0^3 h$

$M_{xy} = 4k \displaystyle\int_0^{\pi/2} \int_0^{r_0} \int_0^{h(r_0-r)/r_0} r^2 z\, dz\, dr\, d\theta$

$= \dfrac{1}{30}k\pi r_0^3 h^2$

$\bar{z} = \dfrac{M_{xy}}{m} = \dfrac{k\pi r_0^3 h^2/30}{k\pi r_0^3 h/6} = \dfrac{h}{5}$

27. $I_z = 4k \displaystyle\int_0^{\pi/2} \int_0^{r_0} \int_0^{h(r_0-r)/r_0} r^3\, dz\, dr\, d\theta$

$= \dfrac{4kh}{r_0} \displaystyle\int_0^{\pi/2} \int_0^{r_0} (r_0 r^3 - r^4)\, dr\, d\theta$

$= \dfrac{4kh}{r_0} \left(\dfrac{r_0^5}{20}\right)\left(\dfrac{\pi}{2}\right)$

$= \dfrac{1}{10}k\pi r_0^4 h$

Since the mass of the core is $m = kV = k\left(\frac{1}{3}\pi r_0^2 h\right)$ from Exercise 23, we have $k = 3m/\pi r_0^2 h$. Thus,

$I_z = \dfrac{1}{10}k\pi r_0^4 h$

$= \dfrac{1}{10}\left(\dfrac{3m}{\pi r_0^2 h}\right)\pi r_0^4 h$

$= \dfrac{3}{10}mr_0^2$

29. $m = k(\pi b^2 h - \pi a^2 h) = k\pi h(b^2 - a^2)$

$I_z = 4k \displaystyle\int_0^{\pi/2} \int_a^{b} \int_0^{h} r^3\, dz\, dr\, d\theta$

$= 4kh \displaystyle\int_0^{\pi/2} \int_a^{b} r^3\, dr\, d\theta$

$= kh \displaystyle\int_0^{\pi/2} (b^4 - a^4)\, d\theta$

$= \dfrac{k\pi(b^4 - a^4)h}{2}$

$= \dfrac{k\pi(b^2 - a^2)(b^2 + a^2)h}{2}$

$= \dfrac{1}{2}m(a^2 + b^2)$

31. $V = \displaystyle\int_0^{2\pi} \int_0^{\pi} \int_0^{4\sin\phi} \rho^2 \sin\phi\, d\rho\, d\phi\, d\theta = 16\pi^2$

33. $m = 8k \displaystyle\int_0^{\pi/2} \int_0^{\pi/2} \int_0^{a} \rho^3 \sin\phi\, d\rho\, d\theta\, d\phi$

$= 2ka^4 \displaystyle\int_0^{\pi/2} \int_0^{\pi/2} \sin\phi\, d\theta\, d\phi$

$= k\pi a^4 \displaystyle\int_0^{\pi/2} \sin\phi\, d\phi$

$= \left[k\pi a^4(-\cos\phi)\right]_0^{\pi/2}$

$= k\pi a^4$

35. $m = \dfrac{2}{3}k\pi r^3$

$\bar{x} = \bar{y} = 0$ by symmetry

$M_{xy} = 4k\displaystyle\int_0^{\pi/2}\int_0^{\pi/2}\int_0^r \rho^3 \cos\phi \sin\phi \, d\rho \, d\theta \, d\phi$

$= \dfrac{1}{2}kr^4\displaystyle\int_0^{\pi/2}\int_0^{\pi/2} \sin 2\phi \, d\theta \, d\phi$

$= \dfrac{kr^4\pi}{4}\displaystyle\int_0^{\pi/2} \sin 2\phi \, d\phi$

$= \left[-\dfrac{1}{8}k\pi r^4 \cos 2\phi\right]_0^{\pi/2} = \dfrac{1}{4}k\pi r^4$

$\bar{z} = \dfrac{M_{xy}}{m} = \dfrac{k\pi r^4/4}{2k\pi r^3/3} = \dfrac{3r}{8}$

37. $I_z = 4k\displaystyle\int_{\pi/4}^{\pi/2}\int_0^{\pi/2}\int_0^{\cos\phi} \rho^4 \sin^3\phi \, d\rho \, d\theta \, d\phi$

$= \dfrac{4}{5}k\displaystyle\int_{\pi/4}^{\pi/2}\int_0^{\pi/2} \cos^5\phi \sin^3\phi \, d\theta \, d\phi$

$= \dfrac{2}{5}k\pi\displaystyle\int_{\pi/4}^{\pi/2} \cos^5\phi(1 - \cos^2\phi)\sin\phi \, d\phi$

$= \left[\dfrac{2}{5}k\pi\left(-\dfrac{1}{6}\cos^6\phi + \dfrac{1}{8}\cos^8\phi\right)\right]_{\pi/4}^{\pi/2}$

$= \dfrac{k\pi}{192}$

39. $x = r\cos\theta \qquad x^2 + y^2 = r^2$

$y = r\sin\theta \qquad \tan\theta = \dfrac{y}{x}$

$z = z \qquad\qquad z = z$

41. $\displaystyle\int_{\theta_1}^{\theta_2}\int_{g_1(\theta)}^{g_2(\theta)}\int_{h_1(r\cos\theta,\, r\sin\theta)}^{h_2(r\cos\theta,\, r\sin\theta)} f(r\cos\theta, r\sin\theta, z)r \, dz \, dr \, d\theta$

43. (a) $r = r_0$: right circular cylinder about z-axis

$\theta = \theta_0$: plane parallel to z-axis

$z = z_0$: plane parallel to xy-plane

(b) $\rho = \rho_0$: sphere of radius ρ_0

$\theta = \theta_0$: plane parallel to z-axis

$\phi = \phi_0$: cone

45. $16\displaystyle\int_0^a\int_0^{\sqrt{a^2-x^2}}\int_0^{\sqrt{a^2-x^2-y^2}}\int_0^{\sqrt{a^2-x^2-y^2-z^2}} dw \, dz \, dy \, dx$

$= 16\displaystyle\int_0^a\int_0^{\sqrt{a^2-x^2}}\int_0^{\sqrt{a^2-x^2-y^2}} \sqrt{a^2 - x^2 - y^2 - z^2} \, dz \, dy \, dx$

$= 16\displaystyle\int_0^{\pi/2}\int_0^a\int_0^{\sqrt{a^2-r^2}} \sqrt{(a^2 - r^2) - z^2} \, dz(r \, dr \, d\theta)$

$= 16\displaystyle\int_0^{\pi/2}\int_0^a \dfrac{1}{2}\left[z\sqrt{(a^2 - r^2) - z^2} + (a^2 - r^2)\arcsin\dfrac{z}{\sqrt{a^2 - r^2}}\right]_0^{\sqrt{a^2-r^2}} r \, dr \, d\theta$

$= 8\displaystyle\int_0^{\pi/2}\int_0^a \dfrac{\pi}{2}(a^2 - r^2)r \, dr \, d\theta$

$= 4\pi\displaystyle\int_0^{\pi/2}\left[\dfrac{a^2r^2}{2} - \dfrac{r^4}{4}\right]_0^a d\theta$

$= a^4\pi\displaystyle\int_0^{\pi/2} d\theta = \dfrac{a^4\pi^2}{2}$

Section 13.8   Change of Variables: Jacobians

1. $x = -\dfrac{1}{2}(u - v)$

$y = \dfrac{1}{2}(u + v)$

$\dfrac{\partial x}{\partial u}\dfrac{\partial y}{\partial v} - \dfrac{\partial y}{\partial u}\dfrac{\partial x}{\partial v} = \left(-\dfrac{1}{2}\right)\left(\dfrac{1}{2}\right) - \left(\dfrac{1}{2}\right)\left(\dfrac{1}{2}\right)$

$\qquad\qquad\qquad = -\dfrac{1}{2}$

3. $x = u - v^2$

$y = u + v$

$\dfrac{\partial x}{\partial u}\dfrac{\partial y}{\partial v} - \dfrac{\partial y}{\partial u}\dfrac{\partial x}{\partial v} = (1)(1) - (1)(-2v) = 1 + 2v$

5. $x = u\cos\theta - v\sin\theta$

$y = u\sin\theta + v\cos\theta$

$\dfrac{\partial x}{\partial u}\dfrac{\partial y}{\partial v} - \dfrac{\partial y}{\partial u}\dfrac{\partial x}{\partial v} = \cos^2\theta + \sin^2\theta = 1$

7. $x = e^u\sin v$

$y = e^u\cos v$

$\dfrac{\partial x}{\partial u}\dfrac{\partial y}{\partial v} - \dfrac{\partial y}{\partial u}\dfrac{\partial x}{\partial v} = (e^u\sin v)(-e^u\sin v) - (e^u\cos v)(e^u\cos v) = -e^{2u}$

9. $x = 3u + 2v$

$y = 3v$

$v = \dfrac{y}{3}$

$u = \dfrac{x - 2v}{3} = \dfrac{x - 2(y/3)}{3}$

$\quad = \dfrac{x}{3} - \dfrac{2y}{9}$

(x, y)	(u, v)
$(0, 0)$	$(0, 0)$
$(3, 0)$	$(1, 0)$
$(2, 3)$	$(0, 1)$

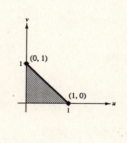

11. $x = \dfrac{1}{2}(u + v)$

$y = \dfrac{1}{2}(u - v)$

$\dfrac{\partial x}{\partial u}\dfrac{\partial y}{\partial v} - \dfrac{\partial y}{\partial u}\dfrac{\partial x}{\partial v} = \left(\dfrac{1}{2}\right)\left(-\dfrac{1}{2}\right) - \left(\dfrac{1}{2}\right)\left(\dfrac{1}{2}\right) = -\dfrac{1}{2}$

$\displaystyle\int_R\!\!\int 4(x^2 + y^2)\, dA = \int_{-1}^{1}\int_{-1}^{1} 4\left[\dfrac{1}{4}(u + v)^2 + \dfrac{1}{4}(u - v)^2\right]\left(\dfrac{1}{2}\right)\, dv\, du$

$\qquad\qquad\qquad = \int_{-1}^{1}\int_{-1}^{1}(u^2 + v^2)\, dv\, du = \int_{-1}^{1} 2\left(u^2 + \dfrac{1}{3}\right)\, du = \left[2\left(\dfrac{u^3}{3} + \dfrac{u}{3}\right)\right]_{-1}^{1} = \dfrac{8}{3}$

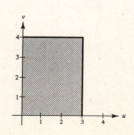

13. $x = u + v$

$y = u$

$\dfrac{\partial x}{\partial u}\dfrac{\partial y}{\partial v} - \dfrac{\partial y}{\partial u}\dfrac{\partial x}{\partial v} = (1)(0) - (1)(1) = -1$

$\displaystyle\int_R\!\!\int y(x - y)\, dA = \int_0^3\int_0^4 uv(1)\, dv\, du = \int_0^3 8u\, du = 36$

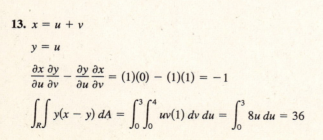

15. $\displaystyle\int_R\!\!\int e^{-xy/2}\,dA$

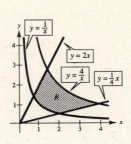

$R:\ y=\dfrac{x}{4},\ y=2x,\ y=\dfrac{1}{x},\ y=\dfrac{4}{x}$

$x=\sqrt{v/u},\ y=\sqrt{uv}\ \Rightarrow\ u=\dfrac{y}{x},\ v=xy$

$\dfrac{\partial(x,y)}{\partial(u,v)}=\begin{vmatrix}\dfrac{\partial x}{\partial u}&\dfrac{\partial x}{\partial v}\\[2mm]\dfrac{\partial y}{\partial u}&\dfrac{\partial y}{\partial v}\end{vmatrix}=\begin{vmatrix}-\dfrac{1}{2}\dfrac{v^{1/2}}{u^{3/2}}&\dfrac{1}{2}\dfrac{1}{u^{1/2}v^{1/2}}\\[3mm]\dfrac{1}{2}\dfrac{v^{1/2}}{u^{1/2}}&\dfrac{1}{2}\dfrac{u^{1/2}}{v^{1/2}}\end{vmatrix}=-\dfrac{1}{4}\!\left(\dfrac{1}{u}+\dfrac{1}{u}\right)=-\dfrac{1}{2u}$

Transformed Region:

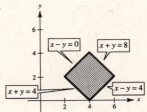

$y=\dfrac{1}{x}\ \Rightarrow\ yx=1\ \Rightarrow\ v=1$

$y=\dfrac{4}{x}\ \Rightarrow\ yx=4\ \Rightarrow\ v=4$

$y=2x\ \Rightarrow\ \dfrac{y}{x}=2\ \Rightarrow\ u=2$

$y=\dfrac{x}{4}\ \Rightarrow\ \dfrac{y}{x}=\dfrac{1}{4}\ \Rightarrow\ u=\dfrac{1}{4}$

$\displaystyle\int_R\!\!\int e^{-xy/2}\,dA=\int_{1/4}^{2}\!\int_{1}^{4}e^{-v/2}\!\left(\dfrac{1}{2u}\right)dv\,du=-\int_{1/4}^{2}\!\left[\dfrac{e^{-v/2}}{u}\right]_1^4 du=-\int_{1/4}^{2}(e^{-2}-e^{-1/2})\dfrac{1}{u}\,du$

$\displaystyle\qquad=-\Big[(e^{-2}-e^{-1/2})\ln u\Big]_{1/4}^{2}=-(e^{-2}-e^{-1/2})\left(\ln 2-\ln\dfrac{1}{4}\right)=(e^{-1/2}-e^{-2})\ln 8\approx 0.9798$

17. $u=x+y=4,\qquad v=x-y=0$

$\ u=x+y=8,\qquad v=x-y=4$

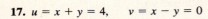

$x=\dfrac{1}{2}(u+v)\qquad y=\dfrac{1}{2}(u-v)$

$\dfrac{\partial(x,y)}{\partial(u,v)}=-\dfrac{1}{2}$

$\displaystyle\int_R\!\!\int (x+y)e^{x-y}\,dA=\int_4^8\!\int_0^4 ue^{v}\!\left(\dfrac{1}{2}\right)dv\,du$

$\displaystyle\qquad=\dfrac{1}{2}\int_4^8 u(e^4-1)\,du=\left[\dfrac{1}{4}u^2(e^4-1)\right]_4^8=12(e^4-1)$

19. $u=x+4y=0,\qquad v=x-y=0$

$\ u=x+4y=5,\qquad v=x-y=5$

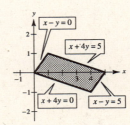

$x=\dfrac{1}{5}(u+4v),\qquad y=\dfrac{1}{5}(u-v)$

$\dfrac{\partial x}{\partial u}\dfrac{\partial y}{\partial v}-\dfrac{\partial y}{\partial u}\dfrac{\partial x}{\partial v}=\left(\dfrac{1}{5}\right)\!\left(-\dfrac{1}{5}\right)-\left(\dfrac{1}{5}\right)\!\left(\dfrac{4}{5}\right)=-\dfrac{1}{5}$

$\displaystyle\int_R\!\!\int \sqrt{(x-y)(x+4y)}\,dA=\int_0^5\!\int_0^5\sqrt{uv}\left(\dfrac{1}{5}\right)du\,dv$

$\displaystyle\qquad=\int_0^5\left[\dfrac{1}{5}\!\left(\dfrac{2}{3}\right)u^{3/2}\sqrt{v}\right]_0^5 dv=\left[\dfrac{2\sqrt5}{3}\!\left(\dfrac{2}{3}\right)v^{3/2}\right]_0^5=\dfrac{100}{9}$

21. $u = x + y, v = x - y, x = \dfrac{1}{2}(u + v), y = \dfrac{1}{2}(u - v)$

$$\frac{\partial x}{\partial u}\frac{\partial y}{\partial v} - \frac{\partial y}{\partial u}\frac{\partial x}{\partial v} = -\frac{1}{2}$$

$$\iint_R \sqrt{x + y}\, dA = \int_0^a \int_{-u}^u \sqrt{u}\left(\frac{1}{2}\right) dv\, du = \int_0^a u\sqrt{u}\, du = \left[\frac{2}{5}u^{5/2}\right]_0^a = \frac{2}{5}a^{5/2}$$

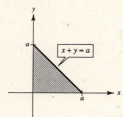

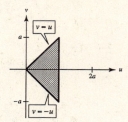

23. $\quad \dfrac{x^2}{a^2} + \dfrac{y^2}{b^2} = 1, x = au, y = bv$

$$\frac{(au)^2}{a^2} + \frac{(bv)^2}{b^2} = 1$$

$$u^2 + v^2 = 1$$

(a) $\dfrac{x^2}{a^2} + \dfrac{y^2}{b^2} = 1 \qquad\qquad u^2 + v^2 = 1$

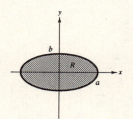

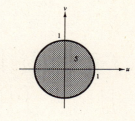

(b) $\dfrac{\partial(x, y)}{\partial(u, v)} = \dfrac{\partial x}{\partial u}\dfrac{\partial y}{\partial v} - \dfrac{\partial y}{\partial u}\dfrac{\partial x}{\partial v}$

$$= (a)(b) - (0)(0) = ab$$

(c) $A = \displaystyle\iint_S ab\, dS$

$$= ab(\pi(1)^2) = \pi ab$$

25. Jacobian $= \dfrac{\partial(x, y)}{\partial(u, v)} = \dfrac{\partial x}{\partial u}\dfrac{\partial y}{\partial v} - \dfrac{\partial y}{\partial u}\dfrac{\partial x}{\partial v}$

27. $x = u(1 - v), \ y = uv(1 - w), \ z = uvw$

$$\frac{\partial(x, y, z)}{\partial(u, v, w)} = \begin{vmatrix} 1 - v & -u & 0 \\ v(1 - w) & u(1 - w) & -uv \\ vw & uw & uv \end{vmatrix} = (1 - v)[u^2v(1 - w) + u^2vw] + u[uv^2(1 - w) + uv^2w]$$

$$= (1 - v)(u^2v) + u(uv^2)$$

$$= u^2v$$

29. $x = \rho \sin\phi \cos\theta, \ y = \rho \sin\phi \sin\theta, \ z = \rho\cos\phi$

$$\frac{\partial(x, y, z)}{\partial(\rho, \theta, \phi)} = \begin{vmatrix} \sin\phi\cos\theta & -\rho\sin\phi\sin\theta & \rho\cos\phi\cos\theta \\ \sin\phi\sin\theta & \rho\sin\phi\cos\theta & \rho\cos\phi\sin\theta \\ \cos\phi & 0 & -\rho\sin\phi \end{vmatrix}$$

$$= \cos\phi[-\rho^2\sin\phi\cos\phi\sin^2\theta - \rho^2\sin\phi\cos\phi\cos^2\theta] - \rho\sin\phi[\rho\sin^2\phi\cos^2\theta + \rho\sin^2\phi\sin^2\theta]$$

$$= \cos\phi[-\rho^2\sin\phi\cos\phi(\sin^2\theta + \cos^2\theta)] - \rho\sin\phi[\rho\sin^2\phi(\cos^2\theta + \sin^2\theta)]$$

$$= -\rho^2\sin\phi\cos^2\phi - \rho^2\sin^3\phi$$

$$= -\rho^2\sin\phi(\cos^2\phi + \sin^2\phi)$$

$$= -\rho^2\sin\phi$$

Review Exercises for Chapter 13

1. $\int_1^{x^2} x \ln y \, dy = \left[xy(-1 + \ln y) \right]_1^{x^2} = x^3(-1 + \ln x^2) + x = x - x^3 + x^3 \ln x^2$

3. $\int_0^1 \int_0^{1+x} (3x + 2y) \, dy \, dx = \int_0^1 \left[3xy + y^2 \right]_0^{1+x} dx = \int_0^1 (4x^2 + 5x + 1) \, dx = \left[\frac{4}{3}x^3 + \frac{5}{2}x^2 + x \right]_0^1 = \frac{29}{6}$

5. $\int_0^3 \int_0^{\sqrt{9-x^2}} 4x \, dy \, dx = \int_0^3 4x\sqrt{9 - x^2} \, dx = \left[-\frac{4}{3}(9 - x^2)^{3/2} \right]_0^3 = 36$

7. $\int_0^3 \int_0^{(3-x)/3} dy \, dx = \int_0^1 \int_0^{3-3y} dx \, dy$

$A = \int_0^1 \int_0^{3-3y} dx \, dy = \int_0^1 (3 - 3y) \, dy = \left[3y - \frac{3}{2}y^2 \right]_0^1 = \frac{3}{2}$

9. $\int_{-5}^3 \int_{-\sqrt{25-x^2}}^{\sqrt{25-x^2}} dy \, dx = \int_{-5}^{-4} \int_{-\sqrt{25-y^2}}^{\sqrt{25-y^2}} dx \, dy + \int_{-4}^4 \int_{-\sqrt{25-y^2}}^3 dx \, dy + \int_4^5 \int_{-\sqrt{25-y^2}}^{\sqrt{25-y^2}} dx \, dy$

$A = 2\int_{-5}^3 \int_0^{\sqrt{25-x^2}} dy \, dx = 2\int_{-5}^3 \sqrt{25 - x^2} \, dx = \left[x\sqrt{25 - x^2} + 25 \arcsin \frac{x}{5} \right]_{-5}^3 = \frac{25\pi}{2} + 12 + 25 \arcsin \frac{3}{5} \approx 67.36$

11. $A = 4\int_0^1 \int_0^{x\sqrt{1-x^2}} dy \, dx = 4\int_0^1 x\sqrt{1 - x^2} \, dx = \left[-\frac{4}{3}(1 - x^2)^{3/2} \right]_0^1 = \frac{4}{3}$

$A = 4\int_0^{1/2} \int_{\sqrt{(1-\sqrt{1-4y^2})/2}}^{\sqrt{(1+\sqrt{1-4y^2})/2}} dx \, dy$

13. $A = \int_2^5 \int_{x-3}^{\sqrt{x-1}} dy \, dx + 2\int_1^2 \int_0^{\sqrt{x-1}} dy \, dx = \int_{-1}^2 \int_{y^2+1}^{y+3} dx \, dy = \frac{9}{2}$

15. Both integrations are over the common region R shown in the figure. Analytically,

$\int_0^1 \int_{2y}^{2\sqrt{2-y^2}} (x + y) \, dx \, dy = \frac{4}{3} + \frac{4}{3}\sqrt{2}$

$\int_0^2 \int_0^{x/2} (x + y) \, dy \, dx + \int_2^{2\sqrt{2}} \int_0^{\sqrt{8-x^2}/2} (x + y) \, dy \, dx = \frac{5}{3} + \left(\frac{4}{3}\sqrt{2} - \frac{1}{3} \right) = \frac{4}{3} + \frac{4}{3}\sqrt{2}$

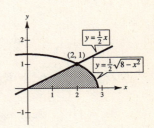

17. $V = \int_0^4 \int_0^{x^2+4} (x^2 - y + 4) \, dy \, dx$

$= \int_0^4 \left[x^2 y - \frac{1}{2}y^2 + 4y \right]_0^{x^2+4} dx$

$= \int_0^4 \left(\frac{1}{2}x^4 + 4x^2 + 8 \right) dx$

$= \left[\frac{1}{10}x^5 + \frac{4}{3}x^3 + 8x \right]_0^4 = \frac{3296}{15}$

19. Volume $\approx$ (base)(height)

$\approx \frac{9}{2}(3) = \frac{27}{2}$

Matches (c)

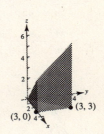

21. $\int_0^\infty \int_0^\infty kxye^{-(x+y)} \, dy \, dx = \int_0^\infty \left[-kxe^{-(x+y)}(y+1) \right]_0^\infty dx = \int_0^\infty kxe^{-x} \, dx = \left[-k(x+1)e^{-x} \right]_0^\infty = k$

Therefore, $k = 1$.

$$P = \int_0^1 \int_0^1 xye^{-(x+y)} \, dy \, dx \approx 0.070$$

23. True

25. True

27. $\int_0^h \int_0^x \sqrt{x^2 + y^2} \, dy \, dx = \int_0^{\pi/4} \int_0^{h \sec \theta} r^2 \, dr \, d\theta$

$$= \frac{h^3}{3} \int_0^{\pi/4} \sec^3 \theta \, d\theta = \frac{h^3}{6} \left[\sec \theta \tan \theta + \ln|\sec \theta + \tan \theta| \right]_0^{\pi/4} = \frac{h^3}{6} \left[\sqrt{2} + \ln\left(\sqrt{2} + 1\right) \right]$$

29. $V = 4 \int_0^h \int_0^{\pi/2} \int_1^{\sqrt{1+z^2}} r \, dr \, d\theta \, dz$

$= 2 \int_0^h \int_0^{\pi/2} (1 + z^2 - 1) \, d\theta \, dz$

$= \pi \int_0^h z^2 \, dz$

$= \left[\pi \left(\frac{1}{3} z^3 \right) \right]_0^h = \frac{\pi h^3}{3}$

31. (a) $(x^2 + y^2)^2 = 9(x^2 - y^2)$

$(r^2)^2 = 9(r^2 \cos^2 \theta - r^2 \sin^2 \theta)$

$r^2 = 9(\cos^2 \theta - \sin^2 \theta) = 9 \cos 2\theta$

$r = 3\sqrt{\cos 2\theta}$

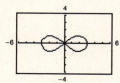

(b) $A = 4 \int_0^{\pi/4} \int_0^{3\sqrt{\cos 2\theta}} r \, dr \, d\theta = 9$

(c) $V = 4 \int_0^{\pi/4} \int_0^{3\sqrt{\cos 2\theta}} \sqrt{9 - r^2} \, r \, dr \, d\theta \approx 20.392$

33. (a) $m = k \int_0^1 \int_{2x^3}^{2x} xy \, dy \, dx = \frac{k}{4}$

$M_x = k \int_0^1 \int_{2x^3}^{2x} xy^2 \, dy \, dx = \frac{16k}{55}$

$M_y = k \int_0^1 \int_{2x^3}^{2x} x^2 y \, dy \, dx = \frac{8k}{45}$

$\bar{x} = \frac{M_y}{m} = \frac{32}{45}$

$\bar{y} = \frac{M_x}{m} = \frac{64}{55}$

(b) $m = k \int_0^1 \int_{2x^3}^{2x} (x^2 + y^2) \, dy \, dx = \frac{17k}{30}$

$M_x = k \int_0^1 \int_{2x^3}^{2x} y(x^2 + y^2) \, dy \, dx = \frac{392k}{585}$

$M_y = k \int_0^1 \int_{2x^3}^{2x} x(x^2 + y^2) \, dy \, dx = \frac{156k}{385}$

$\bar{x} = \frac{M_y}{m} = \frac{936}{1309}$

$\bar{y} = \frac{M_x}{m} = \frac{784}{663}$

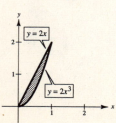

35. $I_x = \int\int_R y^2 \, \rho(x, y) dA = \int_0^a \int_0^b kxy^2 \, dy \, dx = \frac{1}{6}kb^3a^2$

$I_y = \int\int_R x^2 \, \rho(x, y) dA = \int_0^a \int_0^b kx^3 \, dy \, dx = \frac{1}{4}kba^4$

$I_0 = I_x + I_y = \frac{1}{6}kb^3a^2 + \frac{1}{4}kba^4 = \frac{ka^2b}{12}(2b^2 + 3a^2)$

$m = \int\int_R \rho(x, y) dA = \int_0^a \int_0^b kx \, dy \, dx = \frac{1}{2}kba^2$

$\bar{\bar{x}} = \sqrt{\frac{I_y}{m}} = \sqrt{\frac{(1/4)kba^4}{(1/2)kba^2}} = \sqrt{\frac{a^2}{2}} = \frac{a\sqrt{2}}{2}$

$\bar{\bar{y}} = \sqrt{\frac{I_x}{m}} = \sqrt{\frac{(1/6)kb^3a^2}{(1/2)kba^2}} = \sqrt{\frac{b^2}{3}} = \frac{b\sqrt{3}}{3}$

37. $S = \int\int_R \sqrt{1 + (f_x)^2 + (f_y)^2} \, dA$

$= 4\int_0^4 \int_0^{\sqrt{16-x^2}} \sqrt{1 + 4x^2 + 4y^2} \, dy \, dx$

$= 4\int_0^{\pi/2} \int_0^4 \sqrt{1 + 4r^2} \, r \, dr \, d\theta$

$= \left[\frac{1}{3}(65^{3/2} - 1)\theta\right]_0^{\pi/2} = \frac{\pi}{6}\left(65\sqrt{65} - 1\right)$

39. $f(x, y) = 9 - y^2$

$f_x = 0, \ f_y = -2y$

$S = \int\int_R \sqrt{1 + f_x^2 + f_y^2} \, dA$

$= \int_0^3 \int_{-y}^y \sqrt{1 + 4y^2} \, dx \, dy$

$= \int_0^3 \left[\sqrt{1 + 4y^2}\, x\right]_{-y}^y dy$

$= \int_0^3 2y\sqrt{1 + 4y^2} \, dy = \frac{1}{4}\frac{2}{3}(1 + 4y^2)^{3/2}\Big]_0^3 = \frac{1}{6}[(37)^{3/2} - 1]$

41. $\int_{-3}^3 \int_{-\sqrt{9-x^2}}^{\sqrt{9-x^2}} \int_{x^2+y^2}^9 \sqrt{x^2 + y^2} \, dz \, dy \, dx = \int_0^{2\pi} \int_0^3 \int_{r^2}^9 r^2 \, dz \, dr \, d\theta$

$= \int_0^{2\pi} \int_0^3 (9r^2 - r^4) \, dr \, d\theta = \int_0^{2\pi}\left[3r^3 - \frac{r^5}{5}\right]_0^3 d\theta = \frac{162}{5}\int_0^{2\pi} d\theta = \frac{324\pi}{5}$

43. $\int_0^a \int_0^b \int_0^c (x^2 + y^2 + z^2) \, dx \, dy \, dz = \int_0^a \int_0^b \left(\frac{1}{3}c^3 + cy^2 + cz^2\right) dy \, dz$

$= \int_0^a \left(\frac{1}{3}bc^3 + \frac{1}{3}b^3c + bcz^2\right) dz = \frac{1}{3}abc^3 + \frac{1}{3}ab^3c + \frac{1}{3}a^3bc = \frac{1}{3}abc(a^2 + b^2 + c^2)$

45. $\int_{-1}^1 \int_{-\sqrt{1-x^2}}^{\sqrt{1-x^2}} \int_{-\sqrt{1-x^2-y^2}}^{\sqrt{1-x^2-y^2}} (x^2 + y^2) \, dz \, dy \, dx = \int_0^{2\pi} \int_0^1 \int_{-\sqrt{1-r^2}}^{\sqrt{1-r^2}} r^3 \, dz \, dr \, d\theta = \frac{8\pi}{15}$

47. $V = 4\displaystyle\int_0^{\pi/2}\int_0^{2\cos\theta}\int_0^{\sqrt{4-r^2}} r\,dz\,dr\,d\theta$

$= 4\displaystyle\int_0^{\pi/2}\int_0^{2\cos\theta} r\sqrt{4-r^2}\,dr\,d\theta$

$= -\displaystyle\int_0^{\pi/2}\left[\frac{4}{3}(4-r^2)^{3/2}\right]_0^{2\cos\theta} d\theta$

$= \dfrac{32}{3}\displaystyle\int_0^{\pi/2}(1-\sin^3\theta)\,d\theta$

$= \dfrac{32}{3}\left[\theta + \cos\theta - \dfrac{1}{3}\cos^3\theta\right]_0^{\pi/2} = \dfrac{32}{3}\left(\dfrac{\pi}{2}-\dfrac{2}{3}\right)$

49. $m = 4k\displaystyle\int_{\pi/4}^{\pi/2}\int_0^{\pi/2}\int_0^{\cos\phi}\rho^2\sin\phi\,d\rho\,d\theta\,d\phi$

$= \dfrac{4}{3}k\displaystyle\int_{\pi/4}^{\pi/2}\int_0^{\pi/2}\cos^3\phi\sin\phi\,d\theta\,d\phi = \dfrac{2}{3}k\pi\int_{\pi/4}^{\pi/2}\cos^3\phi\sin\phi\,d\phi = \left[-\dfrac{2}{3}k\pi\left(\dfrac{1}{4}\cos^4\phi\right)\right]_{\pi/4}^{\pi/2} = \dfrac{k\pi}{24}$

$M_{xy} = 4k\displaystyle\int_{\pi/4}^{\pi/2}\int_0^{\pi/2}\int_0^{\cos\phi}\rho^3\cos\phi\sin\phi\,d\rho\,d\theta\,d\phi$

$= k\displaystyle\int_{\pi/4}^{\pi/2}\int_0^{\pi/2}\cos^5\phi\sin\phi\,d\theta\,d\phi = \dfrac{1}{2}k\pi\int_{\pi/4}^{\pi/2}\cos^5\phi\sin\phi\,d\phi = \left[-\dfrac{1}{12}k\pi\cos^6\phi\right]_{\pi/4}^{\pi/2} = \dfrac{k\pi}{96}$

$\bar{z} = \dfrac{M_{xy}}{m} = \dfrac{k\pi/96}{k\pi/24} = \dfrac{1}{4}$

$\bar{x} = \bar{y} = 0$ by symmetry

51. $m = k\displaystyle\int_0^{\pi/2}\int_0^{\pi/2}\int_0^a \rho^2\sin\phi\,d\rho\,d\theta\,d\phi = \dfrac{k\pi a^3}{6}$

$M_{xy} = k\displaystyle\int_0^{\pi/2}\int_0^{\pi/2}\int_0^a (\rho\cos\phi)\rho^2\sin\phi\,d\rho\,d\theta\,d\phi = \dfrac{k\pi a^4}{16}$

$\bar{x} = \bar{y} = \bar{z} = \dfrac{M_{xy}}{m} = \dfrac{k\pi a^4}{16}\left(\dfrac{6}{k\pi a^3}\right) = \dfrac{3a}{8}$

53. $I_z = 4k\displaystyle\int_0^{\pi/2}\int_3^4\int_0^{16-r^2} r^3\,dz\,dr\,d\theta$

$= 4k\displaystyle\int_0^{\pi/2}\int_3^4 (16r^3 - r^5)\,dr\,d\theta = \dfrac{833\pi k}{3}$

55. $z = f(x,y) = \sqrt{a^2 - x^2 - y^2}$

$= \sqrt{a^2 - r^2}$

$0 \le r \le \sqrt{2ah - h^2}$

(a) Disc Method

$V = \pi\displaystyle\int_{a-h}^a (a^2 - y^2)\,dy$

$= \pi\left[a^2 y - \dfrac{y^3}{3}\right]_{a-h}^a = \pi\left[\left(a^3 - \dfrac{a^3}{3}\right) - \left(a^2(a-h) - \dfrac{(a-h)^3}{3}\right)\right]$

$= \pi\left[a^3 - \dfrac{a^3}{3} - a^3 + a^2 h + \dfrac{a^3}{3} - a^2 h + ah^2 - \dfrac{h^3}{3}\right] = \pi\left[ah^2 - \dfrac{h^3}{3}\right] = \dfrac{1}{3}\pi h^2[3a - h]$

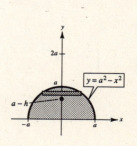

Equivalently, use spherical coordinates

$V = \displaystyle\int_0^{2\pi}\int_0^{\cos^{-1}(a-h/a)}\int_{(a-h)\sec\phi}^a \rho^2\sin\phi\,d\rho\,d\phi\,d\theta$

—CONTINUED—

55. —CONTINUED—

(b) $M_{xy} = \int_0^{2\pi} \int_0^{\cos^{-1}(a-h/a)} \int_{(a-h)\sec\phi}^{a} (\rho\cos\phi)\rho^2\sin\phi\,d\rho\,d\phi\,d\theta$

$= \dfrac{1}{4}h^2\pi(2a-h)^2$

$\bar{z} = \dfrac{M_{xy}}{V} = \dfrac{\dfrac{1}{4}h^2\pi(2a-h)^2}{\dfrac{1}{3}h^2\pi(3a-h)} = \dfrac{3}{4}\dfrac{(2a-h)^2}{3a-h}$

centroid: $\left(0, 0, \dfrac{3(2a-h)^2}{4(3a-h)}\right)$

(c) If $h = a$, $\bar{z} = \dfrac{3(a)^2}{4(2a)} = \dfrac{3}{8}a$

centroid of hemisphere: $\left(0, 0, \dfrac{3}{8}a\right)$

(d) $\lim\limits_{h\to 0}\bar{z} = \lim\limits_{h\to 0}\dfrac{3(2a-h)^2}{4(3a-h)} = \dfrac{3(4a^2)}{12a} = a$

(e) $x^2 + y^2 = \rho^2\sin^2\phi$

$I_z = \int_0^{2\pi}\int_0^{\cos^{-1}(a-h/a)}\int_{(a-h)\sec\phi}^{a}(\rho^2\sin^2\phi)\rho^2\sin\phi\,d\rho\,d\phi\,d\theta$

$= \dfrac{h^3}{30}(20a^2 - 15ah + 3h^2)\pi$

(f) If $h = a$, $I_z = \dfrac{a^3\pi}{30}(20a^2 - 15a^2 + 3a^2) = \dfrac{4}{15}a^5\pi$

57. $\int_0^{2\pi}\int_0^{\pi}\int_0^{6\sin\phi}\rho^2\sin\phi\,d\rho\,d\phi\,d\theta$

Since $\rho = 6\sin\phi$ represents (in the yz-plane) a circle of radius 3 centered at $(0, 3, 0)$, the integral represents the volume of the torus formed by revolving $(0 < \theta < 2\pi)$ this circle about the z-axis.

59. $\dfrac{\partial(x, y)}{\partial(u, v)} = \dfrac{\partial x}{\partial u}\dfrac{\partial y}{\partial v} - \dfrac{\partial y}{\partial u}\dfrac{\partial x}{\partial v}$

$= 1(-3) - 2(3) = -9$

61. $\dfrac{\partial(x, y)}{\partial(u, v)} = \dfrac{\partial x}{\partial u}\dfrac{\partial y}{\partial v} - \dfrac{\partial x}{\partial v}\dfrac{\partial y}{\partial u} = \dfrac{1}{2}\left(-\dfrac{1}{2}\right) - \dfrac{1}{2}\left(\dfrac{1}{2}\right) = -\dfrac{1}{2}$

$x = \dfrac{1}{2}(u + v), y = \dfrac{1}{2}(u - v) \implies u = x + y, v = x - y$

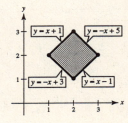

Boundaries in xy-plane	Boundaries in uv-plane
$x + y = 3$	$u = 3$
$x + y = 5$	$u = 5$
$x - y = -1$	$v = -1$
$x - y = 1$	$v = 1$

$\displaystyle\iint_R \ln(x + y)\,dA = \int_3^5\int_{-1}^{1}\ln\left(\dfrac{1}{2}(u+v) + \dfrac{1}{2}(u-v)\right)\left(\dfrac{1}{2}\right)dv\,du = \int_3^5\int_{-1}^{1}\dfrac{1}{2}\ln u\,dv\,du = \int_3^5 \ln u\,du = \Big[u\ln u - u\Big]_3^5$

$= (5\ln 5 - 5) - (3\ln u - 3) = 5\ln 5 - 3\ln 3 - 2 \approx 2.751$

Problem Solving for Chapter 13

1. (a) $V = 16 \displaystyle\int_R\!\!\int \sqrt{1 - x^2}\, dA$

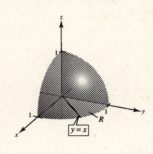

$\qquad = 16 \displaystyle\int_0^{\pi/4}\!\!\int_0^1 \sqrt{1 - r^2\cos^2\theta}\, r\, dr\, d\theta$

$\qquad = -\dfrac{16}{3}\displaystyle\int_0^{\pi/4} \dfrac{1}{\cos^2\theta}\big[(1 - \cos^2\theta)^{3/2} - 1\big]\, d\theta$

$\qquad = -\dfrac{16}{3}\Big[\, \sec\theta + \cos\theta - \tan\theta \,\Big]_0^{\pi/4}$

$\qquad = 8\big(2 - \sqrt{2}\big) \approx 4.6863$

(b) Programs will vary.

3. (a) $\displaystyle\int \dfrac{du}{a^2 + u^2} = \dfrac{1}{a}\arctan\dfrac{u}{a} + c.$ Let $a^2 = 2 - u^2,\, u = v.$

$\qquad$ Then $\displaystyle\int \dfrac{1}{(2 - u^2) + v^2}\, dv = \dfrac{1}{\sqrt{2 - u^2}}\arctan\dfrac{v}{\sqrt{2 - u^2}} + C.$

(b) $I_1 = \displaystyle\int_0^{\sqrt{2}/2}\left[\dfrac{2}{\sqrt{2 - u^2}}\arctan\dfrac{v}{\sqrt{2 - u^2}}\right]_{-u}^{u}\, du$

$\qquad = \displaystyle\int_0^{\sqrt{2}/2} \dfrac{2}{\sqrt{2 - u^2}}\left(\arctan\dfrac{u}{\sqrt{2 - u^2}} - \arctan\dfrac{-u}{\sqrt{2 - u^2}}\right)\, du$

$\qquad = \displaystyle\int_0^{\sqrt{2}/2} \dfrac{4}{\sqrt{2 - u^2}}\arctan\dfrac{u}{\sqrt{2 - u^2}}\, du$

$\qquad$ Let $u = \sqrt{2}\sin\theta,\, du = \sqrt{2}\cos\theta\, d\theta,\, 2 - u^2 = 2 - 2\sin^2\theta = 2\cos^2\theta.$

$\qquad I_1 = 4\displaystyle\int_0^{\pi/6} \dfrac{1}{\sqrt{2}\cos\theta}\arctan\!\left(\dfrac{\sqrt{2}\sin\theta}{\sqrt{2}\cos\theta}\right)\cdot \sqrt{2}\cos\theta\, d\theta$

$\qquad = 4\displaystyle\int_0^{\pi/6}\arctan(\tan\theta)\, d\theta = \dfrac{4\theta^2}{2}\bigg]_0^{\pi/6} = 2\left(\dfrac{\pi}{6}\right)^2 = \dfrac{\pi^2}{18}$

(c) $I_2 = \displaystyle\int_{\sqrt{2}/2}^{\sqrt{2}}\left[\dfrac{2}{\sqrt{2 - u^2}}\arctan\dfrac{v}{\sqrt{2 - u^2}}\right]_{u - \sqrt{2}}^{-u + \sqrt{2}}\, du$

$\qquad = \displaystyle\int_{\sqrt{2}/2}^{\sqrt{2}} \dfrac{2}{\sqrt{2 - u^2}}\left[\arctan\!\left(\dfrac{-u + \sqrt{2}}{\sqrt{2 - u^2}}\right) - \arctan\!\left(\dfrac{u - \sqrt{2}}{\sqrt{2 - u^2}}\right)\right]\, du$

$\qquad = \displaystyle\int_{\sqrt{2}/2}^{\sqrt{2}} \dfrac{4}{\sqrt{2 - u^2}}\arctan\!\left(\dfrac{\sqrt{2} - u}{\sqrt{2 - u^2}}\right)\, du$

$\qquad$ Let $u = \sqrt{2}\sin\theta.$

$\qquad I_2 = 4\displaystyle\int_{\pi/6}^{\pi/2} \dfrac{1}{\sqrt{2}\cos\theta}\arctan\!\left(\dfrac{\sqrt{2} - \sqrt{2}\sin\theta}{\sqrt{2}\cos\theta}\right)\cdot \sqrt{2}\cos\theta\, d\theta$

$\qquad = 4\displaystyle\int_{\pi/6}^{\pi/2}\arctan\!\left(\dfrac{1 - \sin\theta}{\cos\theta}\right)\, d\theta$

—CONTINUED—

3. —CONTINUED—

(d) $\tan\left(\dfrac{1}{2}\left(\dfrac{\pi}{2} - \theta\right)\right) = \sqrt{\dfrac{1 - \cos((\pi/2) - \theta)}{1 + \cos((\pi/2) - \theta)}} = \sqrt{\dfrac{1 - \sin\theta}{1 + \sin\theta}}$

$= \sqrt{\dfrac{(1 - \sin\theta)^2}{(1 + \sin\theta)(1 - \sin\theta)}} = \sqrt{\dfrac{(1 - \sin\theta)^2}{\cos^2\theta}} = \dfrac{1 - \sin\theta}{\cos\theta}$

(e) $I_2 = 4\displaystyle\int_{\pi/6}^{\pi/2} \arctan\left(\dfrac{1 - \sin\theta}{\cos\theta}\right) d\theta = 4\int_{\pi/6}^{\pi/2} \arctan\left(\tan\left(\dfrac{1}{2}\left(\dfrac{\pi}{2} - \theta\right)\right)\right) d\theta$

$= 4\displaystyle\int_{\pi/6}^{\pi/2} \dfrac{1}{2}\left(\dfrac{\pi}{2} - \theta\right) d\theta = 2\int_{\pi/6}^{\pi/2} \left(\dfrac{\pi}{2} - \theta\right) d\theta$

$= 2\left[\dfrac{\pi}{2}\theta - \dfrac{\theta^2}{2}\right]_{\pi/6}^{\pi/2} = 2\left[\left(\dfrac{\pi^2}{4} - \dfrac{\pi^2}{8}\right) - \left(\dfrac{\pi^2}{12} - \dfrac{\pi^2}{72}\right)\right]$

$= 2\left[\dfrac{18 - 9 - 6 + 1}{72}\,\pi^2\right] = \dfrac{4}{36}\,\pi^2 = \dfrac{\pi^2}{9}$

(f) $\dfrac{1}{1 - xy} = 1 + (xy) + (xy)^2 + \cdots \qquad |xy| < 1$

$\displaystyle\int_0^1\int_0^1 \dfrac{1}{1 - xy}\,dx\,dy = \int_0^1\int_0^1 [1 + (xy) + (xy)^2 + \cdots]\,dx\,dy$

$= \displaystyle\int_0^1\int_0^1 \sum_{K=0}^{\infty} (xy)^K\,dx\,dy = \sum_{K=0}^{\infty} \int_0^1 \dfrac{x^{K+1}y^K}{K + 1}\bigg|_0^1\,dy$

$= \displaystyle\sum_{K=0}^{\infty} \int_0^1 \dfrac{y^K}{K + 1}\,dy = \sum_{K=0}^{\infty} \dfrac{y^{K+1}}{(K + 1)^2}\bigg|_0^1$

$= \displaystyle\sum_{K=0}^{\infty} \dfrac{1}{(K + 1)^2} = \sum_{n=1}^{\infty} \dfrac{1}{n^2}$

(g) $u = \dfrac{x + y}{\sqrt{2}},\; v = \dfrac{y - x}{\sqrt{2}}$

$u - v = \dfrac{2x}{\sqrt{2}} \Rightarrow x = \dfrac{u - v}{\sqrt{2}}$

$u + v = \dfrac{2y}{\sqrt{2}} \Rightarrow y = \dfrac{u + v}{\sqrt{2}}$

$\dfrac{\partial(x, y)}{\partial(u, v)} = \begin{vmatrix} 1/\sqrt{2} & -1/\sqrt{2} \\ 1/\sqrt{2} & 1/\sqrt{2} \end{vmatrix} = 1$

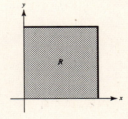

R	S
$(0, 0)$ ↔	$(0, 0)$
$(1, 0)$ ↔	$\left(\dfrac{1}{\sqrt{2}}, -\dfrac{1}{\sqrt{2}}\right)$
$(0, 1)$ ↔	$\left(\dfrac{1}{\sqrt{2}}, \dfrac{1}{\sqrt{2}}\right)$
$(1, 1)$ ↔	$\left(\sqrt{2}, 0\right)$

$\displaystyle\int_0^1\int_0^1 \dfrac{1}{1 - xy}\,dx\,dy = \int_0^{\sqrt{2}/2}\int_{-u}^{u} \dfrac{1}{1 - \dfrac{u^2}{2} + \dfrac{v^2}{2}}\,dv\,du + \int_{\sqrt{2}/2}^{\sqrt{2}}\int_{u - \sqrt{2}}^{-u + \sqrt{2}} \dfrac{1}{1 - \dfrac{u^2}{2} + \dfrac{v^2}{2}}\,dv\,du$

$= I_1 + I_2 = \dfrac{\pi^2}{18} + \dfrac{\pi^2}{9} = \dfrac{\pi^2}{6}$

5. Boundary in *xy*-plane   Boundary in *uv*-plane

$y = \sqrt{x}$ $u = 1$

$y = \sqrt{2x}$ $u = 2$

$y = \frac{1}{3}x^2$ $v = 3$

$y = \frac{1}{4}x^2$ $v = 4$

$$\frac{\partial(x, y)}{\partial(u, v)} = \begin{vmatrix} \frac{1}{3}\left(\frac{v}{u}\right)^{2/3} & \frac{2}{3}\left(\frac{u}{v}\right)^{1/3} \\ \frac{2}{3}\left(\frac{v}{u}\right)^{1/3} & \frac{1}{3}\left(\frac{u}{v}\right)^{2/3} \end{vmatrix} = -\frac{1}{3}$$

$$A = \int\!\!\int_R 1 \, dA = \int\!\!\int_S 1 \left|\frac{\partial(x, y)}{\partial(u, v)}\right| dA = \frac{1}{3}$$

7.

$$V = \int_0^3 \int_0^{2x} \int_x^{6-x} dy \, dz \, dx = 18$$

9. From Exercise 55, Section 13.3,

$$\int_{-\infty}^{\infty} e^{-x^2/2} \, dx = \sqrt{2\pi}$$

Thus, $\displaystyle\int_0^{\infty} e^{-x^2/2} \, dx = \frac{\sqrt{2\pi}}{2}$ and $\displaystyle\int_0^{\infty} e^{-x^2} \, dx = \frac{\sqrt{\pi}}{2}$

$$\int_0^{\infty} x^2 e^{-x^2} \, dx = \left[-\frac{1}{2}xe^{-x^2}\right]_0^{\infty} + \frac{1}{2}\int_0^{\infty} e^{-x^2} \, dx = \frac{1}{2}\frac{\sqrt{\pi}}{2} = \frac{\sqrt{\pi}}{4}$$

11. $f(x, y) = \begin{cases} ke^{-(x+y)/a} & x \geq 0, y \geq 0 \\ 0 & \text{elsewhere} \end{cases}$

$$\int_{-\infty}^{\infty}\int_{-\infty}^{\infty} f(x, y) \, dA = \int_0^{\infty}\int_0^{\infty} ke^{-(x+y)/a} \, dx \, dy$$

$$= k\int_0^{\infty} e^{-x/a} \, dx \cdot \int_0^{\infty} e^{-y/a} \, dy$$

These two integrals are equal to

$$\int_0^{\infty} e^{-x/a} \, dx = \lim_{b\to\infty}\left[(-a)e^{-x/a}\right]_0^b = a.$$

Hence, assuming $a, k > 0$, you obtain

$$1 = ka^2 \quad \text{or} \quad a = \frac{1}{\sqrt{k}}.$$

13. $A = l \cdot w = \left(\dfrac{\Delta x}{\cos \theta}\right)\Delta y = \sec \theta \, \Delta x \, \Delta y$

Area in *xy*-plane: $\Delta x \, \Delta y$

C H A P T E R 14
Vector Analysis

CHAPTER 14
Vector Analysis

Section 14.1   Vector Fields
Solutions to Odd-Numbered Exercises

1. All vectors are parallel to y-axis.

Matches (c)

3. All vectors point outward.

Matches (b)

5. Vectors are parallel to x-axis for $y = n\pi$.

Matches (a)

7. $\mathbf{F}(x, y) = \mathbf{i} + \mathbf{j}$

$\|\mathbf{F}\| = \sqrt{2}$

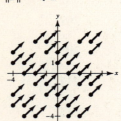

9. $\mathbf{F}(x, y) = x\mathbf{i} + y\mathbf{j}$

$\|\mathbf{F}\| = \sqrt{x^2 + y^2} = c$

$x^2 + y^2 = c^2$

11. $\mathbf{F}(x, y, z) = 3y\mathbf{j}$

$\|\mathbf{F}\| = 3|y| = c$

13. $\mathbf{F}(x, y) = 4x\mathbf{i} + y\mathbf{j}$

$\|\mathbf{F}\| = \sqrt{16x^2 + y^2} = c$

$\dfrac{x^2}{c^2/16} + \dfrac{y^2}{c^2} = 1$

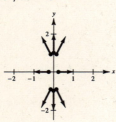

15. $\mathbf{F}(x, y, z) = \mathbf{i} + \mathbf{j} + \mathbf{k}$

$\|\mathbf{F}\| = \sqrt{3}$

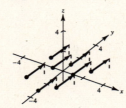

17.

19.

21. $f(x, y) = 5x^2 + 3xy + 10y^2$

$f_x(x, y) = 10x + 3y$

$f_y(x, y) = 3x + 20y$

$\mathbf{F}(x, y) = (10x + 3y)\mathbf{i} + (3x + 20y)\mathbf{j}$

23. $f(x, y, z) = z - ye^{x^2}$

$f_x(x, y, z) = -2xye^{x^2}$

$f_y(x, y, z) = -e^{x^2}$

$f_z = 1$

$\mathbf{F}(x, y, z) = -2xye^{x^2}\mathbf{i} - e^{x^2}\mathbf{j} + \mathbf{k}$

25. $g(x, y, z) = xy \ln(x + y)$

$g_x(x, y, z) = y \ln(x + y) + \dfrac{xy}{x + y}$

$g_y(x, y, z) = x \ln(x + y) + \dfrac{xy}{x + y}$

$g_z(x, y, z) = 0$

$\mathbf{G}(x, y, z) = \left[\dfrac{xy}{x + y} + y \ln(x + y) \right]\mathbf{i} + \left[\dfrac{xy}{x + y} + x \ln(x + y) \right]\mathbf{j}$

27. $\mathbf{F}(x, y) = 12xy\mathbf{i} + 6(x^2 + y)\mathbf{j}$

$M = 12xy$ and $N = 6(x^2 + y)$ have continuous first partial derivatives.

$\dfrac{\partial N}{\partial x} = 12x = \dfrac{\partial M}{\partial y} \Rightarrow \mathbf{F}$ is conservative.

29. $\mathbf{F}(x, y) = \sin y\mathbf{i} + x \cos y\mathbf{j}$

$M = \sin y$ and $N = x \cos y$ have continuous first partial derivatives.

$\dfrac{\partial N}{\partial x} = \cos y = \dfrac{\partial M}{\partial y} \Rightarrow \mathbf{F}$ is conservative.

31. $M = 15y^3, N = -5xy^2$

$\dfrac{\partial N}{\partial x} = -5y^2 \neq \dfrac{\partial M}{\partial y} = 45y^2 \Rightarrow$ Not conservative

33. $M = \dfrac{2}{y}e^{2x/y}, N = \dfrac{-2x}{y^2}e^{2x/y}$

$\dfrac{\partial N}{\partial x} = \dfrac{-2(y + 2x)}{y^3}e^{2x/y} = \dfrac{\partial M}{\partial y} \Rightarrow$ Conservative

35. $\mathbf{F}(x, y) = 2xy\mathbf{i} + x^2\mathbf{j}$

$\dfrac{\partial}{\partial y}[2xy] = 2x$

$\dfrac{\partial}{\partial x}[x^2] = 2x$

Conservative

$f_x(x, y) = 2xy$

$f_y(x, y) = x^2$

$f(x, y) = x^2y + K$

37. $\mathbf{F}(x, y) = xe^{x^2y}(2y\mathbf{i} + x\mathbf{j})$

$\dfrac{\partial}{\partial y}[2xye^{x^2y}] = 2xe^{x^2y} + 2x^3ye^{x^2y}$

$\dfrac{\partial}{\partial x}[x^2e^{x^2y}] = 2xe^{x^2y} + 2x^3ye^{x^2y}$

Conservative

$f_x(x, y) = 2xye^{x^2y}$

$f_y(x, y) = x^2e^{x^2y}$

$f(x, y) = e^{x^2y} + K$

39. $\mathbf{F}(x, y) = \dfrac{x}{x^2 + y^2}\mathbf{i} + \dfrac{y}{x^2 + y^2}\mathbf{j}$

$\dfrac{\partial}{\partial y}\left[\dfrac{x}{x^2 + y^2}\right] = -\dfrac{2xy}{(x^2 + y^2)^2}$

$\dfrac{\partial}{\partial x}\left[\dfrac{y}{x^2 + y^2}\right] = -\dfrac{2xy}{(x^2 + y^2)^2}$

Conservative

$f_x(x, y) = \dfrac{x}{x^2 + y^2}$

$f_y(x, y) = \dfrac{y}{x^2 + y^2}$

$f(x, y) = \dfrac{1}{2}\ln(x^2 + y^2) + K$

41. $\mathbf{F}(x, y) = e^x(\cos y\mathbf{i} + \sin y\mathbf{j})$

$\dfrac{\partial}{\partial y}[e^x \cos y] = -e^x \sin y$

$\dfrac{\partial}{\partial x}[e^x \sin y] = e^x \sin y$

Not conservative

43. $\mathbf{F}(x, y, z) = xyz\mathbf{i} + y\mathbf{j} + z\mathbf{k}, \ (1, 2, 1)$

$\operatorname{curl} \mathbf{F} = \begin{vmatrix} \mathbf{i} & \mathbf{j} & \mathbf{k} \\ \dfrac{\partial}{\partial x} & \dfrac{\partial}{\partial y} & \dfrac{\partial}{\partial z} \\ xyz & y & z \end{vmatrix} = xy\mathbf{j} - xz\mathbf{k}$

$\operatorname{curl} \mathbf{F} (1, 2, 1) = 2\mathbf{j} - \mathbf{k}$

45. $\mathbf{F}(x, y, z) = e^x \sin y\mathbf{i} - e^x \cos y\mathbf{j}, \ (0, 0, 3)$

$\operatorname{curl} \mathbf{F} = \begin{vmatrix} \mathbf{i} & \mathbf{j} & \mathbf{k} \\ \dfrac{\partial}{\partial x} & \dfrac{\partial}{\partial y} & \dfrac{\partial}{\partial z} \\ e^x \sin y & -e^x \cos y & 0 \end{vmatrix} = -2e^x \cos y\mathbf{k}$

$\operatorname{curl} \mathbf{F} (0, 0, 3) = -2\mathbf{k}$

47. $F(x, y, z) = \arctan\left(\dfrac{x}{y}\right)\mathbf{i} + \ln\sqrt{x^2 + y^2}\,\mathbf{j} + \mathbf{k}$

$$\text{curl } \mathbf{F} = \begin{vmatrix} \mathbf{i} & \mathbf{j} & \mathbf{k} \\ \dfrac{\partial}{\partial x} & \dfrac{\partial}{\partial y} & \dfrac{\partial}{\partial z} \\ \arctan\left(\dfrac{x}{y}\right) & \dfrac{1}{2}\ln(x^2 + y^2) & 1 \end{vmatrix} = \left[\dfrac{x}{x^2 + y^2} - \dfrac{(-x/y^2)}{1 + (x/y)^2}\right]\mathbf{k} = \dfrac{2x}{x^2 + y^2}\mathbf{k}$$

49. $F(x, y, z) = \sin(x - y)\mathbf{i} + \sin(y - z)\mathbf{j} + \sin(z - x)\mathbf{k}$

$$\text{curl } \mathbf{F} = \begin{vmatrix} \mathbf{i} & \mathbf{j} & \mathbf{k} \\ \dfrac{\partial}{\partial x} & \dfrac{\partial}{\partial y} & \dfrac{\partial}{\partial z} \\ \sin(x - y) & \sin(y - z) & \sin(z - x) \end{vmatrix} = \cos(y - z)\mathbf{i} + \cos(z - x)\mathbf{j} + \cos(x - y)\mathbf{k}$$

51. $F(x, y, z) = \sin y\,\mathbf{i} - x\cos y\,\mathbf{j} + \mathbf{k}$

$$\text{curl } \mathbf{F} = \begin{vmatrix} \mathbf{i} & \mathbf{j} & \mathbf{k} \\ \dfrac{\partial}{\partial x} & \dfrac{\partial}{\partial y} & \dfrac{\partial}{\partial z} \\ \sin y & -x\cos y & 1 \end{vmatrix} = -2\cos y\,\mathbf{k} \neq 0$$

Not conservative

53. $F(x, y, z) = e^z(y\mathbf{i} + x\mathbf{j} + xy\mathbf{k})$

$$\text{curl } \mathbf{F} = \begin{vmatrix} \mathbf{i} & \mathbf{j} & \mathbf{k} \\ \dfrac{\partial}{\partial x} & \dfrac{\partial}{\partial y} & \dfrac{\partial}{\partial z} \\ ye^z & xe^z & xye^z \end{vmatrix} = 0$$

Conservative

$$f_x(x, y, z) = ye^z$$
$$f_y(x, y, z) = xe^z$$
$$f_z(x, y, z) = xye^z$$
$$f(x, y, z) = xye^z + K$$

55. $F(x, y, z) = \dfrac{1}{y}\mathbf{i} - \dfrac{x}{y^2}\mathbf{j} + (2z - 1)\mathbf{k}$

$$\text{curl } \mathbf{F} = \begin{vmatrix} \mathbf{i} & \mathbf{j} & \mathbf{k} \\ \dfrac{\partial}{\partial x} & \dfrac{\partial}{\partial y} & \dfrac{\partial}{\partial z} \\ \dfrac{1}{y} & -\dfrac{x}{y^2} & 2z - 1 \end{vmatrix} = 0$$

Conservative

$$f_x(x, y, z) = \dfrac{1}{y}$$

$$f_y(x, y, z) = -\dfrac{x}{y^2}$$

$$f_z(x, y, z) = 2z - 1$$

$$f(x, y, z) = \int \dfrac{1}{y}\,dx = \dfrac{x}{y} + g(y, z) + K_1$$

$$f(x, y, z) = \int -\dfrac{x}{y^2}\,dy = \dfrac{x}{y} + h(x, z) + K_2$$

$$f(x, y, z) = \int (2z - 1)\,dz$$

$$= z^2 - z + p(x, y) + K_3$$

$$f(x, y, z) = \dfrac{x}{y} + z^2 - z + K$$

57. $F(x, y, z) = 6x^2\mathbf{i} - xy^2\mathbf{j}$

$$\text{div } \mathbf{F}(x, y, z) = \dfrac{\partial}{\partial x}[6x^2] + \dfrac{\partial}{\partial y}[-xy^2]$$

$$= 12x - 2xy$$

59. $\mathbf{F}(x, y, z) = \sin x \mathbf{i} + \cos y \mathbf{j} + z^2 \mathbf{k}$

$\text{div } \mathbf{F}(x, y, z) = \dfrac{\partial}{\partial x}[\sin x] + \dfrac{\partial}{\partial y}[\cos y] + \dfrac{\partial}{\partial z}[z^2] = \cos x - \sin y + 2z$

61. $\mathbf{F}(x, y, z) = xyz\mathbf{i} + y\mathbf{j} + z\mathbf{k}$

$\text{div } \mathbf{F}(x, y, z) = yz + 1 + 1 = yz + 2$

$\text{div } \mathbf{F}(1, 2, 1) = 4$

63. $\mathbf{F}(x, y, z) = e^x \sin y \mathbf{i} - e^x \cos y \mathbf{j}$

$\text{div } \mathbf{F}(x, y, z) = e^x \sin y + e^x \sin y$

$\text{div } \mathbf{F}(0, 0, 3) = 0$

65. See the definition, page 1008. Examples include velocity fields, gravitational fields and magnetic fields.

67. See the definition on page 1014.

69. $\mathbf{F}(x, y, z) = \mathbf{i} + 2x\mathbf{j} + 3y\mathbf{k}$

$\mathbf{G}(x, y, z) = x\mathbf{i} - y\mathbf{j} + z\mathbf{k}$

$\mathbf{F} \times \mathbf{G} = \begin{vmatrix} \mathbf{i} & \mathbf{j} & \mathbf{k} \\ 1 & 2x & 3y \\ x & -y & z \end{vmatrix} = (2xz + 3y^2)\mathbf{i} - (z - 3xy)\mathbf{j} + (-y - 2x^2)\mathbf{k}$

$\text{curl } (\mathbf{F} \times \mathbf{G}) = \begin{vmatrix} \mathbf{i} & \mathbf{j} & \mathbf{k} \\ \dfrac{\partial}{\partial x} & \dfrac{\partial}{\partial y} & \dfrac{\partial}{\partial z} \\ 2xz + 3y^2 & 3xy - z & -y - 2x^2 \end{vmatrix} = (-1 + 1)\mathbf{i} - (-4x - 2x)\mathbf{j} + (3y - 6y)\mathbf{k} = 6x\mathbf{j} - 3y\mathbf{k}$

71. $\mathbf{F}(x, y, z) = xyz\mathbf{i} + y\mathbf{j} + z\mathbf{k}$

$\text{curl } \mathbf{F} = \begin{vmatrix} \mathbf{i} & \mathbf{j} & \mathbf{k} \\ \dfrac{\partial}{\partial x} & \dfrac{\partial}{\partial y} & \dfrac{\partial}{\partial z} \\ xyz & y & z \end{vmatrix} = xy\mathbf{j} - xz\mathbf{k}$

$\text{curl}(\text{curl } \mathbf{F}) = \begin{vmatrix} \mathbf{i} & \mathbf{j} & \mathbf{k} \\ \dfrac{\partial}{\partial x} & \dfrac{\partial}{\partial y} & \dfrac{\partial}{\partial z} \\ 0 & xy & -xz \end{vmatrix} = z\mathbf{j} + y\mathbf{k}$

73. $\mathbf{F}(x, y, z) = \mathbf{i} + 2x\mathbf{j} + 3y\mathbf{k}$

$\mathbf{G}(x, y, z) = x\mathbf{i} - y\mathbf{j} + z\mathbf{k}$

$\mathbf{F} \times \mathbf{G} = \begin{vmatrix} \mathbf{i} & \mathbf{j} & \mathbf{k} \\ 1 & 2x & 3y \\ x & -y & z \end{vmatrix}$

$\quad = (2xz + 3y^2)\mathbf{i} - (z - 3xy)\mathbf{j} + (-y - 2x^2)\mathbf{k}$

$\text{div}(\mathbf{F} \times \mathbf{G}) = 2z + 3x$

75. $\mathbf{F}(x, y, z) = xyz\mathbf{i} + y\mathbf{j} + z\mathbf{k}$

$\text{curl } \mathbf{F} = \begin{vmatrix} \mathbf{i} & \mathbf{j} & \mathbf{k} \\ \dfrac{\partial}{\partial x} & \dfrac{\partial}{\partial y} & \dfrac{\partial}{\partial z} \\ xyz & y & z \end{vmatrix} = xy\mathbf{j} - xz\mathbf{k}$

$\text{div}(\text{curl } \mathbf{F}) = x - x = 0$

77. Let $\mathbf{F} = M\mathbf{i} + N\mathbf{j} + P\mathbf{k}$ and $\mathbf{G} = Q\mathbf{i} + R\mathbf{j} + S\mathbf{k}$ where $M, N, P, Q, R,$ and S have continuous partial derivatives.

$\mathbf{F} + \mathbf{G} = (M + Q)\mathbf{i} + (N + R)\mathbf{j} + (P + S)\mathbf{k}$

$\text{curl}(\mathbf{F} + \mathbf{G}) = \begin{vmatrix} \mathbf{i} & \mathbf{j} & \mathbf{k} \\ \dfrac{\partial}{\partial x} & \dfrac{\partial}{\partial y} & \dfrac{\partial}{\partial z} \\ M + Q & N + R & P + S \end{vmatrix}$

$= \left[\dfrac{\partial}{\partial y}(P + S) - \dfrac{\partial}{\partial z}(N + R)\right]\mathbf{i} - \left[\dfrac{\partial}{\partial x}(P + S) - \dfrac{\partial}{\partial z}(M + Q)\right]\mathbf{j} + \left[\dfrac{\partial}{\partial x}(N + R) - \dfrac{\partial}{\partial y}(M + Q)\right]\mathbf{k}$

$= \left(\dfrac{\partial P}{\partial y} - \dfrac{\partial N}{\partial z}\right)\mathbf{i} - \left(\dfrac{\partial P}{\partial x} - \dfrac{\partial M}{\partial z}\right)\mathbf{j} + \left(\dfrac{\partial N}{\partial x} - \dfrac{\partial M}{\partial y}\right)\mathbf{k} + \left(\dfrac{\partial S}{\partial y} - \dfrac{\partial R}{\partial z}\right)\mathbf{i} - \left(\dfrac{\partial S}{\partial x} - \dfrac{\partial Q}{\partial z}\right)\mathbf{j} + \left(\dfrac{\partial R}{\partial x} - \dfrac{\partial Q}{\partial y}\right)\mathbf{k}$

$= \text{curl } \mathbf{F} + \text{curl } \mathbf{G}$

79. Let $\mathbf{F} = M\mathbf{i} + N\mathbf{j} + P\mathbf{k}$ and $\mathbf{G} = R\mathbf{i} + S\mathbf{j} + T\mathbf{k}$.

$$\operatorname{div}(\mathbf{F} + \mathbf{G}) = \frac{\partial}{\partial x}(M + R) + \frac{\partial}{\partial y}(N + S) + \frac{\partial}{\partial z}(P + T) = \frac{\partial M}{\partial x} + \frac{\partial R}{\partial x} + \frac{\partial N}{\partial y} + \frac{\partial S}{\partial y} + \frac{\partial P}{\partial z} + \frac{\partial T}{\partial z}$$

$$= \left[\frac{\partial M}{\partial x} + \frac{\partial N}{\partial y} + \frac{\partial P}{\partial z}\right] + \left[\frac{\partial R}{\partial x} + \frac{\partial S}{\partial y} + \frac{\partial T}{\partial z}\right]$$

$$= \operatorname{div} \mathbf{F} + \operatorname{div} \mathbf{G}$$

81. $\mathbf{F} = M\mathbf{i} + N\mathbf{j} + P\mathbf{k}$

$$\nabla \times [\nabla f + (\nabla \times \mathbf{F})] = \mathbf{curl}(\nabla f + (\nabla \times \mathbf{F}))$$

$$= \mathbf{curl}(\nabla f) + \mathbf{curl}(\nabla \times \mathbf{F}) \quad \text{(Exercise 77)}$$

$$= \mathbf{curl}(\nabla \times \mathbf{F}) \quad \text{(Exercise 78)}$$

$$= \nabla \times (\nabla \times \mathbf{F})$$

83. Let $\mathbf{F} = M\mathbf{i} + N\mathbf{j} + P\mathbf{k}$, then $f\mathbf{F} = fM\mathbf{i} + fN\mathbf{j} + fP\mathbf{k}$.

$$\operatorname{div}(f\mathbf{F}) = \frac{\partial}{\partial x}(fM) + \frac{\partial}{\partial y}(fN) + \frac{\partial}{\partial z}(fP) = f\frac{\partial M}{\partial x} + M\frac{\partial f}{\partial x} + f\frac{\partial N}{\partial y} + N\frac{\partial f}{\partial y} + f\frac{\partial P}{\partial z} + P\frac{\partial f}{\partial z}$$

$$= f\left(\frac{\partial M}{\partial x} + \frac{\partial N}{\partial y} + \frac{\partial N}{\partial z}\right) + \left(\frac{\partial f}{\partial x}M + \frac{\partial f}{\partial y}N + \frac{\partial f}{\partial z}P\right)$$

$$= f\operatorname{div} \mathbf{F} + \nabla f \cdot \mathbf{F}$$

In Exercises 85 and 87, $\mathbf{F}(x, y, z) = x\mathbf{i} + y\mathbf{j} + z\mathbf{k}$ and $f(x, y, z) = \|\mathbf{F}(x, y, z)\| = \sqrt{x^2 + y^2 + z^2}$.

85.

$$\ln f = \frac{1}{2}\ln(x^2 + y^2 + z^2)$$

$$\nabla(\ln f) = \frac{x}{x^2 + y^2 + z^2}\mathbf{i} + \frac{y}{x^2 + y^2 + z^2}\mathbf{j} + \frac{z}{x^2 + y^2 + z^2}\mathbf{k} = \frac{x\mathbf{i} + y\mathbf{j} + z\mathbf{k}}{x^2 + y^2 + z^2} = \frac{\mathbf{F}}{f^2}$$

87. $f^n = \left(\sqrt{x^2 + y^2 + z^2}\right)^n$

$$\nabla f^n = n\left(\sqrt{x^2 + y^2 + z^2}\right)^{n-1}\frac{x}{\sqrt{x^2 + y^2 + z^2}}\mathbf{i} + n\left(\sqrt{x^2 + y^2 + z^2}\right)^{n-1}\frac{y}{\sqrt{x^2 + y^2 + z^2}}\mathbf{j}$$

$$+ n\left(\sqrt{x^2 + y^2 + z^2}\right)^{n-1}\frac{z}{\sqrt{x^2 + y^2 + z^2}}\mathbf{k}$$

$$= n\left(\sqrt{x^2 + y^2 + z^2}\right)^{n-2}(x\mathbf{i} + y\mathbf{j} + z\mathbf{k}) = nf^{n-2}\mathbf{F}$$

89. The winds are stronger over Phoenix. Although the winds over both cities are northeasterly, they are more towards the east over Atlanta.

Section 14.2   Line Integrals

1. $x^2 + y^2 = 9$

$\dfrac{x^2}{9} + \dfrac{y^2}{9} = 1$

$\cos^2 t + \sin^2 t = 1$

$\cos^2 t = \dfrac{x^2}{9}$

$\sin^2 t = \dfrac{y^2}{9}$

$x = 3 \cos t$

$y = 3 \sin t$

$\mathbf{r}(t) = 3 \cos t \mathbf{i} + 3 \sin t \mathbf{j}$

$0 \le t \le 2\pi$

3. $\mathbf{r}(t) = \begin{cases} t\mathbf{i}, & 0 \le t \le 3 \\ 3\mathbf{i} + (t-3)\mathbf{j}, & 3 \le t \le 6 \\ (9-t)\mathbf{i} + 3\mathbf{j}, & 6 \le t \le 9 \\ (12-t)\mathbf{j}, & 9 \le t \le 12 \end{cases}$

5. $\mathbf{r}(t) = \begin{cases} t\mathbf{i} + \sqrt{t}\mathbf{j}, & 0 \le t \le 1 \\ (2-t)\mathbf{i} + (2-t)\mathbf{j}, & 1 \le t \le 2 \end{cases}$

7. $\mathbf{r}(t) = 4t\mathbf{i} + 3t\mathbf{j}$, $0 \le t \le 2$; $\mathbf{r}'(t) = 4\mathbf{i} + 3\mathbf{j}$

$\displaystyle \int_C (x - y)\, ds = \int_0^2 (4t - 3t)\sqrt{(4)^2 + (3)^2}\, dt = \int_0^2 5t\, dt = \left[\dfrac{5t^2}{2}\right]_0^2 = 10$

9. $\mathbf{r}(t) = \sin t \mathbf{i} + \cos t \mathbf{j} + 8t\mathbf{k}$, $0 \le t \le \dfrac{\pi}{2}$; $\mathbf{r}'(t) = \cos t \mathbf{i} - \sin t \mathbf{j} + 8\mathbf{k}$

$\displaystyle \int_C (x^2 + y^2 + z^2)\, ds = \int_0^{\pi/2} (\sin^2 t + \cos^2 t + 64t^2)\sqrt{(\cos t)^2 + (-\sin t)^2 + 64}\, dt$

$\displaystyle = \int_0^{\pi/2} \sqrt{65}(1 + 64t^2)\, dt = \left[\sqrt{65}\left(t + \dfrac{64t^3}{3}\right)\right]_0^{\pi/2} = \sqrt{65}\left(\dfrac{\pi}{2} + \dfrac{8\pi^3}{3}\right) = \dfrac{\sqrt{65}\,\pi}{6}(3 + 16\pi^2)$

11. $\mathbf{r}(t) = t\mathbf{i}$, $0 \le t \le 3$

$\displaystyle \int_C (x^2 + y^2)\, ds = \int_0^3 [t^2 + 0^2]\sqrt{1 + 0}\, dt$

$\displaystyle = \int_0^3 t^2\, dt$

$\displaystyle = \left[\dfrac{1}{3}t^3\right]_0^3 = 9$

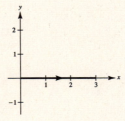

13. $\mathbf{r}(t) = \cos t \mathbf{i} + \sin t \mathbf{j}$, $0 \le t \le \dfrac{\pi}{2}$

$\displaystyle \int_C (x^2 + y^2)\, ds = \int_0^{\pi/2} [\cos^2 t + \sin^2 t]\sqrt{(-\sin t)^2 + (\cos t)^2}\, dt$

$\displaystyle = \int_0^{\pi/2} dt = \dfrac{\pi}{2}$

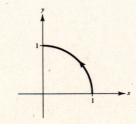

15. $\mathbf{r}(t) = t\mathbf{i} + t\mathbf{j},\ 0 \le t \le 1$

$$\int_C \left(x + 4\sqrt{y} \right) ds = \int_0^1 \left(t + 4\sqrt{t} \right)\sqrt{1+1}\ dt$$

$$= \left[\sqrt{2}\left(\frac{t^2}{2} + \frac{8}{3}t^{3/2} \right) \right]_0^1 = \frac{19\sqrt{2}}{6}$$

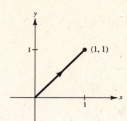

17. $\mathbf{r}(t) = \begin{cases} t\mathbf{i}, & 0 \le t \le 1 \\ (2-t)\mathbf{i} + (t-1)\mathbf{j}, & 1 \le t \le 2 \\ (3-t)\mathbf{j}, & 2 \le t \le 3 \end{cases}$

$$\int_{C_1} \left(x + 4\sqrt{y} \right) ds = \int_0^1 t\ dt = \frac{1}{2}$$

$$\int_{C_2} \left(x + 4\sqrt{y} \right) ds = \int_1^2 \left[(2-t) + 4\sqrt{t-1} \right]\sqrt{1+1}\ dt$$

$$= \sqrt{2}\left[2t - \frac{t^2}{2} + \frac{8}{3}(t-1)^{3/2} \right]_1^2 = \frac{19\sqrt{2}}{6}$$

$$\int_{C_3} \left(x + 4\sqrt{y} \right) ds = \int_2^3 4\sqrt{3-t}\ dt = \left[-\frac{8}{3}(3-t)^{3/2} \right]_2^3 = \frac{8}{3}$$

$$\int_C \left(x + 4\sqrt{y} \right) ds = \frac{1}{2} + \frac{19\sqrt{2}}{6} + \frac{8}{3} = \frac{19 + 19\sqrt{2}}{6} = \frac{19(1 + \sqrt{2})}{6}$$

19. $\rho(x, y, z) = \dfrac{1}{2}(x^2 + y^2 + z^2)$

$\quad \mathbf{r}(t) = 3\cos t\mathbf{i} + 3\sin t\mathbf{j} + 2t\mathbf{k},\ 0 \le t \le 4\pi$

$\quad \mathbf{r}'(t) = -3\sin t\mathbf{i} + 3\cos t\mathbf{j} + 2\mathbf{k}$

$\quad \|\mathbf{r}'(t)\| = \sqrt{(-3\sin t)^2 + (3\cos t)^2 + (2)^2} = \sqrt{13}$

$\quad \text{Mass} = \int_C \rho(x, y, z)\ ds = \int_0^{4\pi} \frac{1}{2}[(3\cos t)^2 + (3\sin t)^2 + (2t)^2]\sqrt{13}\ dt$

$$= \frac{\sqrt{13}}{2}\int_0^{4\pi} (9 + 4t^2)\ dt = \left[\frac{\sqrt{13}}{2}\left(9t + \frac{4t^3}{3} \right) \right]_0^{4\pi}$$

$$= \frac{2\sqrt{13}\pi}{3}(27 + 64\pi^2) \approx 4973.8$$

21. $\mathbf{F}(x, y) = xy\mathbf{i} + y\mathbf{j}$

$\quad C:\ \mathbf{r}(t) = 4t\mathbf{i} + t\mathbf{j},\ 0 \le t \le 1$

$\quad \mathbf{F}(t) = 4t^2\mathbf{i} + t\mathbf{j}$

$\quad \mathbf{r}'(t) = 4\mathbf{i} + \mathbf{j}$

$$\int_C \mathbf{F} \cdot d\mathbf{r} = \int_0^1 (16t^2 + t)\ dt$$

$$= \left[\frac{16}{3}t^3 + \frac{1}{2}t^2 \right]_0^1 = \frac{35}{6}$$

23. $\mathbf{F}(x, y) = 3x\mathbf{i} + 4y\mathbf{i}$

$\quad C:\ \mathbf{r}(t) = 2\cos t\mathbf{i} + 2\sin t\mathbf{j},\ 0 \le t \le \dfrac{\pi}{2}$

$\quad \mathbf{F}(t) = 6\cos t\mathbf{i} + 8\sin t\mathbf{j}$

$\quad \mathbf{r}'(t) = -2\sin t\mathbf{i} + 2\cos t\mathbf{j}$

$$\int_C \mathbf{F} \cdot d\mathbf{r} = \int_0^{\pi/2} (-12\sin t\cos t + 16\sin t\cos t)\ dt$$

$$= \left[2\sin^2 t \right]_0^{\pi/2} = 2$$

25. $\mathbf{F}(x, y, z) = x^2 y\mathbf{i} + (x - z)\mathbf{j} + xyz\mathbf{k}$

    C: $\mathbf{r}(t) = t\mathbf{i} + t^2\mathbf{j} + 2\mathbf{k}$, $0 \le t \le 1$

    $\mathbf{F}(t) = t^4\mathbf{i} + (t - 2)\mathbf{j} + 2t^3\mathbf{k}$

    $\mathbf{r}'(t) = \mathbf{i} + 2t\mathbf{j}$

$$\int_C \mathbf{F} \cdot d\mathbf{r} = \int_0^1 [t^4 + 2t(t - 2)]\, dt$$

$$= \left[\frac{t^5}{5} + \frac{2t^3}{3} - 2t^2 \right]_0^1 = -\frac{17}{15}$$

27. $\mathbf{F}(x, y, z) = x^2 z\mathbf{i} + 6y\mathbf{j} + yz^2\mathbf{k}$

    $\mathbf{r}(t) = t\mathbf{i} + t^2\mathbf{j} + \ln t\mathbf{k}$, $1 \le t \le 3$

    $\mathbf{F}(t) = t^2 \ln t\mathbf{i} + 6t^2\mathbf{j} + t^2 \ln^2 t\mathbf{k}$

$$d\mathbf{r} = \left(\mathbf{i} + 2t\mathbf{j} + \frac{1}{t}\mathbf{k} \right) dt$$

$$\int_C \mathbf{F} \cdot d\mathbf{r} = \int_1^3 [t^2 \ln t + 12t^3 + t(\ln t)^2]\, dt$$

$$\approx 249.49$$

29. $\mathbf{F}(x, y) = -x\mathbf{i} - 2y\mathbf{j}$

    C: $y = x^3$ from $(0, 0)$ to $(2, 8)$

    $\mathbf{r}(t) = t\mathbf{i} + t^3\mathbf{j}$, $0 \le t \le 2$

    $\mathbf{r}'(t) = \mathbf{i} + 3t^2\mathbf{j}$

    $\mathbf{F}(t) = -t\mathbf{i} - 2t^3\mathbf{j}$

$\mathbf{F} \cdot \mathbf{r}' = -t - 6t^5$

$$\text{Work} = \int_C \mathbf{F} \cdot d\mathbf{r} = \int_0^2 (-t - 6t^5)\, dt = \left[-\frac{1}{2}t^2 - t^6 \right]_0^2 = -66$$

31. $\mathbf{F}(x, y) = 2x\mathbf{i} + y\mathbf{j}$

    C: counterclockwise around the triangle whose vertices are $(0, 0)$, $(1, 0)$, $(1, 1)$

$$\mathbf{r}(t) = \begin{cases} t\mathbf{i}, & 0 \le t \le 1 \\ \mathbf{i} + (t - 1)\mathbf{j}, & 1 \le t \le 2 \\ (3 - t)\mathbf{i} + (3 - t)\mathbf{j}, & 2 \le t \le 3 \end{cases}$$

On C_1: $\mathbf{F}(t) = 2t\mathbf{i}$, $\mathbf{r}'(t) = \mathbf{i}$

$$\text{Work} = \int_{C_1} \mathbf{F} \cdot d\mathbf{r} = \int_0^1 2t\, dt = 1$$

On C_2: $\mathbf{F}(t) = 2\mathbf{i} + (t - 1)\mathbf{j}$, $\mathbf{r}'(t) = \mathbf{j}$

$$\text{Work} = \int_{C_2} \mathbf{F} \cdot d\mathbf{r} = \int_1^2 (t - 1)\, dt = \frac{1}{2}$$

On C_3: $\mathbf{F}(t) = 2(3 - t)\mathbf{i} + (3 - t)\mathbf{j}$, $\mathbf{r}'(t) = -\mathbf{i} - \mathbf{j}$

$$\text{Work} = \int_{C_3} \mathbf{F} \cdot d\mathbf{r} = \int_2^3 [-2(3 - t) - (3 - t)]\, dt = -\frac{3}{2}$$

$$\text{Total work} = \int_C \mathbf{F} \cdot d\mathbf{r} = 1 + \frac{1}{2} - \frac{3}{2} = 0$$

33. $\mathbf{F}(x, y, z) = x\mathbf{i} + y\mathbf{j} - 5z\mathbf{k}$

    C: $\mathbf{r}(t) = 2 \cos t\mathbf{i} + 2 \sin t\mathbf{j} + t\mathbf{k}$, $0 \le t \le 2\pi$

    $\mathbf{r}'(t) = -2 \sin t\mathbf{i} + 2 \cos t\mathbf{j} + \mathbf{k}$

    $\mathbf{F}(t) = 2 \cos t\mathbf{i} + 2 \sin t\mathbf{j} - 5t\mathbf{k}$

    $\mathbf{F} \cdot \mathbf{r}' = -5t$

$$\text{Work} = \int_C \mathbf{F} \cdot d\mathbf{r} = \int_0^{2\pi} -5t\, dt = -10\pi^2$$

35. $\mathbf{r}(t) = 3 \sin t\mathbf{i} + 3 \cos t\mathbf{j} + \dfrac{10}{2\pi}t\mathbf{k}$, $0 \le t \le 2\pi$

    $\mathbf{F} = 150\mathbf{k}$

$$d\mathbf{r} = \left(3 \cos t\mathbf{i} - 3 \sin t\mathbf{j} + \frac{10}{2\pi}\mathbf{k} \right) dt$$

$$\int_C \mathbf{F} \cdot d\mathbf{r} = \int_0^{2\pi} \frac{1500}{2\pi}\, dt = \left[\frac{1500}{2\pi}t \right]_0^{2\pi} = 1500 \text{ ft} \cdot \text{lb}$$

37. $\mathbf{F}(x, y) = x^2\mathbf{i} + xy\mathbf{j}$

   (a) $\mathbf{r}_1(t) = 2t\mathbf{i} + (t - 1)\mathbf{j}, 1 \le t \le 3$

          $\mathbf{r}_1{}'(t) = 2\mathbf{i} + \mathbf{j}$

          $\mathbf{F}(t) = 4t^2\mathbf{i} + 2t(t - 1)\mathbf{j}$

$$\int_{C_1} \mathbf{F} \cdot d\mathbf{r} = \int_1^3 (8t^2 + 2t(t - 1))\, dt = \frac{236}{3}$$

Both paths join $(2, 0)$ and $(6, 2)$. The integrals are negatives of each other because the orientations are different.

   (b) $\mathbf{r}_2(t) = 2(3 - t)\mathbf{i} + (2 - t)\mathbf{j}, 0 \le t \le 2$

          $\mathbf{r}_2{}'(t) = -2\mathbf{i} - \mathbf{j}$

          $\mathbf{F}(t) = 4(3 - t)^2\mathbf{i} + 2(3 - t)(2 - t)\mathbf{j}$

$$\int_{C_2} \mathbf{F} \cdot d\mathbf{r} = \int_0^2 [-8(3 - t)^2 - 2(3 - t)(2 - t)]\, dt$$

$$= -\frac{236}{3}$$

39. $\mathbf{F}(x, y) = y\mathbf{i} - x\mathbf{j}$

   $C:\ \mathbf{r}(t) = t\mathbf{i} - 2t\mathbf{j}$

     $\mathbf{r}'(t) = \mathbf{i} - 2\mathbf{j}$

     $\mathbf{F}(t) = -2t\mathbf{i} - t\mathbf{j}$

     $\mathbf{F} \cdot \mathbf{r}' = -2t + 2t = 0$

Thus, $\displaystyle\int_C \mathbf{F} \cdot d\mathbf{r} = 0.$

41. $\mathbf{F}(x, y) = (x^3 - 2x^2)\mathbf{i} + \left(x - \dfrac{y}{2}\right)\mathbf{j}$

   $C:\ \mathbf{r}(t) = t\mathbf{i} + t^2\mathbf{j}$

     $\mathbf{r}'(t) = \mathbf{i} + 2t\mathbf{j}$

     $\mathbf{F}(t) = (t^3 - 2t^2)\mathbf{i} + \left(t - \dfrac{t^2}{2}\right)\mathbf{j}$

     $\mathbf{F} \cdot \mathbf{r}' = (t^3 - 2t^2) + 2t\left(t - \dfrac{t^2}{2}\right) = 0$

Thus, $\displaystyle\int_C \mathbf{F} \cdot d\mathbf{r} = 0.$

43. $x = 2t,\ y = 10t,\ 0 \le t \le 1 \implies y = 5x$ or $x = \dfrac{y}{5},\ 0 \le y \le 10$

$$\int_C (x + 3y^2)\, dy = \int_0^{10} \left(\frac{y}{5} + 3y^2\right) dy = \left[\frac{y^2}{10} + y^3\right]_0^{10} = 1010$$

45. $x = 2t,\ y = 10t,\ 0 \le t \le 1 \implies x = \dfrac{y}{5},\ 0 \le y \le 10,\ dx = \dfrac{1}{5}\, dy$

$$\int_C xy\, dx + y\, dy = \int_0^{10} \left(\frac{y^2}{25} + y\right) dy = \left[\frac{y^3}{75} + \frac{y^2}{2}\right]_0^{10} = \frac{190}{3} \quad \text{OR}$$

$y = 5x,\ dy = 5\, dx,\ 0 \le x \le 2$

$$\int_C xy\, dx + y\, dy = \int_0^2 (5x^2 + 25x)\, dx = \left[\frac{5x^3}{3} + \frac{25x^2}{2}\right]_0^2 = \frac{190}{3}$$

47. $\mathbf{r}(t) = t\mathbf{i},\ 0 \le t \le 5$

   $x(t) = t,\quad y(t) = 0$

   $dx = dt,\quad dy = 0$

$$\int_C (2x - y)\, dx + (x + 3y)\, dy = \int_0^5 2t\, dt = 25$$

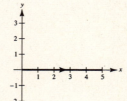

49. $\mathbf{r}(t) = \begin{cases} t\mathbf{i}, & 0 \le t \le 3 \\ 3\mathbf{i} + (t-3)\mathbf{j}, & 3 \le t \le 6 \end{cases}$

C_1: $x(t) = t,\ y(t) = 0,$

  $dx = dt,\ dy = 0$

$$\int_{C_1} (2x - y)\, dx + (x + 3y)\, dy = \int_0^3 2t\, dt = 9$$

C_2: $x(t) = 3,\ y(t) = t - 3$

  $dx = 0,\ dy = dt$

$$\int_{C_2} (2x - y)\, dx + (x + 3y)\, dy = \int_3^6 [3 + 3(t-3)]\, dt = \left[\frac{3t^2}{2} - 6t \right]_3^6 = \frac{45}{2}$$

$$\int_C (2x - y)\, dx + (x + 3y)\, dy = 9 + \frac{45}{2} = \frac{63}{2}$$

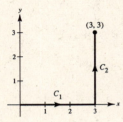

51. $x(t) = t,\ y(t) = 1 - t^2,\ 0 \le t \le 1,\ dx = dt,\ dy = -2t\, dt$

$$\int_C (2x - y)\, dx + (x + 3y)\, dy = \int_0^1 [(2t - 1 + t^2) + (t + 3 - 3t^2)(-2t)]\, dt$$

$$= \int_0^1 (6t^3 - t^2 - 4t - 1)\, dt = \left[\frac{3t^4}{2} - \frac{t^3}{3} - 2t^2 - t \right]_0^1 = -\frac{11}{6}$$

53. $x(t) = t,\ y(t) = 2t^2,\ 0 \le t \le 2$

  $dx = dt,\ dy = 4t\, dt$

$$\int_C (2x - y)\, dx + (x + 3y)\, dy = \int_0^2 (2t - 2t^2)\, dt + (t + 6t^2)4t\, dt$$

$$= \int_0^2 (24t^3 + 2t^2 + 2t)\, dt = \left[6t^4 + \frac{2}{3}t^3 + t^2 \right]_0^2 = \frac{316}{3}$$

55. $f(x, y) = h$

  C: line from $(0, 0)$ to $(3, 4)$

$\mathbf{r} = 3t\mathbf{i} + 4t\mathbf{j},\ 0 \le t \le 1$

$\mathbf{r}'(t) = 3\mathbf{i} + 4\mathbf{j}$

$\|\mathbf{r}'(t)\| = 5$

Lateral surface area:

$$\int_C f(x, y)\, ds = \int_0^1 5h\, dt = 5h$$

57. $f(x, y) = xy$

  C: $x^2 + y^2 = 1$ from $(1, 0)$ to $(0, 1)$

$\mathbf{r}(t) = \cos t\mathbf{i} + \sin t\mathbf{j},\ 0 \le t \le \dfrac{\pi}{2}$

$\mathbf{r}'(t) = -\sin t\mathbf{i} + \cos t\mathbf{j}$

$\|\mathbf{r}'(t)\| = 1$

Lateral surface area:

$$\int_C f(x, y)\, ds = \int_0^{\pi/2} \cos t \sin t\, dt$$

$$= \left[\frac{\sin^2 t}{2} \right]_0^{\pi/2} = \frac{1}{2}$$

59. $f(x, y) = h$

C: $y = 1 - x^2$ from $(1, 0)$ to $(0, 1)$

$$\mathbf{r}(t) = (1 - t)\mathbf{i} + [1 - (1 - t)^2]\mathbf{j}, \ 0 \le t \le 1$$

$$\mathbf{r}'(t) = -\mathbf{i} + 2(1 - t)\mathbf{j}$$

$$\|\mathbf{r}'(t)\| = \sqrt{1 + 4(1 - t)^2}$$

Lateral surface area:

$$\int_C f(x, y) \, ds = \int_0^1 h\sqrt{1 + 4(1 - t)^2} \, dt$$

$$= -\frac{h}{4}\left[2(1 - t)\sqrt{1 + 4(1 - t)^2} + \ln|2(1 - t) + \sqrt{1 + 4(1 - t)^2}\,|\right]_0^1$$

$$= \frac{h}{4}\left[2\sqrt{5} + \ln\left(2 + \sqrt{5}\right)\right] \approx 1.4789h$$

61. $f(x, y) = xy$

C: $y = 1 - x^2$ from $(1, 0)$ to $(0, 1)$

You could parameterize the curve C as in Exercises 59 and 60. Alternatively, let $x = \cos t$, then:

$$y = 1 - \cos^2 t = \sin^2 t$$

$$\mathbf{r}(t) = \cos t\mathbf{i} + \sin^2 t\mathbf{j}, \ 0 \le t \le \frac{\pi}{2}$$

$$\mathbf{r}'(t) = -\sin t\mathbf{i} + 2 \sin t \cos t\mathbf{j}$$

$$\|\mathbf{r}'(t)\| = \sqrt{\sin^2 t + 4 \sin^2 t \cos^2 t} = \sin t\sqrt{1 + 4 \cos^2 t}$$

Lateral surface area:

$$\int_C f(x, y) \, ds = \int_0^{\pi/2} \cos t \sin^2 t\left(\sin t\sqrt{1 + 4 \cos^2 t}\right) dt = \int_0^{\pi/2} \sin^2 t[(1 + 4 \cos^2 t)^{1/2} \sin t \cos t] \, dt$$

Let $u = \sin^2 t$ and $dv = (1 + 4 \cos^2 t)^{1/2} \sin t \cos t$, then $du = 2 \sin t \cos t \, dt$ and $v = -\frac{1}{12}(1 + 4 \cos^2 t)^{3/2}$.

$$\int_C f(x, y) \, ds = \left[-\frac{1}{12} \sin^2 t(1 + 4 \cos^2 t)^{3/2}\right]_0^{\pi/2} + \frac{1}{6}\int_0^{\pi/2} (1 + 4 \cos^2 t)^{3/2} \sin t \cos t \, dt$$

$$= \left[-\frac{1}{12} \sin^2 t(1 + 4 \cos^2 t)^{3/2} - \frac{1}{120}(1 + 4 \cos^2 t)^{5/2}\right]_0^{\pi/2}$$

$$= \left(-\frac{1}{12} - \frac{1}{120}\right) + \frac{1}{120}(5)^{5/2} = \frac{1}{120}\left(25\sqrt{5} - 11\right) \approx 0.3742$$

63. (a) $f(x, y) = 1 + y^2$

$$\mathbf{r}(t) = 2 \cos t\mathbf{i} + 2 \sin t\mathbf{j}, \ 0 \le t \le 2\pi$$

$$\mathbf{r}'(t) = -2 \sin t\mathbf{i} + 2 \cos t\mathbf{j}$$

$$\|\mathbf{r}'(t)\| = 2$$

$$S = \int_C f(x, y) \, ds = \int_0^{2\pi} (1 + 4 \sin^2 t)(2) \, dt$$

$$= \left[2t + 4(t - \sin t \cos t)\right]_0^{2\pi} = 12\pi \approx 37.70 \text{ cm}^2$$

(b) $0.2(12\pi) = \dfrac{12\pi}{5} \approx 7.54 \text{ cm}^3$

(c)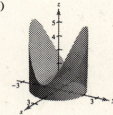

65. $S \approx 25$

Matches b

67. (a) Graph of: $\mathbf{r}(t) = 3 \cos t \mathbf{i} + 3 \sin t \mathbf{j} + (1 + \sin^2 2t)\mathbf{k}, \ 0 \leq t \leq 2\pi$

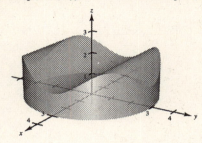

(b) Consider the portion of the surface in the first quadrant. The curve $z = 1 + \sin^2 2t$ is over the curve $\mathbf{r}_1(t) = 3 \cos t \mathbf{i} + 3 \sin t \mathbf{j}, \ 0 \leq t \leq \pi/2$. Hence, the total lateral surface area is

$$4 \int_C f(x, y) \, ds = 4 \int_0^{\pi/2} (1 + \sin^2 2t)3 \, dt = 12\left(\frac{3\pi}{4}\right) = 9\pi \text{ sq. cm}$$

(c) The cross sections parallel to the xz-plane are rectangles of height $1 + 4(y/3)^2(1 - y^2/9)$ and base $2\sqrt{9 - y^2}$. Hence,

$$\text{Volume} = 2 \int_0^3 2\sqrt{9 - y^2}\left(1 + 4\frac{y^2}{9}\left(1 - \frac{y^2}{9}\right)\right) dy = \frac{27\pi}{2} \approx 42.412 \text{ cm}^3$$

69. See the definition of Line Integral, page 1020.

See Theorem 14.4.

71. The greater the height of the surface over the curve, the greater the lateral surface area. Hence,

$$z_3 < z_1 < z_2 < z_4.$$

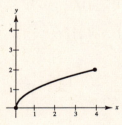

73. False

$$\int_C xy \, ds = \sqrt{2} \int_0^1 t^2 \, dt$$

75. False, the orientations are different.

Section 14.3 Conservative Vector Fields and Independence of Path

1. $\mathbf{F}(x, y) = x^2\mathbf{i} + xy\mathbf{j}$

(a) $\mathbf{r}_1(t) = t\mathbf{i} + t^2\mathbf{j}$, $0 \le t \le 1$

$\mathbf{r}_1'(t) = \mathbf{i} + 2t\mathbf{j}$

$\mathbf{F}(t) = t^2\mathbf{i} + t^3\mathbf{j}$

$\displaystyle\int_C \mathbf{F} \cdot d\mathbf{r} = \int_0^1 (t^2 + 2t^4)\, dt = \frac{11}{15}$

(b) $\mathbf{r}_2(\theta) = \sin\theta\mathbf{i} + \sin^2\theta\mathbf{j}$, $0 \le \theta \le \dfrac{\pi}{2}$

$\mathbf{r}_2'(\theta) = \cos\theta\mathbf{i} + 2\sin\theta\cos\theta\mathbf{j}$

$\mathbf{F}(t) = \sin^2\theta\mathbf{i} + \sin^3\theta\mathbf{j}$

$\displaystyle\int_C \mathbf{F} \cdot d\mathbf{r} = \int_0^{\pi/2} (\sin^2\theta\cos\theta + 2\sin^4\theta\cos\theta)\, d\theta$

$\displaystyle = \left[\frac{\sin^3\theta}{3} + \frac{2\sin^5\theta}{5}\right]_0^{\pi/2} = \frac{11}{15}$

3. $\mathbf{F}(x, y) = y\mathbf{i} - x\mathbf{j}$

(a) $\mathbf{r}_1(\theta) = \sec\theta\mathbf{i} + \tan\theta\mathbf{j}$, $0 \le \theta \le \dfrac{\pi}{3}$

$\mathbf{r}_1'(\theta) = \sec\theta\tan\theta\mathbf{i} + \sec^2\theta\mathbf{j}$

$\mathbf{F}(\theta) = \tan\theta\mathbf{i} - \sec\theta\mathbf{j}$

$\displaystyle\int_C \mathbf{F} \cdot d\mathbf{r} = \int_0^{\pi/3} (\sec\theta\tan^2\theta - \sec^3\theta)\, d\theta = \int_0^{\pi/3} [\sec\theta(\sec^2\theta - 1) - \sec^3\theta]\, d\theta$

$\displaystyle = -\int_0^{\pi/3} \sec\theta\, d\theta = \left[-\ln|\sec\theta + \tan\theta|\right]_0^{\pi/3} = -\ln(2 + \sqrt{3}) \approx -1.317$

(b) $\mathbf{r}_2(t) = \sqrt{t+1}\,\mathbf{i} + \sqrt{t}\,\mathbf{j}$, $0 \le t \le 3$

$\mathbf{r}_2'(t) = \dfrac{1}{2\sqrt{t+1}}\mathbf{i} + \dfrac{1}{2\sqrt{t}}\mathbf{j}$

$\mathbf{F}(t) = \sqrt{t}\,\mathbf{i} - \sqrt{t+1}\,\mathbf{j}$

$\displaystyle\int_C \mathbf{F} \cdot d\mathbf{r} = \int_0^3 \left[\frac{\sqrt{t}}{2\sqrt{t+1}} - \frac{\sqrt{t+1}}{2\sqrt{t}}\right] dt = -\frac{1}{2}\int_0^3 \frac{1}{\sqrt{t}\sqrt{t+1}}\, dt = -\frac{1}{2}\int_0^3 \frac{1}{\sqrt{t^2 + t + (1/4) - (1/4)}}\, dt$

$\displaystyle = -\frac{1}{2}\int_0^3 \frac{1}{\sqrt{[t + (1/2)]^2 - (1/4)}}\, dt = \left[-\frac{1}{2}\ln\left|\left(t + \frac{1}{2}\right) + \sqrt{t^2 + t}\right|\right]_0^3$

$\displaystyle = -\frac{1}{2}\left[\ln\left(\frac{7}{2} + 2\sqrt{3}\right) - \ln\left(\frac{1}{2}\right)\right] = -\frac{1}{2}\ln(7 + 4\sqrt{3}) \approx -1.317$

5. $\mathbf{F}(x, y) = e^x\sin y\mathbf{i} + e^x\cos y\mathbf{j}$

$\dfrac{\partial N}{\partial x} = e^x\cos y \qquad \dfrac{\partial M}{\partial y} = e^x\cos y$

Since $\dfrac{\partial N}{\partial x} = \dfrac{\partial M}{\partial y}$, $\mathbf{F}$ is conservative.

7. $\mathbf{F}(x, y) = \dfrac{1}{y}\mathbf{i} + \dfrac{x}{y^2}\mathbf{j}$

$\dfrac{\partial N}{\partial x} = \dfrac{1}{y^2} \qquad \dfrac{\partial M}{\partial y} = -\dfrac{1}{y^2}$

Since $\dfrac{\partial N}{\partial x} \ne \dfrac{\partial M}{\partial y}$, $\mathbf{F}$ is not conservative.

9. $\mathbf{F}(x, y, z) = y^2z\mathbf{i} + 2xyz\mathbf{j} + xy^2\mathbf{k}$

$\operatorname{curl}\mathbf{F} = \mathbf{0} \implies \mathbf{F}$ is conservative.

11. $F(x, y) = 2xy\mathbf{i} + x^2\mathbf{j}$

(a) $r_1(t) = t\mathbf{i} + t^2\mathbf{j},\ 0 \le t \le 1$

$r_1'(t) = \mathbf{i} + 2t\mathbf{j}$

$F(t) = 2t^3\mathbf{i} + t^2\mathbf{j}$

$\displaystyle\int_C F \cdot d\mathbf{r} = \int_0^1 4t^3\,dt = 1$

(b) $r_2(t) = t\mathbf{i} + t^3\mathbf{j},\ 0 \le t \le 1$

$r_2'(t) = \mathbf{i} + 3t^2\mathbf{j}$

$F(t) = 2t^4\mathbf{i} + t^2\mathbf{j}$

$\displaystyle\int_C F \cdot d\mathbf{r} = \int_0^1 5t^4\,dt = 1$

13. $F(x, y) = y\mathbf{i} - x\mathbf{j}$

(a) $r_1(t) = t\mathbf{i} + t\mathbf{j},\ 0 \le t \le 1$

$r_1'(t) = \mathbf{i} + \mathbf{j}$

$F(t) = t\mathbf{i} - t\mathbf{j}$

$\displaystyle\int_C F \cdot d\mathbf{r} = 0$

(b) $r_2(t) = t\mathbf{i} + t^2\mathbf{j},\ 0 \le t \le 1$

$r_2'(t) = \mathbf{i} + 2t\mathbf{j}$

$F(t) = t^2\mathbf{i} - t\mathbf{j}$

$\displaystyle\int_C F \cdot d\mathbf{r} = \int_0^1 -t^2\,dt = -\frac{1}{3}$

(c) $r_3(t) = t\mathbf{i} + t^3\mathbf{j},\ 0 \le t \le 1$

$r_3'(t) = \mathbf{i} + 3t^2\mathbf{j}$

$F(t) = t^3\mathbf{i} - t\mathbf{j}$

$\displaystyle\int_C F \cdot d\mathbf{r} = \int_0^1 -2t^3\,dt = -\frac{1}{2}$

15. $\displaystyle\int_C y^2\,dx + 2xy\,dy$

Since $\partial M/\partial y = \partial N/\partial x = 2y$, $F(x, y) = y^2\mathbf{i} + 2xy\mathbf{j}$ is conservative. The potential function is $f(x, y) = xy^2 + k$. Therefore, we can use the Fundamental Theorem of Line Integrals.

(a) $\displaystyle\int_C y^2\,dx + 2xy\,dy = \left[x^2y\right]_{(0,0)}^{(4,4)} = 64$

(b) $\displaystyle\int_C y^2\,dx + 2xy\,dy = \left[x^2y\right]_{(-1,0)}^{(1,0)} = 0$

(c) and (d) Since C is a closed curve, $\displaystyle\int_C y^2\,dx + 2xy\,dy = 0$.

17. $\displaystyle\int_C 2xy\,dx + (x^2 + y^2)\,dy$

Since $\partial M/\partial y = \partial N/\partial x = 2x$,

$\qquad F(x, y) = 2xy\mathbf{i} + (x^2 + y^2)\mathbf{j}$ is conservative.

The potential function is $f(x, y) = x^2y + \dfrac{y^3}{3} + k$.

(a) $\displaystyle\int_C 2xy\,dx + (x^2 + y^2)\,dy = \left[x^2y + \frac{y^3}{3}\right]_{(5,0)}^{(0,4)} = \frac{64}{3}$

(b) $\displaystyle\int_C 2xy\,dx + (x^2 + y^2)\,dy = \left[x^2y + \frac{y^3}{3}\right]_{(2,0)}^{(0,4)} = \frac{64}{3}$

19. $F(x, y, z) = yz\mathbf{i} + xz\mathbf{j} + xy\mathbf{k}$

Since **curl** $F = 0$, $F(x, y, z)$ is conservative. The potential function is $f(x, y, z) = xyz + k$.

(a) $r_1(t) = t\mathbf{i} + 2\mathbf{j} + t\mathbf{k},\ 0 \le t \le 4$

$\displaystyle\int_C F \cdot d\mathbf{r} = \left[xyz\right]_{(0,2,0)}^{(4,2,4)} = 32$

(b) $r_2(t) = t^2\mathbf{i} + t\mathbf{j} + t^2\mathbf{k},\ 0 \le t \le 2$

$\displaystyle\int_C F \cdot d\mathbf{r} = \left[xyz\right]_{(0,0,0)}^{(4,2,4)} = 32$

21. $F(x, y, z) = (2y + x)\mathbf{i} + (x^2 - z)\mathbf{j} + (2y - 4z)\mathbf{k}$

$F(x, y, z)$ is not conservative.

(a) $r_1(t) = t\mathbf{i} + t^2\mathbf{j} + \mathbf{k},\ 0 \le t \le 1$

$r_1'(t) = \mathbf{i} + 2t\mathbf{j}$

$F(t) = (2t^2 + t)\mathbf{i} + (t^2 - 1)\mathbf{j} + (2t^2 - 4)\mathbf{k}$

$\displaystyle\int_C F \cdot d\mathbf{r} = \int_0^1 (2t^3 + 2t^2 - t)\,dt = \frac{2}{3}$

—CONTINUED—

21. **—CONTINUED—**

(b) $\mathbf{r}_2(t) = t\mathbf{i} + t\mathbf{j} + (2t - 1)^2\mathbf{k},\ 0 \le t \le 1$

$\mathbf{r}_2'(t) = \mathbf{i} + \mathbf{j} + 4(2t - 1)\mathbf{k}$

$\mathbf{F}(t) = 3t\mathbf{i} + [t^2 - (2t - 1)^2]\mathbf{j} + [2t - 4(2t - 1)^2]\mathbf{k}$

$$\int_C \mathbf{F} \cdot d\mathbf{r} = \int_0^1 [3t + t^2 - (2t - 1)^2 + 8t(2t - 1) - 16(2t - 1)^3]\,dt$$

$$= \int_0^1 [17t^2 - 5t - (2t - 1)^2 - 16(2t - 1)^3]\,dt = \left[\frac{17t^3}{3} - \frac{5t^2}{2} - \frac{(2t - 1)^3}{6} - 2(2t - 1)^4\right]_0^1 = \frac{17}{6}$$

23. $\mathbf{F}(x, y, z) = e^z(y\mathbf{i} + x\mathbf{j} + xy\mathbf{k})$

$\mathbf{F}(x, y, z)$ is conservative. The potential function is
$f(x, y, z) = xye^z + k.$

(a) $\mathbf{r}_1(t) = 4\cos t\mathbf{i} + 4\sin t\mathbf{j} + 3\mathbf{k},\ 0 \le t \le \pi$

$$\int_C \mathbf{F} \cdot d\mathbf{r} = \left[xye^z\right]_{(4,\,0,\,3)}^{(-4,\,0,\,3)} = 0$$

(b) $\mathbf{r}_2(t) = (4 - 8t)\mathbf{i} + 3\mathbf{k},\ 0 \le t \le 1$

$$\int_C \mathbf{F} \cdot d\mathbf{r} = \left[xye^z\right]_{(4,\,0,\,3)}^{(-4,\,0,\,3)} = 0$$

25. $\displaystyle\int_C (y\mathbf{i} + x\mathbf{j}) \cdot d\mathbf{r} = \left[xy\right]_{(0,\,0)}^{(3,\,8)} = 24$

27. $\displaystyle\int_C \cos x \sin y\,dx + \sin x \cos y\,dy = \left[\sin x \sin y\right]_{(0,\,-\pi)}^{(3\pi/2,\,\pi/2)} = -1$

29. $\displaystyle\int_C e^x \sin y\,dx + e^x \cos y\,dy = \left[e^x \sin y\right]_{(0,\,0)}^{(2\pi,\,0)} = 0$

31. $\displaystyle\int_C (y + 2z)\,dx + (x - 3z)\,dy + (2x - 3y)\,dz$

$\mathbf{F}(x, y, z)$ is conservative and the potential function is $f(x, y, z) = xy - 3yz + 2xz.$

(a) $\left[xy - 3yz + 2xz\right]_{(0,\,0,\,0)}^{(1,\,1,\,1)} = 0 - 0 = 0$

(b) $\left[xy - 3yz + 2xz\right]_{(0,\,0,\,0)}^{(0,\,0,\,1)} + \left[xy - 3yz + 2xz\right]_{(0,\,0,\,1)}^{(1,\,1,\,1)} = 0 + 0 = 0$

(c) $\left[xy - 3yz + 2xz\right]_{(0,\,0,\,0)}^{(1,\,0,\,0)} + \left[xy - 3yz + 2xz\right]_{(1,\,0,\,0)}^{(1,\,1,\,0)} + \left[xy - 3yz + 2xz\right]_{(1,\,1,\,0)}^{(1,\,1,\,1)} = 0 + 1 + (-1) = 0$

33. $\displaystyle\int_C -\sin x\,dx + z\,dy + y\,dz = \left[\cos x + yz\right]_{(0,\,0,\,0)}^{(\pi/2,\,3,\,4)} = 12 - 1 = 11$

35. $\mathbf{F}(x, y) = 9x^2y^2\mathbf{i} + (6x^3y - 1)\mathbf{j}$ is conservative.

Work $= \left[3x^3y^2 - y\right]_{(0,\,0)}^{(5,\,9)} = 30{,}366$

37. $\mathbf{r}(t) = 2 \cos 2\pi t \mathbf{i} + 2 \sin 2\pi t \mathbf{j}$

$\mathbf{r}'(t) = -4\pi \sin 2\pi t \mathbf{i} + 4\pi \cos 2\pi t \mathbf{j}$

$\mathbf{a}(t) = -8\pi^2 \cos 2\pi t \mathbf{i} - 8\pi^2 \sin 2\pi t \mathbf{j}$

$\mathbf{F}(t) = m\mathbf{a}(t) = \dfrac{1}{32}\mathbf{a}(t) = -\dfrac{\pi^2}{4}(\cos 2\pi t \mathbf{i} + \sin 2\pi t \mathbf{j})$

$W = \displaystyle\int_C \mathbf{F} \cdot d\mathbf{r} = \int_C -\dfrac{\pi^2}{4}(\cos 2\pi t \mathbf{i} + \sin 2\pi t \mathbf{j}) \cdot 4\pi(-\sin 2\pi t \mathbf{i} + \cos 2\pi t \mathbf{j})\, dt = -\pi^3 \int_C 0\, dt = 0$

39. Since the sum of the potential and kinetic energies remains constant from point to point, if the kinetic energy is decreasing at a rate of 10 units per minute, then the potential energy is increasing at a rate of 10 units per minute.

41. No. The force field is conservative.   **43.** See Theorem 14.5, page 1033.

45. (a) The direct path along the line segment joining $(-4, 0)$ to $(3, 4)$ requires less work than the path going from $(-4, 0)$ to $(-4, 4)$ and then to $(3, 4)$.

(b) The closed curve given by the line segments joining $(-4, 0)$, $(-4, 4)$, $(3, 4)$, and $(-4, 0)$ satisfies $\displaystyle\int_C \mathbf{F} \cdot d\mathbf{r} \neq 0$.

47. False, it would be true if $\mathbf{F}$ were conservative.   **49.** True

51. Let

$\mathbf{F} = M\mathbf{i} + N\mathbf{j} = \dfrac{\partial f}{\partial y}\mathbf{i} - \dfrac{\partial f}{\partial x}\mathbf{j}.$

Then $\dfrac{\partial M}{\partial y} = \dfrac{\partial}{\partial y}\left(\dfrac{\partial f}{\partial y}\right) = \dfrac{\partial^2 f}{\partial y^2}$ and $\dfrac{\partial N}{\partial x} = \dfrac{\partial}{\partial x}\left(-\dfrac{\partial f}{\partial x}\right) = -\dfrac{\partial^2 f}{\partial x^2}.$ Since

$\dfrac{\partial^2 f}{\partial x^2} + \dfrac{\partial^2 f}{\partial y^2} = 0$ we have $\dfrac{\partial M}{\partial y} = \dfrac{\partial N}{\partial x}.$

Thus, $\mathbf{F}$ is conservative. Therefore, by Theorem 14.7, we have

$\displaystyle\int_C \left(\dfrac{\partial f}{\partial y}\, dx - \dfrac{\partial f}{\partial x}\, dy\right) = \int_C (M\, dx + N\, dy) = \int_C \mathbf{F} \cdot d\mathbf{r} = 0$

for every closed curve in the plane.

Section 14.4   Green's Theorem

1. $\mathbf{r}(t) = \begin{cases} t\mathbf{i}, & 0 \leq t \leq 4 \\ 4\mathbf{i} + (t-4)\mathbf{j}, & 4 \leq t \leq 8 \\ (12-t)\mathbf{i} + 4\mathbf{j}, & 8 \leq t \leq 12 \\ (16-t)\mathbf{j}, & 12 \leq t \leq 16 \end{cases}$

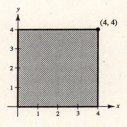

$\displaystyle\int_C y^2\, dx + x^2\, dy = \int_0^4 [0\, dt + t^2(0)] + \int_4^8 [(t-4)^2(0) + 16\, dt]$

$+ \displaystyle\int_8^{12} [16(-dt) + (12-t)^2(0)] + \int_{12}^{16} [(16-t)^2(0) + 0(-dt)]$

$= 0 + 64 - 64 + 0 = 0$

By Green's Theorem, $\displaystyle\iint_R \left(\dfrac{\partial N}{\partial x} - \dfrac{\partial M}{\partial y}\right) dA = \int_0^4 \int_0^4 (2x - 2y)\, dy\, dx = \int_0^4 (8x - 16)\, dx = 0.$

3. $\mathbf{r}(t) = \begin{cases} t\mathbf{i} + t^2/4\mathbf{j}, & 0 \le t \le 4 \\ (8-t)\mathbf{i} + (8-t)\mathbf{j}, & 4 \le t \le 8 \end{cases}$

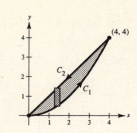

$$\int_C y^2\,dx + x^2\,dy = \int_0^4 \left[\frac{t^4}{16}(dt) + t^2\left(\frac{t}{2}\,dt\right)\right] + \int_4^8 \left[(8-t)^2(-dt) + (8-t)^2(-dt)\right]$$

$$= \int_0^4 \left[\frac{t^4}{16} + \frac{t^3}{2}\right]dt + \int_4^8 -2(8-t)^2\,dt = \frac{224}{5} - \frac{128}{3} = \frac{32}{15}$$

By Green's Theorem,

$$\iint_R \left(\frac{\partial N}{\partial x} - \frac{\partial M}{\partial y}\right)dA = \int_0^4\int_{x^2/4}^x (2x - 2y)\,dy\,dx = \int_0^4 \left(x^2 - \frac{x^3}{2} + \frac{x^4}{16}\right)dx = \frac{32}{15}.$$

5. $C: x^2 + y^2 = 4$

Let $x = 2\cos t$ and $y = 2\sin t$, $0 \le t \le 2\pi$.

$$\int_C xe^y\,dx + e^x\,dy = \int_0^{2\pi} \left[2\cos t\, e^{2\sin t}(-2\sin t) + e^{2\cos t}(2\cos t)\right]dt \approx 19.99$$

$$\iint_R \left(\frac{\partial N}{\partial x} - \frac{\partial M}{\partial y}\right)dA = \int_{-2}^2\int_{-\sqrt{4-x^2}}^{\sqrt{4-x^2}} (e^x - xe^y)\,dy\,dx = \int_{-2}^2 \left[2\sqrt{4-x^2}\,e^x - xe^{\sqrt{4-x^2}} + xe^{-\sqrt{4-x^2}}\right]dx \approx 19.99$$

In Exercises 7 and 9, $\dfrac{\partial N}{\partial x} - \dfrac{\partial M}{\partial y} = 1$.

7. $\displaystyle\int_C (y-x)\,dx + (2x-y)\,dy = \int_0^2\int_{x^2-x}^x dy\,dx$

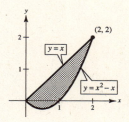

$$= \int_0^2 (2x - x^2)\,dx$$

$$= \frac{4}{3}$$

9. From the accompanying figure, we see that R is the shaded region. Thus, Green's Theorem yields

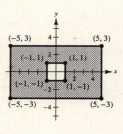

$$\int_C (y-x)\,dx + (2x-y)\,dy = \iint_R 1\,dA$$

$$= \text{Area of } R$$

$$= 6(10) - 2(2)$$

$$= 56.$$

11. Since the curves $y = 0$ and $y = 4 - x^2$ intersect at $(-2, 0)$ and $(2, 0)$, Green's Theorem yields

$$\int_C 2xy\,dx + (x+y)\,dy = \iint_R (1 - 2x)\,dA = \int_{-2}^2\int_0^{4-x^2} (1 - 2x)\,dy\,dx$$

$$= \int_{-2}^2 \left[y - 2xy\right]_0^{4-x^2} dx$$

$$= \int_{-2}^2 (4 - 8x - x^2 + 2x^3)\,dx$$

$$= \left[4x - 4x^2 - \frac{x^3}{3} + \frac{x^4}{2}\right]_{-2}^2$$

$$= -\frac{8}{3} - \frac{8}{3} + 16 = \frac{32}{3}.$$

13. Since R is the interior of the circle $x^2 + y^2 = a^2$, Green's Theorem yields

$$\int_C (x^2 - y^2)\, dx + 2xy\, dy = \iint_R (2y + 2y)\, dA$$

$$= \int_{-a}^{a} \int_{-\sqrt{a^2 - x^2}}^{\sqrt{a^2 - x^2}} 4y\, dy\, dx = 4 \int_{-a}^{a} 0\, dx = 0.$$

15. Since $\dfrac{\partial M}{\partial y} = \dfrac{2x}{x^2 + y^2} = \dfrac{\partial N}{\partial x}$,

we have path independence and

$$\iint_R \left(\frac{\partial N}{\partial x} - \frac{\partial M}{\partial y} \right) dA = 0.$$

17. By Green's Theorem,

$$\int_C \sin x \cos y\, dx + (xy + \cos x \sin y)\, dy = \iint_R [(y - \sin x \sin y) - (-\sin x \sin y)]\, dA$$

$$= \int_0^1 \int_x^{\sqrt{x}} y\, dy\, dx = \frac{1}{2} \int_0^1 (x - x^2)\, dx = \frac{1}{2} \left[\frac{x^2}{2} - \frac{x^3}{3} \right]_0^1 = \frac{1}{12}.$$

19. By Green's Theorem,

$$\int_C xy\, dx + (x + y)\, dy = \iint_R (1 - x)\, dA$$

$$= \int_0^{2\pi} \int_1^3 (1 - r \cos \theta)r\, dr\, d\theta = \int_0^{2\pi} \left(4 - \frac{26}{3} \cos \theta \right) d\theta = 8\pi.$$

21. $\mathbf{F}(x, y) = xy\mathbf{i} + (x + y)\mathbf{j}$

$C\colon x^2 + y^2 = 4$

$$\text{Work} = \int_C xy\, dx + (x + y)\, dy = \iint_R (1 - x)\, dA = \int_0^{2\pi} \int_0^2 (1 - r \cos \theta)r\, dr\, d\theta = \int_0^{2\pi} \left(2 - \frac{8}{3} \cos \theta \right) d\theta = 4\pi$$

23. $\mathbf{F}(x, y) = (x^{3/2} - 3y)\mathbf{i} + (6x + 5\sqrt{y})\mathbf{j}$

$C\colon$ boundary of the triangle with vertices $(0, 0)$, $(5, 0)$, $(0, 5)$

$$\text{Work} = \int_C (x^{3/2} - 3y)\, dx + \left(6x + 5\sqrt{y} \right) dy = \iint_R 9\, dA = 9\left(\tfrac{1}{2}\right)(5)(5) = \tfrac{225}{2}$$

25. $C\colon$ let $x = a \cos t$, $y = a \sin t$, $0 \le t \le 2\pi$. By Theorem 14.9, we have

$$A = \frac{1}{2} \int_C x\, dy - y\, dx = \frac{1}{2} \int_0^{2\pi} [a \cos t(a \cos t) - a \sin t(-a \sin t)]\, dt = \frac{1}{2} \int_0^{2\pi} a^2\, dt = \left[\frac{a^2}{2} t \right]_0^{2\pi} = \pi a^2.$$

27. From the accompanying figure we see that

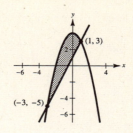

C_1: $y = 2x + 1$, $dy = 2\,dx$

C_2: $y = 4 - x^2$, $dy = -2x\,dx$.

Thus, by Theorem 14.9, we have

$$A = \frac{1}{2}\int_{-3}^{1} [x(2) - (2x + 1)]\,dx + \frac{1}{2}\int_{1}^{-3} [x(-2x) - (4 - x^2)]\,dx$$

$$= \frac{1}{2}\int_{-3}^{1} (-1)\,dx + \frac{1}{2}\int_{1}^{-3} (-x^2 - 4)\,dx$$

$$= \frac{1}{2}\int_{-3}^{1} (-1)\,dx + \frac{1}{2}\int_{-3}^{1} (x^2 + 4)\,dx = \frac{1}{2}\int_{-3}^{1} (3 + x^2)\,dx = \frac{1}{2}\left[3x + \frac{x^3}{3}\right]_{-3}^{1} = \frac{32}{3}.$$

29. See Theorem 14.8, page 1042.

31. Answers will vary.

$$\mathbf{F}_1(x, y) = y\mathbf{i} + x\mathbf{j}$$

$$\mathbf{F}_2(x, y) = x^2\mathbf{i} + y^2\mathbf{j}$$

$$\mathbf{F}_3(x, y) = 2xy\mathbf{i} + x^2\mathbf{j}$$

33. $A = \displaystyle\int_{-2}^{2} (4 - x^2)\,dx = \left[4x - \frac{x^3}{3}\right]_{-2}^{2} = \frac{32}{3}$

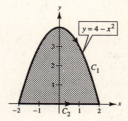

$$\bar{x} = \frac{1}{2A}\int_{C_1} x^2\,dy + \frac{1}{2A}\int_{C_2} x^2\,dy$$

For C_1, $dy = -2x\,dx$ and for C_2, $dy = 0$. Thus,

$$\bar{x} = \frac{1}{2(32/3)}\int_{2}^{-2} x^2(-2x\,dx) = \left[\frac{3}{64}\left(-\frac{x^4}{2}\right)\right]_{2}^{-2} = 0.$$

To calculate $\bar{y}$, note that $y = 0$ along C_2. Thus,

$$\bar{y} = \frac{-1}{2(32/3)}\int_{2}^{-2} (4 - x^2)^2\,dx = \frac{3}{64}\int_{-2}^{2} (16 - 8x^2 + x^4)\,dx = \frac{3}{64}\left[16x - \frac{8x^3}{3} + \frac{x^5}{5}\right]_{-2}^{2} = \frac{8}{5}.$$

$$(\bar{x}, \bar{y}) = \left(0, \frac{8}{5}\right)$$

35. Since $A = \displaystyle\int_{0}^{1} (x - x^3)\,dx = \left[\frac{x^2}{2} - \frac{x^4}{4}\right]_{0}^{1} = \frac{1}{4}$, we have $\dfrac{1}{2A} = 2$. On C_1 we have $y = x^3$, $dy = 3x^2\,dx$ and on C_2 we have

$y = x$, $dy = dx$. Thus,

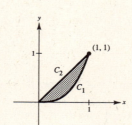

$$\bar{x} = 2\int_{C} x^2\,dy = 2\int_{C_1} x^2(3x^2\,dx) + 2\int_{C_2} x^2\,dx$$

$$= 6\int_{0}^{1} x^4\,dx + 2\int_{1}^{0} x^2\,dx = \frac{6}{5} - \frac{2}{3} = \frac{8}{15}$$

$$\bar{y} = -2\int_{C} y^2\,dx$$

$$= -2\int_{0}^{1} x^6\,dx - 2\int_{1}^{0} x^2\,dx = -\frac{2}{7} + \frac{2}{3} = \frac{8}{21}.$$

$$(\bar{x}, \bar{y}) = \left(\frac{8}{15}, \frac{8}{21}\right)$$

37. $A = \dfrac{1}{2}\displaystyle\int_0^{2\pi} a^2(1 - \cos\theta)^2\,d\theta$

$= \dfrac{a^2}{2}\displaystyle\int_0^{2\pi}\left(1 - 2\cos\theta + \dfrac{1}{2} + \dfrac{\cos 2\theta}{2}\right)d\theta = \dfrac{a^2}{2}\left[\dfrac{3\theta}{2} - 2\sin\theta + \dfrac{1}{4}\sin 2\theta\right]_0^{2\pi} = \dfrac{a^2}{2}(3\pi) = \dfrac{3\pi a^2}{2}$

39. In this case the inner loop has domain $\dfrac{2\pi}{3} \le \theta \le \dfrac{4\pi}{3}$. Thus,

$A = \dfrac{1}{2}\displaystyle\int_{2\pi/3}^{4\pi/3}(1 + 4\cos\theta + 4\cos^2\theta)\,d\theta$

$= \dfrac{1}{2}\displaystyle\int_{2\pi/3}^{4\pi/3}(3 + 4\cos\theta + 2\cos 2\theta)\,d\theta = \dfrac{1}{2}\left[3\theta + 4\sin\theta + \sin 2\theta\right]_{2\pi/3}^{4\pi/3} = \pi - \dfrac{3\sqrt{3}}{2}.$

41. $I = \displaystyle\int_C \dfrac{y\,dx - x\,dy}{x^2 + y^2}$

(a) Let $\mathbf{F} = \dfrac{y}{x^2 + y^2}\mathbf{i} - \dfrac{x}{x^2 + y^2}\mathbf{j}.$

$\mathbf{F}$ is conservative since $\dfrac{\partial N}{\partial x} = \dfrac{\partial M}{\partial y} = \dfrac{x^2 - y^2}{(x^2 + y^2)^2}.$

$\mathbf{F}$ is defined and has continuous first partials everywhere except at the origin. If C is a circle (a closed path) that does not contain the origin, then

$\displaystyle\int_C \mathbf{F}\cdot d\mathbf{r} = \int_C M\,dx + N\,dy = \iint_R\left(\dfrac{\partial N}{\partial x} - \dfrac{\partial M}{\partial y}\right)dA = 0.$

(b) Let $\mathbf{r} = a\cos t\,\mathbf{i} - a\sin t\,\mathbf{j},\ 0 \le t \le 2\pi$ be a circle C_1 oriented clockwise inside C (see figure). Introduce line segments C_2 and C_3 as illustrated in Example 6 of this section in the text. For the region inside C and outside C_1, Green's Theorem applies. Note that since C_2 and C_3 have opposite orientations, the line integrals over them cancel. Thus, $C_4 = C_1 + C_2 + C + C_3$ and

$\displaystyle\int_{C_4}\mathbf{F}\cdot d\mathbf{r} = \int_{C_1}\mathbf{F}\cdot d\mathbf{r} + \int_C \mathbf{F}\cdot d\mathbf{r} = 0.$

But,

$\displaystyle\int_{C_1}\mathbf{F}\cdot d\mathbf{r} = \int_0^{2\pi}\left[\dfrac{(-a\sin t)(-a\sin t)}{a^2\cos^2 t + a^2\sin^2 t} + \dfrac{(-a\cos t)(-a\cos t)}{a^2\cos^2 t + a^2\sin^2 t}\right]dt$

$= \displaystyle\int_0^{2\pi}(\sin^2 t + \cos^2 t)\,dt = \left[t\right]_0^{2\pi} = 2\pi.$

Finally, $\displaystyle\int_C \mathbf{F}\cdot d\mathbf{r} = -\int_{C_1}\mathbf{F}\cdot d\mathbf{r} = -2\pi.$

Note: If C were orientated clockwise, then the answer would have been 2π.

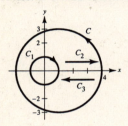

43. Pentagon: $(0, 0), (2, 0), (3, 2), (1, 4), (-1, 1)$

$A = \frac{1}{2}[(0 - 0) + (4 - 0) + (12 - 2) + (1 + 4) + (0 - 0)] = \frac{19}{2}$

45. $\int_C y^n \, dx + x^n \, dy = \int\int_R \left(\frac{\partial N}{\partial x} - \frac{\partial M}{\partial y} \right) dA$

For the line integral, use the two paths

C_1: $\mathbf{r}_1(x) = x\mathbf{i}, \quad -a \le x \le a$

C_2: $\mathbf{r}_2(x) = x\mathbf{i} + \sqrt{a^2 - x^2} \, \mathbf{j}, \quad x = a \text{ to } x = -a$

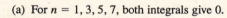

$\int_{C_1} y^n \, dx + x^n \, dy = 0$

$\int_{C_2} y^n \, dx + x^n \, dy = \int_a^{-a} \left[(a^2 - x^2)^{n/2} + x^n \frac{-x}{\sqrt{a^2 - x^2}} \right] dx$

$\int\int_R \left(\frac{\partial N}{\partial x} - \frac{\partial M}{\partial y} \right) dA = \int_{-a}^a \int_0^{\sqrt{a^2 - x^2}} [nx^{n-1} - ny^{n-1}] \, dy \, dx$

(a) For $n = 1, 3, 5, 7$, both integrals give 0.

(b) For n even, you obtain

$n = 2: -\frac{4}{3}a^3 \qquad n = 4: -\frac{16}{15}a^5 \qquad n = 6: -\frac{32}{35}a^7 \qquad n = 8: -\frac{256}{315}a^9$

(c) If n is odd then the integral equals 0.

47. $\int_C (f D_N g - g D_N f) \, ds = \int_C f D_N g \, ds - \int_C g D_N f \, ds$

$= \int\int_R (f\nabla^2 g + \nabla f \cdot \nabla g) \, dA - \int\int_R (g\nabla^2 f + \nabla g \cdot \nabla f) \, dA = \int\int_R (f\nabla^2 g - g\nabla^2 f) \, dA$

49. $\mathbf{F} = M\mathbf{i} + N\mathbf{j}$

$\frac{\partial N}{\partial x} = \frac{\partial M}{\partial y} \implies \frac{\partial N}{\partial x} - \frac{\partial M}{\partial y} = 0.$

$\int_C \mathbf{F} \cdot d\mathbf{r} = \int_C M \, dx + N \, dy = \int\int_R \left(\frac{\partial N}{\partial x} - \frac{\partial M}{\partial y} \right) dA = \int\int_R (0) \, dA = 0$

Section 14.5   Parametric Surfaces

1. $\mathbf{r}(u, v) = u\mathbf{i} + v\mathbf{j} + uv\mathbf{k}$

$z = xy$

Matches c.

3. $\mathbf{r}(u, v) = 2 \cos v \cos u\mathbf{i} + 2 \cos v \sin u\mathbf{j} + 2 \sin v\mathbf{k}$

$x^2 + y^2 + z^2 = 4$

Matches b.

5. $\mathbf{r}(u, v) = u\mathbf{i} + v\mathbf{j} + \frac{v}{2}\mathbf{k}$

$y - 2z = 0$

Plane

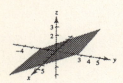

7. $\mathbf{r}(u, v) = 2 \cos u\mathbf{i} + v\mathbf{j} + 2 \sin u\mathbf{k}$

$x^2 + z^2 = 4$

Cylinder

For Exercises 9 and 11,

$$\mathbf{r}(u, v) = u \cos v\mathbf{i} + u \sin v\mathbf{j} + u^2\mathbf{k}, \ 0 \le u \le 2, \ 0 \le v \le 2\pi.$$

Eliminating the parameter yields

$$z = x^2 + y^2, \ 0 \le z \le 4.$$

9. $\mathbf{s}(u, v) = u \cos v\mathbf{i} + u \sin v\mathbf{j} - u^2\mathbf{k}, \ 0 \le u \le 2, \ 0 \le v \le 2\pi$

$z = -(x^2 + y^2)$

The paraboloid is reflected (inverted) through the *xy*-plane.

11. $\mathbf{s}(u, v) = u \cos v\mathbf{i} + u \sin v\mathbf{j} + u^2\mathbf{k}, \ 0 \le u \le 3, \ 0 \le v \le 2\pi$

The height of the paraboloid is increased from 4 to 9.

13. $\mathbf{r}(u, v) = 2u \cos v\mathbf{i} + 2u \sin v\mathbf{j} + u^4\mathbf{k},$

$0 \le u \le 1, \ 0 \le v \le 2\pi$

$z = \dfrac{(x^2 + y^2)^2}{16}$

15. $\mathbf{r}(u, v) = 2 \sinh u \cos v\mathbf{i} + \sinh u \sin v\mathbf{j} + \cosh u\mathbf{k},$

$0 \le u \le 2, \ 0 \le v \le 2\pi$

$\dfrac{z^2}{1} - \dfrac{x^2}{4} - \dfrac{y^2}{1} = 1$

17. $\mathbf{r}(u, v) = (u - \sin u) \cos v\mathbf{i} + (1 - \cos u) \sin v\mathbf{j} + u\mathbf{k},$

$0 \le u \le \pi, \ 0 \le v \le 2\pi$

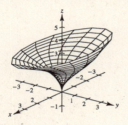

19. $z = y$

$\mathbf{r}(u, v) = u\mathbf{i} + v\mathbf{j} + v\mathbf{k}$

21. $x^2 + y^2 = 16$

$\mathbf{r}(u, v) = 4 \cos u\mathbf{i} + 4 \sin u\mathbf{j} + v\mathbf{k}$

23. $z = x^2$

$\mathbf{r}(u, v) = u\mathbf{i} + v\mathbf{j} + u^2\mathbf{k}$

25. $z = 4$ inside $x^2 + y^2 = 9.$

$\mathbf{r}(u, v) = v \cos u\mathbf{i} + v \sin u\mathbf{j} + 4\mathbf{k}, \ 0 \le v \le 3$

27. Function: $y = \dfrac{x}{2}, \ 0 \le x \le 6$

Axis of revolution: *x*-axis

$x = u, \ y = \dfrac{u}{2} \cos v, \ z = \dfrac{u}{2} \sin v$

$0 \le u \le 6, \ 0 \le v \le 2\pi$

29. Function: $x = \sin z, \ 0 \le z \le \pi$

Axis of revolution: *z*-axis

$x = \sin u \cos v, \ y = \sin u \sin v, \ z = u$

$0 \le u \le \pi, \ 0 \le v \le 2\pi$

31. $\mathbf{r}(u, v) = (u + v)\mathbf{i} + (u - v)\mathbf{j} + v\mathbf{k}$, $(1, -1, 1)$

$\mathbf{r}_u(u, v) = \mathbf{i} + \mathbf{j}$, $\mathbf{r}_v(u, v) = \mathbf{i} - \mathbf{j} + \mathbf{k}$

At $(1, -1, 1)$, $u = 0$ and $v = 1$.

$\mathbf{r}_u(0, 1) = \mathbf{i} + \mathbf{j}$, $\mathbf{r}_v(0, 1) = \mathbf{i} - \mathbf{j} + \mathbf{k}$

$\mathbf{N} = \mathbf{r}_u(0, 1) \times \mathbf{r}_v(0, 1) = \begin{vmatrix} \mathbf{i} & \mathbf{j} & \mathbf{k} \\ 1 & 1 & 0 \\ 1 & -1 & 1 \end{vmatrix} = \mathbf{i} - \mathbf{j} - 2\mathbf{k}$

Tangent plane: $(x - 1) - (y + 1) - 2(z - 1) = 0$

$$x - y - 2z = 0$$

(The original plane!)

33. $\mathbf{r}(u, v) = 2u \cos v\mathbf{i} + 3u \sin v\mathbf{j} + u^2\mathbf{k}$, $(0, 6, 4)$

$\mathbf{r}_u(u, v) = 2 \cos v\mathbf{i} + 3 \sin v\mathbf{j} + 2u\mathbf{k}$

$\mathbf{r}_v(u, v) = -2u \sin v\mathbf{i} + 3u \cos v\mathbf{j}$

At $(0, 6, 4)$, $u = 2$ and $v = \pi/2$.

$\mathbf{r}_u\left(2, \dfrac{\pi}{2}\right) = 3\mathbf{j} + 4\mathbf{k}$, $\mathbf{r}_v\left(2, \dfrac{\pi}{2}\right) = -4\mathbf{i}$

$\mathbf{N} = \mathbf{r}_u\left(2, \dfrac{\pi}{2}\right) \times \mathbf{r}_v\left(2, \dfrac{\pi}{2}\right)$

$\quad = \begin{vmatrix} \mathbf{i} & \mathbf{j} & \mathbf{k} \\ 0 & 3 & 4 \\ -4 & 0 & 0 \end{vmatrix} = -16\mathbf{j} + 12\mathbf{k}$

Direction numbers: $0, 4, -3$

Tangent plane: $4(y - 6) - 3(z - 4) = 0$

$$4y - 3z = 12$$

35. $\mathbf{r}(u, v) = 2u\mathbf{i} - \dfrac{v}{2}\mathbf{j} + \dfrac{v}{2}\mathbf{k}$, $0 \le u \le 2$, $0 \le v \le 1$

$\mathbf{r}_u(u, v) = 2\mathbf{i}$, $\mathbf{r}_v(u, v) = -\dfrac{1}{2}\mathbf{j} + \dfrac{1}{2}\mathbf{k}$

$\mathbf{r}_u \times \mathbf{r}_v = \begin{vmatrix} \mathbf{i} & \mathbf{j} & \mathbf{k} \\ 2 & 0 & 0 \\ 0 & -\frac{1}{2} & \frac{1}{2} \end{vmatrix} = -\mathbf{j} - \mathbf{k}$

$\|\mathbf{r}_u \times \mathbf{r}_v\| = \sqrt{2}$

$A = \displaystyle\int_0^1 \int_0^2 \sqrt{2}\, du\, dv = 2\sqrt{2}$

37. $\mathbf{r}(u, v) = a \cos u\mathbf{i} + a \sin u\mathbf{j} + v\mathbf{k}$, $0 \le u \le 2\pi$, $0 \le v \le b$

$\mathbf{r}_u(u, v) = -a \sin u\mathbf{i} + a \cos u\mathbf{j}$

$\mathbf{r}_v(u, v) = \mathbf{k}$

$\mathbf{r}_u \times \mathbf{r}_v = \begin{vmatrix} \mathbf{i} & \mathbf{j} & \mathbf{k} \\ -a \sin u & a \cos u & 0 \\ 0 & 0 & 1 \end{vmatrix} = a \cos u\mathbf{i} + a \sin u\mathbf{j}$

$\|\mathbf{r}_u \times \mathbf{r}_v\| = a$

$A = \displaystyle\int_0^b \int_0^{2\pi} a\, du\, dv = 2\pi ab$

39. $\mathbf{r}(u, v) = au \cos v\mathbf{i} + au \sin v\mathbf{j} + u\mathbf{k}$, $0 \le u \le b$, $0 \le v \le 2\pi$

$\mathbf{r}_u(u, v) = a \cos v\mathbf{i} + a \sin v\mathbf{j} + \mathbf{k}$

$\mathbf{r}_v(u, v) = -au \sin v\mathbf{i} + au \cos v\mathbf{j}$

$\mathbf{r}_u \times \mathbf{r}_v = \begin{vmatrix} \mathbf{i} & \mathbf{j} & \mathbf{k} \\ a \cos v & a \sin v & 1 \\ -au \sin v & au \cos v & 0 \end{vmatrix} = -au \cos v\mathbf{i} - au \sin v\mathbf{j} + a^2 u\mathbf{k}$

$\|\mathbf{r}_u \times \mathbf{r}_v\| = au\sqrt{1 + a^2}$

$A = \displaystyle\int_0^{2\pi} \int_0^b a\sqrt{1 + a^2}\, u\, du\, dv = \pi ab^2\sqrt{1 + a^2}$

41. $\mathbf{r}(u, v) = \sqrt{u} \cos v \mathbf{i} + \sqrt{u} \sin v \mathbf{j} + u \mathbf{k}, \ 0 \le u \le 4, \ 0 \le v \le 2\pi$

$\mathbf{r}_u(u, v) = \dfrac{\cos v}{2\sqrt{u}}\mathbf{i} + \dfrac{\sin v}{2\sqrt{u}}\mathbf{j} + \mathbf{k}$

$\mathbf{r}_v(u, v) = -\sqrt{u} \sin v \mathbf{i} + \sqrt{u} \cos v \mathbf{j}$

$\mathbf{r}_u \times \mathbf{r}_v = \begin{vmatrix} \mathbf{i} & \mathbf{j} & \mathbf{k} \\ \dfrac{\cos v}{2\sqrt{u}} & \dfrac{\sin v}{2\sqrt{u}} & 1 \\ -\sqrt{u} \sin v & \sqrt{u} \cos v & 0 \end{vmatrix} = -\sqrt{u} \cos v \mathbf{i} - \sqrt{u} \sin v \mathbf{j} + \dfrac{1}{2}\mathbf{k}$

$\|\mathbf{r}_u \times \mathbf{r}_v\| = \sqrt{u + \dfrac{1}{4}}$

$A = \displaystyle\int_0^{2\pi} \int_0^4 \sqrt{u + \dfrac{1}{4}} \, du \, dv = \dfrac{\pi}{6}\left(17\sqrt{17} - 1\right) \approx 36.177$

43. See the definition, page 1051.

45. (a) From $(-10, 10, 0)$

    (b) From $(10, 10, 10)$

    (c) From $(0, 10, 0)$

    (d) From $(10, 0, 0)$

47. (a) $\mathbf{r}(u, v) = (4 + \cos v) \cos u \mathbf{i} +$
$(4 + \cos v) \sin u \mathbf{j} + \sin v \mathbf{k},$
$0 \le u \le 2\pi, \ 0 \le v \le 2\pi$

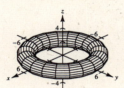

    (b) $\mathbf{r}(u, v) = (4 + 2\cos v) \cos u \mathbf{i} +$
$(4 + 2\cos v) \sin u \mathbf{j} + 2\sin v \mathbf{k},$
$0 \le u \le 2\pi, \ 0 \le v \le 2\pi$

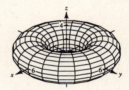

    (c) $\mathbf{r}(u, v) = (8 + \cos v) \cos u \mathbf{i} +$
$(8 + \cos v) \sin u \mathbf{j} + \sin v \mathbf{k},$
$0 \le u \le 2\pi, \ 0 \le v \le 2\pi$

    (d) $\mathbf{r}(u, v) = (8 + 3\cos v) \cos u \mathbf{i} +$
$(8 + 3\cos v) \sin u \mathbf{j} + 3\sin v \mathbf{k},$
$0 \le u \le 2\pi, \ 0 \le v \le 2\pi$

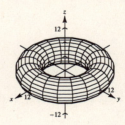

The radius of the generating circle that is revolved about the z-axis is b, and its center is a units from the axis of revolution.

49. $\mathbf{r}(u, v) = 20 \sin u \cos v \mathbf{i} + 20 \sin u \sin v \mathbf{j} + 20 \cos u \mathbf{k}$ $0 \le u \le \pi/3, \ 0 \le v \le 2\pi$

$\mathbf{r}_u = 20 \cos u \cos v \mathbf{i} + 20 \cos u \sin v \mathbf{j} - 20 \sin u \mathbf{k}$

$\mathbf{r}_v = -20 \sin u \sin v \mathbf{i} + 20 \sin u \cos v \mathbf{j}$

$$\mathbf{r}_u \times \mathbf{r}_v = \begin{vmatrix} \mathbf{i} & \mathbf{j} & \mathbf{k} \\ 20 \cos u \cos v & 20 \cos u \sin v & -20 \sin u \\ -20 \sin u \sin v & 20 \sin u \cos v & 0 \end{vmatrix}$$

$\quad = 400 \sin^2 u \cos v \mathbf{i} + 400 \sin^2 u \sin v \mathbf{j} + 400(\cos u \sin u \cos^2 v + \cos u \sin u \sin^2 v)\mathbf{k}$

$\quad = 400[\sin^2 u \cos v \mathbf{i} + \sin^2 u \sin v \mathbf{j} + \cos u \sin u \mathbf{k}]$

$\|\mathbf{r}_u \times \mathbf{r}_v\| = 400\sqrt{\sin^4 u \cos^2 v + \sin^4 u \sin^2 v + \cos^2 u \sin^2 u}$

$\quad = 400\sqrt{\sin^4 u + \cos^2 u \sin^2 u}$

$\quad = 400\sqrt{\sin^2 u} = 400 \sin u$

$S = \displaystyle\int_S\!\!\int dS = \int_0^{2\pi} \int_0^{\pi/3} 400 \sin u \, du \, dv = \int_0^{2\pi} \Big[-400 \cos u \Big]_0^{\pi/3} dv$

$\quad = \displaystyle\int_0^{2\pi} 200 \, dv = 400\pi \, \text{m}^2$

51. $\mathbf{r}(u, v) = u \cos v \mathbf{i} + u \sin v \mathbf{j} + 2v\mathbf{k}, \ 0 \le u \le 3, \ 0 \le v \le 2\pi$

$\mathbf{r}_u(u, v) = \cos v \mathbf{i} + \sin v \mathbf{j}$

$\mathbf{r}_v(u, v) = -u \sin v \mathbf{i} + u \cos v \mathbf{j} + 2\mathbf{k}$

$$\mathbf{r}_u \times \mathbf{r}_v = \begin{vmatrix} \mathbf{i} & \mathbf{j} & \mathbf{k} \\ \cos v & \sin v & 0 \\ -u \sin v & u \cos v & 2 \end{vmatrix} = 2 \sin v \mathbf{i} - 2 \cos v \mathbf{j} + u\mathbf{k}$$

$\|\mathbf{r}_u \times \mathbf{r}_v\| = \sqrt{4 + u^2}$

$A = \displaystyle\int_0^{2\pi} \int_0^3 \sqrt{4 + u^2} \, du \, dv = \pi \left[3\sqrt{13} + 4 \ln\left(\frac{3 + \sqrt{13}}{2} \right) \right]$

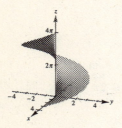

53. Essay

Section 14.6   Surface Integrals

1. $S: z = 4 - x, \ 0 \le x \le 4, \ 0 \le y \le 4, \ \dfrac{\partial z}{\partial x} = -1, \ \dfrac{\partial z}{\partial y} = 0$

$\displaystyle\int_S\!\!\int (x - 2y + z) \, dS = \int_0^4 \int_0^4 (x - 2y + 4 - x)\sqrt{1 + (-1)^2 + (0)^2} \, dy \, dx$

$\quad = \sqrt{2} \displaystyle\int_0^4 \int_0^4 (4 - 2y) \, dy \, dx = 0$

3. S: $z = 10$, $x^2 + y^2 \le 1$, $\dfrac{\partial z}{\partial x} = \dfrac{\partial z}{\partial y} = 0$

$$\iint_S (x - 2y + z)\, dS = \int_{-1}^{1} \int_{-\sqrt{1-x^2}}^{\sqrt{1-x^2}} (x - 2y + 10)\sqrt{1 + (0)^2 + (0)^2}\, dy\, dx$$

$$= \int_0^{2\pi} \int_0^1 (r\cos\theta - 2r\sin\theta + 10)r\, dr\, d\theta$$

$$= \int_0^{2\pi} \left(\frac{1}{3}\cos\theta - \frac{2}{3}\sin\theta + 5\right) d\theta$$

$$= \left[\frac{1}{3}\sin\theta + \frac{2}{3}\cos\theta + 5\theta\right]_0^{2\pi} = 10\pi$$

5. S: $z = 6 - x - 2y$, (first octant) $\dfrac{\partial z}{\partial x} = -1$, $\dfrac{\partial z}{\partial y} = -2$

$$\iint_S xy\, dS = \int_0^6 \int_0^{3-(x/2)} xy\sqrt{1 + (-1)^2 + (-2)^2}\, dy\, dx$$

$$= \sqrt{6} \int_0^6 \left[\frac{xy^2}{2}\right]_0^{3-(x/2)} dx$$

$$= \frac{\sqrt{6}}{2} \int_0^6 x\left(9 - 3x + \frac{1}{4}x^2\right) dx$$

$$= \frac{\sqrt{6}}{2} \left[\frac{9x^2}{2} - x^3 + \frac{x^4}{16}\right]_0^6 = \frac{27\sqrt{6}}{2}$$

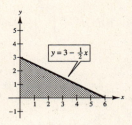

$y = 3 - \frac{1}{2}x$

7. S: $z = 9 - x^2$, $0 \le x \le 2$, $0 \le y \le x$,

$$\frac{\partial z}{\partial x} = -2x, \quad \frac{\partial z}{\partial y} = 0$$

$$\iint_S xy\, dS = \int_0^2 \int_y^2 xy\sqrt{1 + 4x^2}\, dx\, dy = \frac{391\sqrt{17} + 1}{240}$$

9. S: $z = 10 - x^2 - y^2$, $0 \le x \le 2$, $0 \le y \le 2$

$$\iint_S (x^2 - 2xy)\, dS = \int_0^2 \int_0^2 (x^2 - 2xy)\sqrt{1 + 4x^2 + 4y^2}\, dy\, dx \approx -11.47$$

11. S: $2x + 3y + 6z = 12$ (first octant) $\Rightarrow z = 2 - \dfrac{1}{3}x - \dfrac{1}{2}y$

$\rho(x, y, z) = x^2 + y^2$

$$m = \iint_R (x^2 + y^2)\sqrt{1 + \left(-\frac{1}{3}\right)^2 + \left(-\frac{1}{2}\right)^2}\, dA$$

$$= \frac{7}{6} \int_0^6 \int_0^{4-(2x/3)} (x^2 + y^2)\, dy\, dx$$

$$= \frac{7}{6} \int_0^6 \left[x^2\left(4 - \frac{2}{3}x\right) + \frac{1}{3}\left(4 - \frac{2}{3}x\right)^3\right] dx = \frac{7}{6}\left[\frac{4}{3}x^3 - \frac{1}{6}x^4 - \frac{1}{8}\left(4 - \frac{2}{3}x\right)^4\right]_0^6 = \frac{364}{3}$$

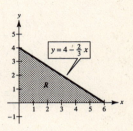

$y = 4 - \frac{2}{3}x$

R

13. S: $\mathbf{r}(u, v) = u\mathbf{i} + v\mathbf{j} + \dfrac{v}{2}\mathbf{k}$, $0 \leq u \leq 1$, $0 \leq v \leq 2$

$$\|\mathbf{r}_u \times \mathbf{r}_v\| = \left\| -\frac{1}{2}\mathbf{j} + \mathbf{k} \right\| = \frac{\sqrt{5}}{2}$$

$$\iint_S (y + 5)\, dS = \int_0^2 \int_0^1 (v + 5)\frac{\sqrt{5}}{2}\, du\, dv = 6\sqrt{5}$$

15. S: $\mathbf{r}(u, v) = 2\cos u\, \mathbf{i} + 2\sin u\, \mathbf{j} + v\mathbf{k}$, $0 \leq u \leq \dfrac{\pi}{2}$, $0 \leq v \leq 2$

$$\|\mathbf{r}_u \times \mathbf{r}_v\| = \|2\cos u\, \mathbf{i} + 2\sin u\, \mathbf{j}\| = 2$$

$$\iint_S xy\, dS = \int_0^2 \int_0^{\pi/2} 8\cos u \sin u\, du\, dv = 8$$

17. $f(x, y, z) = x^2 + y^2 + z^2$

    S: $z = x + 2$, $x^2 + y^2 \leq 1$

$$\iint_S f(x, y, z)\, dS = \int_{-1}^1 \int_{-\sqrt{1-x^2}}^{\sqrt{1-x^2}} [x^2 + y^2 + (x + 2)^2]\sqrt{1 + (1)^2 + (0)^2}\, dy\, dx$$

$$= \sqrt{2}\int_0^{2\pi} \int_0^1 [r^2 + (r\cos\theta + 2)^2]r\, dr\, d\theta$$

$$= \sqrt{2}\int_0^{2\pi} \int_0^1 [r^2 + r^2\cos^2\theta + 4r\cos\theta + 4]r\, dr\, d\theta$$

$$= \sqrt{2}\int_0^{2\pi} \left[\frac{r^4}{4} + \frac{r^4}{4}\cos^2\theta + \frac{4r^3}{3}\cos\theta + 2r^2\right]_0^1 d\theta$$

$$= \sqrt{2}\int_0^{2\pi} \left[\frac{9}{4} + \left(\frac{1}{4}\right)\frac{1 + \cos 2\theta}{2} + \frac{4}{3}\cos\theta\right] d\theta$$

$$= \sqrt{2}\left[\frac{9}{4}\theta + \frac{1}{8}\left(\theta + \frac{1}{2}\sin 2\theta\right) + \frac{4}{3}\sin\theta\right]_0^{2\pi} = \sqrt{2}\left[\frac{18\pi}{4} + \frac{\pi}{4}\right] = \frac{19\sqrt{2}\pi}{4}$$

19. $f(x, y, z) = \sqrt{x^2 + y^2 + z^2}$

    S: $z = \sqrt{x^2 + y^2}$, $x^2 + y^2 \leq 4$

$$\iint_S f(x, y, z)\, dS = \int_{-2}^2 \int_{-\sqrt{4-x^2}}^{\sqrt{4-x^2}} \sqrt{x^2 + y^2 + \left(\sqrt{x^2 + y^2}\right)^2} \sqrt{1 + \left(\frac{x}{\sqrt{x^2 + y^2}}\right)^2 + \left(\frac{y}{\sqrt{x^2 + y^2}}\right)^2}\, dy\, dx$$

$$= \sqrt{2}\int_{-2}^2 \int_{-\sqrt{4-x^2}}^{\sqrt{4-x^2}} \sqrt{x^2 + y^2}\sqrt{\frac{x^2 + y^2 + x^2 + y^2}{x^2 + y^2}}\, dy\, dx$$

$$= 2\int_{-2}^2 \int_{-\sqrt{4-x^2}}^{\sqrt{4-x^2}} \sqrt{x^2 + y^2}\, dy\, dx$$

$$= 2\int_0^{2\pi} \int_0^2 r^2\, dr\, d\theta = 2\int_0^{2\pi} \left[\frac{r^3}{3}\right]_0^2 d\theta = \left[\frac{16}{3}\theta\right]_0^{2\pi} = \frac{32\pi}{3}$$

21. $f(x, y, z) = x^2 + y^2 + z^2$

$S: x^2 + y^2 = 9, \ 0 \le x \le 3, \ 0 \le y \le 3, 0 \le z \le 9$

Project the solid onto the yz-plane; $x = \sqrt{9 - y^2}, \ 0 \le y \le 3, \ 0 \le z \le 9$.

$$\iint_S f(x, y, z) \, dS = \int_0^3 \int_0^9 [(9 - y^2) + y^2 + z^2] \sqrt{1 + \left(\frac{-y}{\sqrt{9 - y^2}}\right)^2 + (0)^2} \, dz \, dy$$

$$= \int_0^3 \int_0^9 (9 + z^2) \frac{3}{\sqrt{9 - y^2}} \, dz \, dy = \int_0^3 \left[\frac{3}{\sqrt{9 - y^2}}\left(9z + \frac{z^3}{3}\right)\right]_0^9 \, dy$$

$$= 324 \int_0^3 \frac{3}{\sqrt{9 - y^2}} \, dy = \left[972 \arcsin\left(\frac{y}{3}\right)\right]_0^3 = 972\left(\frac{\pi}{2} - 0\right) = 486\pi$$

23. $\mathbf{F}(x, y, z) = 3z\mathbf{i} - 4\mathbf{j} + y\mathbf{k}$

$S: x + y + z = 1$ (first octant)

$G(x, y, z) = x + y + z - 1$

$\nabla G(x, y, z) = \mathbf{i} + \mathbf{j} + \mathbf{k}$

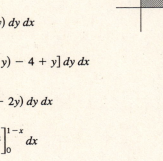

$$\iint_S \mathbf{F} \cdot \mathbf{N} \, dS = \iint_R \mathbf{F} \cdot \nabla G \, dA = \int_0^1 \int_0^{1-x} (3z - 4 + y) \, dy \, dx$$

$$= \int_0^1 \int_0^{1-x} [3(1 - x - y) - 4 + y] \, dy \, dx$$

$$= \int_0^1 \int_0^{1-x} (-1 - 3x - 2y) \, dy \, dx$$

$$= \int_0^1 \left[-y - 3xy - y^2\right]_0^{1-x} \, dx$$

$$= -\int_0^1 [(1 - x) + 3x(1 - x) + (1 - x)^2] \, dx$$

$$= -\int_0^1 (2 - 2x^2) \, dx = -\frac{4}{3}$$

25. $\mathbf{F}(x, y, z) = x\mathbf{i} + y\mathbf{j} + z\mathbf{k}$

$S: z = 9 - x^2 - y^2, \ 0 \le z$

$G(x, y, z) = x^2 + y^2 + z - 9$

$\nabla G(x, y, z) = 2x\mathbf{i} + 2y\mathbf{j} + \mathbf{k}$

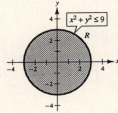

$$\iint_S \mathbf{F} \cdot \mathbf{N} \, dS = \iint_R \mathbf{F} \cdot \nabla G \, dA = \iint_R (2x^2 + 2y^2 + z) \, dA$$

$$= \iint_R [2x^2 + 2y^2 + (9 - x^2 - y^2)] \, dA$$

$$= \iint_R (x^2 + y^2 + 9) \, dA$$

$$= \int_0^{2\pi} \int_0^3 (r^2 + 9)r \, dr \, d\theta$$

$$= \int_0^{2\pi} \left[\frac{r^4}{4} + \frac{9r^2}{2}\right]_0^3 \, d\theta = \frac{243\pi}{2}$$

27. $\mathbf{F}(x, y, z) = 4\mathbf{i} - 3\mathbf{j} + 5\mathbf{k}$

$S: \ z = x^2 + y^2, \ x^2 + y^2 \le 4$

$G(x, y, z) = -x^2 - y^2 + z$

$\nabla G(x, y, z) = -2x\mathbf{i} - 2y\mathbf{j} + \mathbf{k}$

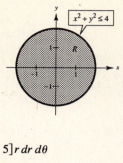

$$\iint_S \mathbf{F} \cdot \mathbf{N} \, dS = \iint_R \mathbf{F} \cdot \nabla G \, dA = \iint_R (-8x + 6y + 5) \, dA$$

$$= \int_0^{2\pi} \int_0^2 [-8r \cos\theta + 6r \sin\theta + 5] r \, dr \, d\theta$$

$$= \int_0^{2\pi} \left[-\frac{8}{3} r^3 \cos\theta + 2r^3 \sin\theta + \frac{5}{2} r^2 \right]_0^2 d\theta$$

$$= \int_0^{2\pi} \left[-\frac{64}{3} \cos\theta + 16 \sin\theta + 10 \right] d\theta$$

$$= \left[-\frac{64}{3} \sin\theta - 16 \cos\theta + 10\theta \right]_0^{2\pi} = 20\pi$$

29. $\mathbf{F}(x, y, z) = 4xy\mathbf{i} + z^2\mathbf{j} + yz\mathbf{k}$

$S:$ unit cube bounded by $x = 0, \ x = 1, \ y = 0, \ y = 1, \ z = 0, \ z = 1$

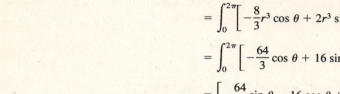

$S_1:$ The top of the cube

$\mathbf{N} = \mathbf{k}, \ z = 1$

$$\iint_{S_1} \mathbf{F} \cdot \mathbf{N} \, dS = \int_0^1 \int_0^1 y(1) \, dy \, dx = \frac{1}{2}$$

$S_2:$ The bottom of the cube

$\mathbf{N} = -\mathbf{k}, \ z = 0$

$$\iint_{S_2} \mathbf{F} \cdot \mathbf{N} \, dS = \int_0^1 \int_0^1 -y(0) \, dy \, dx = 0$$

$S_4:$ The back of the cube

$\mathbf{N} = -\mathbf{i}, \ x = 0$

$$\iint_{S_4} \mathbf{F} \cdot \mathbf{N} \, dS = \int_0^1 \int_0^1 -4(0)y \, dy \, dx = 0$$

$S_6:$ The left side of the cube

$\mathbf{N} = -\mathbf{j}, \ y = 0$

$$\iint_{S_6} \mathbf{F} \cdot \mathbf{N} \, dS = \int_0^1 \int_0^1 -z^2 \, dz \, dx = -\frac{1}{3}$$

$S_3:$ The front of the cube

$\mathbf{N} = \mathbf{i}, \ x = 1$

$$\iint_{S_3} \mathbf{F} \cdot \mathbf{N} \, dS = \int_0^1 \int_0^1 4(1)y \, dy \, dz = 2$$

$S_5:$ The right side of the cube

$\mathbf{N} = \mathbf{j}, \ y = 1$

$$\iint_{S_5} \mathbf{F} \cdot \mathbf{N} \, dS = \int_0^1 \int_0^1 z^2 \, dz \, dx = \frac{1}{3}$$

Therefore,

$$\iint_S \mathbf{F} \cdot \mathbf{N} \, dS = \frac{1}{2} + 0 + 2 + 0 + \frac{1}{3} - \frac{1}{3} = \frac{5}{2}.$$

31. The surface integral of f over a surface S, where S is given by $z = g(x, y)$, is defined as

$$\iint_S f(x, y, z) \, dS = \lim_{\|\Delta\| \to 0} \sum_{i=1}^{n} f(x_i, y_i, z_i) \Delta S_i. \ \text{(page 1061)}$$

See Theorem 14.10, page 1061.

33. See the definition, page 1067.

See Theorem 14.11, page 1067.

35. (a)

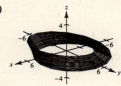

(b) If a normal vector at a point P on the surface is moved around the Möbius strip once, it will point in the opposite direction.

(c) $\mathbf{r}(u, 0) = 4\cos(2u)\mathbf{i} + 4\sin(2u)\mathbf{j}$

This is a circle.

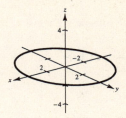

(d) (construction)

(e) You obtain a strip with a double twist and twice as long as the original Möbius strip.

37. $z = \sqrt{x^2 + y^2}$, $0 \le z \le a$

$$m = \iint_S k \, dS = k \iint_R \sqrt{1 + \left(\frac{x}{\sqrt{x^2 + y^2}}\right)^2 + \left(\frac{y}{\sqrt{x^2 + y^2}}\right)^2} \, dA = k \iint_R \sqrt{2} \, dA = \sqrt{2}\, k\pi a^2$$

$$I_z = \iint_S k(x^2 + y^2) \, dS = \iint_R k(x^2 + y^2)\sqrt{2} \, dA$$

$$= \sqrt{2}k \int_0^{2\pi} \int_0^a r^3 \, dr \, d\theta = \frac{\sqrt{2}ka^4}{4}(2\pi)$$

$$= \frac{\sqrt{2}k\pi a^4}{2} = \frac{a^2}{2}\left(\sqrt{2}k\pi a^2\right) = \frac{a^2 m}{2}$$

39. $x^2 + y^2 = a^2$, $0 \le z \le h$

$\rho(x, y, z) = 1$

$y = \pm\sqrt{a^2 - x^2}$

Project the solid onto the xz-plane.

$$I_z = 4\iint_S (x^2 + y^2)(1) \, dS$$

$$= 4\int_0^h \int_0^a [x^2 + (a^2 - x^2)]\sqrt{1 + \left(\frac{-x}{\sqrt{a^2 - x^2}}\right)^2 + (0)^2} \, dx \, dz$$

$$= 4a^3 \int_0^h \int_0^a \frac{1}{\sqrt{a^2 - x^2}} \, dx \, dz$$

$$= 4a^3 \int_0^h \left[\arcsin\frac{x}{a}\right]_0^a dz = 4a^3\left(\frac{\pi}{2}\right)(h) = 2\pi a^3 h$$

41. S: $z = 16 - x^2 - y^2$, $z \ge 0$

$\mathbf{F}(x, y, z) = 0.5z\mathbf{k}$

$$\iint_S \rho \mathbf{F} \cdot \mathbf{N} \, dS = \iint_R \rho \mathbf{F} \cdot (-g_x(x, y)\mathbf{i} - g_y(x, y)\mathbf{j} + \mathbf{k}) \, dA = \iint_R 0.5\rho z \mathbf{k} \cdot (2x\mathbf{i} + 2y\mathbf{j} + \mathbf{k}) \, dA$$

$$= \iint_R 0.5\rho z \, dA = \iint_R 0.5\rho(16 - x^2 - y^2) \, dA$$

$$= 0.5\rho \int_0^{2\pi} \int_0^4 (16 - r^2)r \, dr \, d\theta = 0.5\rho \int_0^{2\pi} 64 \, d\theta = 64\pi\rho$$

Section 14.7   Divergence Theorem

1. Surface Integral: There are six surfaces to the cube, each with $dS = \sqrt{1}\, dA$.

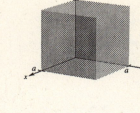

$$z = 0, \quad \mathbf{N} = -\mathbf{k}, \quad \mathbf{F} \cdot \mathbf{N} = -z^2, \quad \int_{S_1}\!\!\int 0\, dA = 0$$

$$z = a, \quad \mathbf{N} = \mathbf{k}, \quad \mathbf{F} \cdot \mathbf{N} = z^2, \quad \int_{S_2}\!\!\int a^2\, dA = \int_0^a \int_0^a a^2\, dx\, dy = a^4$$

$$x = 0, \quad \mathbf{N} = -\mathbf{i}, \quad \mathbf{F} \cdot \mathbf{N} = -2x, \quad \int_{S_3}\!\!\int 0\, dA = 0$$

$$x = a, \quad \mathbf{N} = \mathbf{i}, \quad \mathbf{F} \cdot \mathbf{N} = 2x, \quad \int_{S_4}\!\!\int 2a\, dy\, dz = \int_0^a \int_0^a 2a\, dy\, dz = 2a^3$$

$$y = 0, \quad \mathbf{N} = -\mathbf{j}, \quad \mathbf{F} \cdot \mathbf{N} = 2y, \quad \int_{S_5}\!\!\int 0\, dA = 0$$

$$y = a, \quad \mathbf{N} = \mathbf{j}, \quad \mathbf{F} \cdot \mathbf{N} = -2y, \quad \int_{S_6}\!\!\int -2a\, dA = \int_0^a \int_0^a -2a\, dz\, dx = -2a^3$$

Therefore, $\int_S\!\!\int \mathbf{F} \cdot \mathbf{N}\, dS = a^4 + 2a^3 - 2a^3 = a^4$.

Divergence Theorem: Since div $\mathbf{F} = 2z$, the Divergence Theorem yields

$$\iiint_Q \text{div } \mathbf{F}\, dV = \int_0^a \int_0^a \int_0^a 2z\, dz\, dy\, dx = \int_0^a \int_0^a a^2\, dy\, dx = a^4.$$

3. Surface Integral: There are four surfaces to this solid.

$z = 0, \mathbf{N} = -\mathbf{k}, \mathbf{F} \cdot \mathbf{N} = -z$

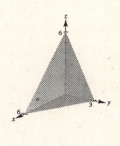

$$\int_{S_1}\!\!\int 0\, dS = 0$$

$y = 0, \mathbf{N} = -\mathbf{j}, \mathbf{F} \cdot \mathbf{N} = 2y - z, \ dS = dA = dx\, dz$

$$\int_{S_2}\!\!\int -z\, dS = \int_0^6 \int_0^{6-z} -z\, dx\, dz = \int_0^6 (z^2 - 6z)\, dz = -36$$

$x = 0, \mathbf{N} = -\mathbf{i}, \mathbf{F} \cdot \mathbf{N} = y - 2x, \ dS = dA = dz\, dy$

$$\int_{S_3}\!\!\int y\, dS = \int_0^3 \int_0^{6-2y} y\, dz\, dy = \int_0^3 (6y - 2y^2)\, dy = 9$$

$x + 2y + z = 6, \ \mathbf{N} = \dfrac{\mathbf{i} + 2\mathbf{j} + \mathbf{k}}{\sqrt{6}}, \ \mathbf{F} \cdot \mathbf{N} = \dfrac{2x - 5y + 3z}{\sqrt{6}}, \ dS = \sqrt{6}\, dA$

$$\int_{S_4}\!\!\int (2x - 5y + 3z)\, dz\, dy = \int_0^3 \int_0^{6-2y} (18 - x - 11y)\, dx\, dy = \int_0^3 (90 - 90y + 20y^2)\, dy = 45$$

Therefore, $\int_S\!\!\int \mathbf{F} \cdot \mathbf{N}\, dS = 0 - 36 + 9 + 45 = 18$.

Divergence Theorem: Since div $\mathbf{F} = 1$, we have

$$\iiint_Q dV = (\text{Volume of solid}) = \frac{1}{3}(\text{Area of base}) \times (\text{Height}) = \frac{1}{3}(9)(6) = 18.$$

5. Since div $\mathbf{F} = 2x + 2y + 2z$, we have

$$\iiint_Q \text{div } \mathbf{F} \, dV = \int_0^a \int_0^a \int_0^a (2x + 2y + 2z) \, dz \, dy \, dx$$

$$= \int_0^a \int_0^a (2ax + 2ay + a^2) \, dy \, dx = \int_0^a (2a^2x + 2a^3) \, dx = \left[a^2x^2 + 2a^3x \right]_0^a = 3a^4.$$

7. Since div $\mathbf{F} = 2x - 2x + 2xyz = 2xyz$

$$\iiint_Q \text{div } \mathbf{F} \, dV = \iiint_Q 2xyz \, dV = \int_0^a \int_0^{2\pi} \int_0^{\pi/2} 2(\rho \sin \phi \cos \theta)(\rho \sin \phi \sin \theta)(\rho \cos \phi)\rho^2 \sin \phi \, d\phi \, d\theta \, d\rho$$

$$= \int_0^a \int_0^{2\pi} \int_0^{\pi/2} 2\rho^5 (\sin \theta \cos \theta)(\sin^3 \phi \cos \phi) \, d\phi \, d\theta \, d\rho$$

$$= \int_0^a \int_0^{2\pi} \frac{1}{2}\rho^5 \sin \theta \cos \theta \, d\theta \, d\rho = \int_0^a \left[\left(\frac{\rho^5}{2} \right) \frac{\sin^2 \theta}{2} \right]_0^{2\pi} d\rho = 0.$$

9. Since div $\mathbf{F} = 3$, we have

$$\iiint_Q 3 \, dV = 3(\text{Volume of sphere}) = 3\left[\frac{4}{3}\pi(2)^3 \right] = 32\pi.$$

11. Since div $\mathbf{F} = 1 + 2y - 1 = 2y$, we have

$$\iiint_Q 2y \, dV = \int_0^4 \int_{-3}^3 \int_{-\sqrt{9-y^2}}^{\sqrt{9-y^2}} 2y \, dx \, dy \, dz = \int_0^4 \int_{-3}^3 4y\sqrt{9 - y^2} \, dy \, dz = \int_0^4 \left[-\frac{4}{3}(9 - y^2)^{3/2} \right]_{-3}^3 dz = 0.$$

13. Since div $\mathbf{F} = 3x^2 + x^2 + 0 = 4x^2$, we have

$$\iiint_Q 4x^2 \, dV = \int_0^6 \int_0^4 \int_0^{4-y} 4x^2 \, dz \, dy \, dx = \int_0^6 \int_0^4 4x^2(4 - y) \, dy \, dx = \int_0^6 32x^2 \, dx = 2304.$$

15. $\mathbf{F}(x, y, z) = xy\mathbf{i} + 4y\mathbf{j} + xz\mathbf{k}$

div $\mathbf{F} = y + 4 + x$

$$\iint_S \mathbf{F} \cdot \mathbf{N} \, dS = \iiint_Q \text{div } \mathbf{F} \, dV = \iiint_Q (y + x + 4) \, dV$$

$$= \int_0^3 \int_0^\pi \int_0^{2\pi} (\rho \sin \phi \sin \theta + \rho \sin \phi \cos \theta + 4)\rho^2 \sin \phi \, d\theta \, d\phi \, d\rho$$

$$= \int_0^3 \int_0^\pi \int_0^{2\pi} [\rho^3 \sin^2 \phi \sin \theta + \rho^3 \sin^2 \phi \cos \theta + 4\rho^2 \sin \phi] \, d\theta \, d\phi \, d\rho$$

$$= \int_0^3 \int_0^\pi \left[-\rho^3 \sin^2 \phi \cos \theta + \rho^3 \sin^2 \phi \sin \theta + 4\rho^2 \sin \phi \cdot \theta \right]_0^{2\pi} d\phi \, d\rho$$

$$= \int_0^3 \int_0^\pi 8\pi\rho^2 \sin \phi \, d\phi \, d\rho$$

$$= \int_0^3 \left[-8\pi\rho^2 \cos \phi \right]_0^\pi d\rho$$

$$= \int_0^3 16\pi\rho^2 \, d\rho = \left[\frac{16\pi\rho^3}{3} \right]_0^3 = 144\pi.$$

17. Using the Divergence Theorem, we have

$$\iint_S \text{curl } \mathbf{F} \cdot \mathbf{N} \, dS = \iiint_Q \text{div (curlF)} \, dV$$

$$\text{curl } \mathbf{F}(x, y, z) = \begin{vmatrix} \mathbf{i} & \mathbf{j} & \mathbf{k} \\ \dfrac{\partial}{\partial x} & \dfrac{\partial}{\partial y} & \dfrac{\partial}{\partial z} \\ 4xy + z^2 & 2x^2 + 6yz & 2xz \end{vmatrix} = -6y\mathbf{i} - (2z - 2z)\mathbf{j} + (4x - 4x)\mathbf{k} = -6y\mathbf{i}$$

$$\text{div (curl } \mathbf{F}) = 0.$$

Therefore, $\displaystyle\iiint_Q \text{div (curl } \mathbf{F}) \, dV = 0.$

19. See Theorem 14.12, page 1073.

21. Using the triple integral to find volume, we need $\mathbf{F}$ so that

$$\text{div } \mathbf{F} = \frac{\partial M}{\partial x} + \frac{\partial N}{\partial y} + \frac{\partial P}{\partial z} = 1.$$

Hence, we could have $\mathbf{F} = x\mathbf{i}$, $\mathbf{F} = y\mathbf{j}$, or $\mathbf{F} = z\mathbf{k}$.

For $dA = dy \, dz$ consider $\mathbf{F} = x\mathbf{i}$, $x = f(y, z)$, then $\mathbf{N} = \dfrac{\mathbf{i} + f_y\mathbf{j} + f_z\mathbf{k}}{\sqrt{1 + f_y^2 + f_z^2}}$ and $dS = \sqrt{1 + f_y^2 + f_z^2} \, dy \, dz$.

For $dA = dz \, dx$ consider $\mathbf{F} = y\mathbf{j}$, $y = f(x, z)$, then $\mathbf{N} = \dfrac{f_x\mathbf{i} + \mathbf{j} + f_z\mathbf{k}}{\sqrt{1 + f_x^2 + f_z^2}}$ and $dS = \sqrt{1 + f_x^2 + f_z^2} \, dz \, dx$.

For $dA = dx \, dy$ consider $\mathbf{F} = z\mathbf{k}$, $z = f(x, y)$, then $\mathbf{N} = \dfrac{f_x\mathbf{i} + f_y\mathbf{j} + \mathbf{k}}{\sqrt{1 + f_x^2 + f_y^2}}$ and $dS = \sqrt{1 + f_x^2 + f_y^2} \, dx \, dy$.

Correspondingly, we then have $V = \displaystyle\iint_S \mathbf{F} \cdot \mathbf{N} \, dS = \iint_S x \, dy \, dz = \iint_S y \, dz \, dx = \iint_S z \, dx \, dy$.

23. Using the Divergence Theorem, we have $\displaystyle\iint_S \text{curl } \mathbf{F} \cdot \mathbf{N} \, dS = \iiint_Q \text{div (curl } \mathbf{F}) \, dV$. Let

$$\mathbf{F}(x, y, z) = M\mathbf{i} + N\mathbf{j} + P\mathbf{k}$$

$$\text{curl } \mathbf{F} = \left(\frac{\partial P}{\partial y} - \frac{\partial N}{\partial z}\right)\mathbf{i} - \left(\frac{\partial P}{\partial x} - \frac{\partial M}{\partial z}\right)\mathbf{j} + \left(\frac{\partial N}{\partial x} - \frac{\partial M}{\partial y}\right)\mathbf{k}$$

$$\text{div (curl } \mathbf{F}) = \frac{\partial^2 P}{\partial x \partial y} - \frac{\partial^2 N}{\partial x \partial z} - \frac{\partial^2 P}{\partial y \partial x} + \frac{\partial^2 M}{\partial y \partial z} + \frac{\partial^2 N}{\partial z \partial x} - \frac{\partial^2 M}{\partial z \partial y} = 0.$$

Therefore, $\displaystyle\iint_S \text{curl } \mathbf{F} \cdot \mathbf{N} \, dS = \iiint_Q 0 \, dV = 0.$

25. If $\mathbf{F}(x, y, z) = x\mathbf{i} + y\mathbf{j} + z\mathbf{k}$, then div $\mathbf{F} = 3$.

$$\iint_S \mathbf{F} \cdot \mathbf{N} \, dS = \iiint_Q \text{div } \mathbf{F} \, dV = \iiint_Q 3 \, dV = 3V.$$

27. $\displaystyle\iint_S f D_{\mathbf{N}} g \, dS = \iint_S f \nabla g \cdot \mathbf{N} \, dS$

$$= \iiint_Q \text{div } (f \nabla g) \, dV = \iiint_Q (f \, \text{div } \nabla g + \nabla f \cdot \nabla g) \, dV = \iiint_Q (f \nabla^2 g + \nabla f \cdot \nabla g) \, dV$$

Section 14.8   Stokes's Theorem

1. $F(x, y, z) = (2y - z)\mathbf{i} + xyz\mathbf{j} + e^z\mathbf{k}$

$$\mathbf{curl\ F} = \begin{vmatrix} \mathbf{i} & \mathbf{j} & \mathbf{k} \\ \frac{\partial}{\partial x} & \frac{\partial}{\partial y} & \frac{\partial}{\partial z} \\ 2y - z & xyz & e^z \end{vmatrix} = -xy\mathbf{i} - \mathbf{j} + (yz - 2)\mathbf{k}$$

3. $F(x, y, z) = 2z\mathbf{i} - 4x^2\mathbf{j} + \arctan x\mathbf{k}$

$$\mathbf{curl\ F} = \begin{vmatrix} \mathbf{i} & \mathbf{j} & \mathbf{k} \\ \frac{\partial}{\partial x} & \frac{\partial}{\partial y} & \frac{\partial}{\partial z} \\ 2z & -4x^2 & \arctan x \end{vmatrix} = \left(2 - \frac{1}{1 + x^2}\right)\mathbf{j} - 8x\mathbf{k}$$

5. $F(x, y, z) = e^{x^2 + y^2}\mathbf{i} + e^{y^2 + z^2}\mathbf{j} + xyz\mathbf{k}$

$$\mathbf{curl\ F} = \begin{vmatrix} \mathbf{i} & \mathbf{j} & \mathbf{k} \\ \frac{\partial}{\partial x} & \frac{\partial}{\partial y} & \frac{\partial}{\partial z} \\ e^{x^2 + y^2} & e^{y^2 + z^2} & xyz \end{vmatrix}$$

$$= (xz - 2ze^{y^2 + z^2})\mathbf{i} - yz\mathbf{j} - 2ye^{x^2 + y^2}\mathbf{k}$$

$$= z(x - 2e^{y^2 + z^2})\mathbf{i} - yz\mathbf{j} - 2ye^{x^2 + y^2}\mathbf{k}$$

7. In this case, $M = -y + z$, $N = x - z$, $P = x - y$ and C is the circle $x^2 + y^2 = 1$, $z = 0$, $dz = 0$.

Line Integral: $\displaystyle\int_C \mathbf{F} \cdot d\mathbf{r} = \int_C (-y + z)\, dx + (x - z)\, dy + (x - y)\, dz = \int_C -y\, dx + x\, dy$

Letting $x = \cos t$, $y = \sin t$, we have $dx = -\sin t\, dt$, $dy = \cos t\, dt$ and

$$\int_C -y\, dx + x\, dy = \int_0^{2\pi} (\sin^2 t + \cos^2 t)\, dt = 2\pi.$$

Double Integral: Consider $F(x, y, z) = x^2 + y^2 + z^2 - 1$.

Then

$$\mathbf{N} = \frac{\nabla F}{\|\nabla F\|} = \frac{2x\mathbf{i} + 2y\mathbf{j} + 2z\mathbf{k}}{2\sqrt{x^2 + y^2 + z^2}} = x\mathbf{i} + y\mathbf{j} + z\mathbf{k}.$$

Since

$$z^2 = 1 - x^2 - y^2, \ z_x = \frac{-2x}{2z} = \frac{-x}{z}, \text{ and } z_y = \frac{-y}{z}, \ dS = \sqrt{1 + \frac{x^2}{z^2} + \frac{y^2}{z^2}}\, dA = \frac{1}{z}dA.$$

Now, since **curl F** $= 2\mathbf{k}$, we have

$$\int_S\int (\mathbf{curl\ F}) \cdot \mathbf{N}\, dS = \int_R\int 2z\left(\frac{1}{z}\right) dA = \int_R\int 2\, dA = 2(\text{Area of circle of radius } 1) = 2\pi.$$

9. Line Integral: From the accompanying figure we see that for

$C_1: z = 0, \; dz = 0$

$C_2: x = 0, \; dx = 0$

$C_3: y = 0, \; dy = 0.$

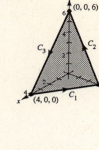

Hence, $\displaystyle \int_C \mathbf{F} \cdot d\mathbf{r} = \int_C xyz \, dx + y \, dy + z \, dz$

$$= \int_{C_1} y \, dy + \int_{C_2} y \, dy + z \, dz + \int_{C_3} z \, dz$$

$$= \int_0^3 y \, dy + \int_3^0 y \, dy + \int_0^6 z \, dz + \int_6^0 z \, dz = 0.$$

Double Integral: $\mathbf{curl} \, \mathbf{F} = xy\mathbf{j} - xz\mathbf{k}$

Considering $F(x, y, z) = 3x + 4y + 2z - 12$, then

$$\mathbf{N} = \frac{\nabla F}{\|\nabla F\|} = \frac{3\mathbf{i} + 4\mathbf{j} + 2\mathbf{k}}{\sqrt{29}} \text{ and } dS = \sqrt{29} \, dA.$$

Thus,

$$\int_S \int (\mathbf{curl} \, \mathbf{F}) \cdot \mathbf{N} \, dS = \int_R \int (4xy - 2xz) \, dA$$

$$= \int_0^4 \int_0^{(-3x+12)/4} \left[4xy - 2x\left(6 - 2y - \frac{3}{2}x \right) \right] dy \, dx$$

$$= \int_0^4 \int_0^{(12-3x)/4} (8xy + 3x^2 - 12x) \, dy \, dx$$

$$= \int_0^4 0 \, dx = 0.$$

11. Let $A = (0, 0, 0)$, $B = (1, 1, 1)$ and $C = (0, 2, 0)$. Then $\mathbf{U} = \overrightarrow{AB} = \mathbf{i} + \mathbf{j} + \mathbf{k}$ and $\mathbf{V} = \overrightarrow{AC} = 2\mathbf{j}$. Thus,

$$\mathbf{N} = \frac{\mathbf{U} \times \mathbf{V}}{\|\mathbf{U} \times \mathbf{V}\|} = \frac{-2\mathbf{i} + 2\mathbf{k}}{2\sqrt{2}} = \frac{-\mathbf{i} + \mathbf{k}}{\sqrt{2}}.$$

Surface S has direction numbers $-1, 0, 1$, with equation $z - x = 0$ and $dS = \sqrt{2} \, dA$. Since $\mathbf{curl} \, \mathbf{F} = -3\mathbf{i} - \mathbf{j} - 2\mathbf{k}$, we have

$$\int_S \int (\mathbf{curl} \, \mathbf{F}) \cdot \mathbf{N} \, dS = \int_R \int \frac{1}{\sqrt{2}}(\sqrt{2}) \, dA = \int_R \int dA = (\text{Area of triangle with } h = 1, b = 2) = 1.$$

13. $\mathbf{F}(x, y, z) = z^2\mathbf{i} + x^2\mathbf{j} + y^2\mathbf{k}$, $S: z = 4 - x^2 - y^2, \; 0 \le z$

$$\mathbf{curl} \, \mathbf{F} = \begin{vmatrix} \mathbf{i} & \mathbf{j} & \mathbf{k} \\ \dfrac{\partial}{\partial x} & \dfrac{\partial}{\partial y} & \dfrac{\partial}{\partial z} \\ z^2 & x^2 & y^2 \end{vmatrix} = 2y\mathbf{i} + 2z\mathbf{j} + 2x\mathbf{k}$$

$G(x, y, z) = x^2 + y^2 + z - 4$

$\nabla G(x, y, z) = 2x\mathbf{i} + 2y\mathbf{j} + \mathbf{k}$

$$\int_S \int (\mathbf{curl} \, \mathbf{F}) \cdot \mathbf{N} \, dS = \int_R \int (4xy + 4yz + 2x) \, dA = \int_{-2}^2 \int_{-\sqrt{4-x^2}}^{\sqrt{4-x^2}} [4xy + 4y(4 - x^2 - y^2) + 2x] \, dy \, dx$$

$$= \int_{-2}^2 \int_{-\sqrt{4-x^2}}^{\sqrt{4-x^2}} [4xy + 16y - 4x^2y - 4y^3 + 2x] \, dy \, dx$$

$$= \int_{-2}^2 4x\sqrt{4 - x^2} \, dx = 0$$

15. $\mathbf{F}(x, y, z) = z^2\mathbf{i} + y\mathbf{j} + xz\mathbf{k}$, $S: z = \sqrt{4 - x^2 - y^2}$

$$\mathbf{curl\ F} = \begin{vmatrix} \mathbf{i} & \mathbf{j} & \mathbf{k} \\ \dfrac{\partial}{\partial x} & \dfrac{\partial}{\partial y} & \dfrac{\partial}{\partial z} \\ z^2 & y & xz \end{vmatrix} = z\mathbf{j}$$

$$G(x, y, z) = z - \sqrt{4 - x^2 - y^2}$$

$$\nabla G(x, y, z) = \frac{x}{\sqrt{4 - x^2 - y^2}}\mathbf{i} + \frac{y}{\sqrt{4 - x^2 - y^2}}\mathbf{j} + \mathbf{k}$$

$$\iint_S (\mathbf{curl\ F}) \cdot \mathbf{N}\, dS = \iint_R \frac{yz}{\sqrt{4 - x^2 - y^2}}\, dA = \iint_R \frac{y\sqrt{4 - x^2 - y^2}}{\sqrt{4 - x^2 - y^2}}\, dA = \int_{-2}^{2}\int_{-\sqrt{4-x^2}}^{\sqrt{4-x^2}} y\, dy\, dx = 0$$

17. $\mathbf{F}(x, y, z) = -\ln\sqrt{x^2 + y^2}\,\mathbf{i} + \arctan\dfrac{x}{y}\mathbf{j} + \mathbf{k}$

$$\mathbf{curl\ F} = \begin{vmatrix} \mathbf{i} & \mathbf{j} & \mathbf{k} \\ \dfrac{\partial}{\partial x} & \dfrac{\partial}{\partial y} & \dfrac{\partial}{\partial z} \\ -1/2 \ln(x^2 + y^2) & \arctan x/y & 1 \end{vmatrix} = \left[\frac{(1/y)}{1 + (x^2/y^2)} + \frac{y}{x^2 + y^2}\right]\mathbf{k} = \left[\frac{2y}{x^2 + y^2}\right]\mathbf{k}$$

$S: z = 9 - 2x - 3y$ over one petal of $r = 2 \sin 2\theta$ in the first octant.

$$G(x, y, z) = 2x + 3y + z - 9$$

$$\nabla G(x, y, z) = 2\mathbf{i} + 3\mathbf{j} + \mathbf{k}$$

$$\iint_S (\mathbf{curl\ F}) \cdot \mathbf{N}\, dS = \iint_R \frac{2y}{x^2 + y^2}\, dA$$

$$= \int_0^{\pi/2}\int_0^{2 \sin 2\theta} \frac{2r \sin\theta}{r^2} r\, dr\, d\theta$$

$$= \int_0^{\pi/2}\int_0^{4 \sin\theta \cos\theta} 2 \sin\theta\, dr\, d\theta$$

$$= \int_0^{\pi/2} 8 \sin^2\theta \cos\theta\, d\theta = \left[\frac{8 \sin^3\theta}{3}\right]_0^{\pi/2} = \frac{8}{3}$$

19. From Exercise 10, we have $\mathbf{N} = \dfrac{2x\mathbf{i} - \mathbf{k}}{\sqrt{1 + 4x^2}}$ and $dS = \sqrt{1 + 4x^2}\, dA$. Since $\mathbf{curl\ F} = xy\mathbf{j} - xz\mathbf{k}$, we have

$$\iint_S (\mathbf{curl\ F}) \cdot \mathbf{N}\, dS = \iint_R xz\, dA = \int_0^a\int_0^a x^3\, dy\, dx = \int_0^a ax^3\, dx = \left[\frac{ax^4}{4}\right]_0^a = \frac{a^5}{4}.$$

21. $\mathbf{F}(x, y, z) = \mathbf{i} + \mathbf{j} - 2\mathbf{k}$　　　　　　　　　**23.** See Theorem 14.13, page 1081.

$$\mathbf{curl\ F} = \begin{vmatrix} \mathbf{i} & \mathbf{j} & \mathbf{k} \\ \dfrac{\partial}{\partial x} & \dfrac{\partial}{\partial y} & \dfrac{\partial}{\partial z} \\ 1 & 1 & -2 \end{vmatrix} = \mathbf{0}$$

Letting $\mathbf{N} = \mathbf{k}$, we have $\displaystyle\iint_S (\mathbf{curl\ F}) \cdot \mathbf{N}\, dS = 0$.

25. (a) $\displaystyle\int_C f\nabla g \cdot d\mathbf{r} = \int\!\!\!\int_S \mathbf{curl}[f\nabla g] \cdot \mathbf{N}\, dS$ (Stokes's Theorem)

$$f\nabla g = f\frac{\partial g}{\partial x}\mathbf{i} + f\frac{\partial g}{\partial y}\mathbf{j} + f\frac{\partial g}{\partial z}\mathbf{k}$$

$$\mathbf{curl}\,(f\nabla g) = \begin{vmatrix} \mathbf{i} & \mathbf{j} & \mathbf{k} \\ \frac{\partial}{\partial x} & \frac{\partial}{\partial y} & \frac{\partial}{\partial z} \\ f(\partial g/\partial x) & f(\partial g/\partial y) & f(\partial g/\partial z) \end{vmatrix}$$

$$= \left[\left[f\left(\frac{\partial^2 g}{\partial y\partial z}\right) + \left(\frac{\partial f}{\partial y}\right)\left(\frac{\partial g}{\partial z}\right)\right] - \left[f\left(\frac{\partial^2 g}{\partial z\partial y}\right) + \left(\frac{\partial f}{\partial z}\right)\left(\frac{\partial g}{\partial y}\right)\right]\right]\mathbf{i}$$

$$- \left[\left[f\left(\frac{\partial^2 g}{\partial x\partial z}\right) + \left(\frac{\partial f}{\partial x}\right)\left(\frac{\partial g}{\partial z}\right)\right] - \left[f\left(\frac{\partial^2 g}{\partial z\partial x}\right) + \left(\frac{\partial f}{\partial z}\right)\left(\frac{\partial g}{\partial x}\right)\right]\right]\mathbf{j}$$

$$+ \left[\left[f\left(\frac{\partial^2 g}{\partial x\partial y}\right) + \left(\frac{\partial f}{\partial x}\right)\left(\frac{\partial g}{\partial y}\right)\right] - \left[f\left(\frac{\partial^2 g}{\partial y\partial x}\right) + \left(\frac{\partial f}{\partial y}\right)\left(\frac{\partial g}{\partial x}\right)\right]\right]\mathbf{k}$$

$$= \left[\left(\frac{\partial f}{\partial y}\right)\left(\frac{\partial g}{\partial z}\right) - \left(\frac{\partial f}{\partial z}\right)\left(\frac{\partial g}{\partial y}\right)\right]\mathbf{i} - \left[\left(\frac{\partial f}{\partial x}\right)\left(\frac{\partial g}{\partial z}\right) - \left(\frac{\partial f}{\partial z}\right)\left(\frac{\partial g}{\partial x}\right)\right]\mathbf{j} + \left[\left(\frac{\partial f}{\partial x}\right)\left(\frac{\partial g}{\partial y}\right) - \left(\frac{\partial f}{\partial y}\right)\left(\frac{\partial g}{\partial x}\right)\right]\mathbf{k}$$

$$= \begin{vmatrix} \mathbf{i} & \mathbf{j} & \mathbf{k} \\ \frac{\partial f}{\partial x} & \frac{\partial f}{\partial y} & \frac{\partial f}{\partial z} \\ \frac{\partial g}{\partial x} & \frac{\partial g}{\partial y} & \frac{\partial g}{\partial z} \end{vmatrix} = \nabla f \times \nabla g$$

Therefore, $\displaystyle\int_C f\nabla g \cdot d\mathbf{r} = \int\!\!\!\int_S \mathbf{curl}[f\nabla g] \cdot \mathbf{N}\, dS = \int\!\!\!\int_S [\nabla f \times \nabla g] \cdot \mathbf{N}\, dS.$

(b) $\displaystyle\int_C (f\nabla f) \cdot d\mathbf{r} = \int\!\!\!\int_S (\nabla f \times \nabla f) \cdot \mathbf{N}\, dS$ (using part a.)

$\qquad\qquad = 0$ since $\nabla f \times \nabla f = \mathbf{0}$.

(c) $\displaystyle\int_C (f\nabla g + g\nabla f) \cdot d\mathbf{r} = \int_C (f\nabla g) \cdot d\mathbf{r} + \int_C (g\nabla f) \cdot d\mathbf{r}$

$$= \int\!\!\!\int_S (\nabla f \times \nabla g) \cdot \mathbf{N}\, dS + \int\!\!\!\int_S (\nabla g \times \nabla f) \cdot \mathbf{N}\, dS \quad \text{(using part a.)}$$

$$= \int\!\!\!\int_S (\nabla f \times \nabla g) \cdot \mathbf{N}\, dS + \int\!\!\!\int_S -(\nabla f \times \nabla g) \cdot \mathbf{N}\, dS = 0$$

27. Let $\mathbf{C} = a\mathbf{i} + b\mathbf{j} + c\mathbf{k}$, then

$$\frac{1}{2}\int_C (\mathbf{C} \times \mathbf{r}) \cdot d\mathbf{r} = \frac{1}{2}\int\!\!\!\int_S \mathbf{curl}\,(\mathbf{C} \times \mathbf{r}) \cdot \mathbf{N}\, dS = \frac{1}{2}\int\!\!\!\int_S 2\mathbf{C} \cdot \mathbf{N}\, dS = \int\!\!\!\int_S \mathbf{C} \cdot \mathbf{N}\, dS$$

since

$$\mathbf{C} \times \mathbf{r} = \begin{vmatrix} \mathbf{i} & \mathbf{j} & \mathbf{k} \\ a & b & c \\ x & y & z \end{vmatrix} = (bz - cy)\mathbf{i} - (az - cx)\mathbf{j} + (ay - bx)\mathbf{k}$$

and

$$\mathbf{curl}(\mathbf{C} \times \mathbf{r}) = \begin{vmatrix} \mathbf{i} & \mathbf{j} & \mathbf{k} \\ \frac{\partial}{\partial x} & \frac{\partial}{\partial y} & \frac{\partial}{\partial z} \\ bz - cy & cx - az & ay - bx \end{vmatrix} = 2(a\mathbf{i} + b\mathbf{j} + c\mathbf{k}) = 2\mathbf{C}.$$

Review Exercises for Chapter 14

1. $\mathbf{F}(x, y, z) = x\mathbf{i} + \mathbf{j} + 2\mathbf{k}$

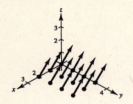

3. $f(x, y, z) = 8x^2 + xy + z^2$

$$\mathbf{F}(x, y, z) = (16x + y)\mathbf{i} + x\mathbf{j} + 2z\mathbf{k}$$

5. Since $\partial M/\partial y = -1/y^2 \neq \partial N/\partial x$, $\mathbf{F}$ is not conservative.

7. Since $\partial M/\partial y = 12xy = \partial N/\partial x$, $\mathbf{F}$ is conservative. From $M = \partial U/\partial x = 6xy^2 - 3x^2$ and $N = \partial U/\partial y = 6x^2y + 3y^2 - 7$, partial integration yields $U = 3x^2y^2 - x^3 + h(y)$ and $U = 3x^2y^2 + y^3 - 7y + g(x)$ which suggests $h(y) = y^3 - 7y$, $g(x) = -x^3$, and $U(x, y) = 3x^2y^2 - x^3 + y^3 - 7y + C$.

9. Since

$$\frac{\partial M}{\partial y} = 4x = \frac{\partial N}{\partial x},$$

$$\frac{\partial M}{\partial z} = 1 \neq \frac{\partial P}{\partial x}.$$

$\mathbf{F}$ is not conservative.

11. Since

$$\frac{\partial M}{\partial y} = \frac{-1}{y^2 z} = \frac{\partial N}{\partial x}, \quad \frac{\partial M}{\partial z} = \frac{-1}{yz^2} = \frac{\partial P}{\partial x}, \quad \frac{\partial N}{\partial z} = \frac{x}{y^2 z^2} = \frac{\partial P}{\partial y},$$

$\mathbf{F}$ is conservative. From

$$M = \frac{\partial U}{\partial x} = \frac{1}{yz}, \quad N = \frac{\partial U}{\partial y} = \frac{-x}{y^2 z}, \quad P = \frac{\partial U}{\partial z} = \frac{-x}{yz^2}$$

we obtain

$$U = \frac{x}{yz} + f(y, z), \quad U = \frac{x}{yz} + g(x, z), \quad U = \frac{x}{yz} + h(x, y) \implies f(x, y, z) = \frac{x}{yz} + K$$

13. Since $\mathbf{F} = x^2\mathbf{i} + y^2\mathbf{j} + z^2\mathbf{k}$:

(a) div $\mathbf{F} = 2x + 2y + 2z$

(b) **curl** $\mathbf{F} = \left(\dfrac{\partial P}{\partial y} - \dfrac{\partial N}{\partial z}\right)\mathbf{i} - \left(\dfrac{\partial P}{\partial x} - \dfrac{\partial M}{\partial z}\right)\mathbf{j} + \left(\dfrac{\partial N}{\partial x} - \dfrac{\partial M}{\partial y}\right)\mathbf{k} = 0\mathbf{i} - 0\mathbf{j} + 0\mathbf{k} = \mathbf{0}$

15. Since $\mathbf{F} = (\cos y + y \cos x)\mathbf{i} + (\sin x - x \sin y)\mathbf{j} + xyz\mathbf{k}$:

(a) div $\mathbf{F} = -y \sin x - x \cos y + xy$

(b) **curl** $\mathbf{F} = xz\mathbf{i} - yz\mathbf{j} + (\cos x - \sin y + \sin y - \cos x)\mathbf{k} = xz\mathbf{i} - yz\mathbf{j}$

17. Since $F = \arcsin x\mathbf{i} + xy^2\mathbf{j} + yz^2\mathbf{k}$:

    (a) $\operatorname{div} \mathbf{F} = \dfrac{1}{\sqrt{1-x^2}} + 2xy + 2yz$

    (b) $\operatorname{\mathbf{curl}} \mathbf{F} = z^2\mathbf{i} + y^2\mathbf{k}$

19. Since $F = \ln(x^2 + y^2)\mathbf{i} + \ln(x^2 + y^2)\mathbf{j} + z\mathbf{k}$:

    (a) $\operatorname{div} \mathbf{F} = \dfrac{2x}{x^2 + y^2} + \dfrac{2y}{x^2 + y^2} + 1$

            $= \dfrac{2x + 2y}{x^2 + y^2} + 1$

    (b) $\operatorname{\mathbf{curl}} \mathbf{F} = \dfrac{2x - 2y}{x^2 + y^2}\mathbf{k}$

21. (a) Let $x = t, y = t, -1 \le t \le 2$, then $ds = \sqrt{2}\, dt$.

$$\int_C (x^2 + y^2)\, ds = \int_{-1}^{2} 2t^2\sqrt{2}\, dt = \left[2\sqrt{2}\left(\frac{t^3}{3}\right)\right]_{-1}^{2} = 6\sqrt{2}$$

    (b) Let $x = 4\cos t, y = 4\sin t, 0 \le t \le 2\pi$, then $ds = 4\, dt$.

$$\int_C (x^2 + y^2)\, ds = \int_0^{2\pi} 16(4\, dt) = 128\pi$$

23. $x = \cos t + t\sin t, y = \sin t - t\cos t, 0 \le t \le 2\pi, \dfrac{dx}{dt} = t\cos t, \dfrac{dy}{dt} = t\sin t$

$$\int_C (x^2 + y^2)\, ds = \int_0^{2\pi} [(\cos t + t\sin t)^2 + (\sin t - t\cos t)^2]\sqrt{t^2\cos^2 t + t^2\sin^2 t}\, dt = \int_0^{2\pi} [t^3 + t]\, dt$$

$$= 2\pi^2(1 + 2\pi^2)$$

25. (a) Let $x = 2t, y = -3t, 0 \le t \le 1$

$$\int_C (2x - y)\, dx + (x + 3y)\, dy = \int_0^1 [7t(2) + (-7t)(-3)]\, dt = \int_0^1 35t\, dt = \frac{35}{2}$$

    (b) $x = 3\cos t, y = 3\sin t, dx = -3\sin t\, dt, dy = 3\cos t\, dt, 0 \le t \le 2\pi$

$$\int_C (2x - y)\, dx + (x + 3y)\, dy = \int_0^{2\pi} (9 + 9\sin t\cos t)\, dt = 18\pi$$

27. $\displaystyle\int_C (2x + y)\, ds, \mathbf{r}(t) = a\cos^3 t\mathbf{i} + a\sin^3 t\mathbf{j}, 0 \le t \le \dfrac{\pi}{2}$

$x'(t) = -3a \cdot \cos^2 t\sin t$

$y'(t) = 3a \cdot \sin^2 t\cos t$

$$\int_C (2x + y)\, ds = \int_0^{\pi/2} (2(a \cdot \cos^3 t) + a \cdot \sin^3 t)\sqrt{x'(t)^2 + y'(t)^2}\, dt = \frac{9a^2}{5}$$

29. $f(x, y) = 5 + \sin(x + y)$

    $C\colon y = 3x$ from $(0, 0)$ to $(2, 6)$

    $\mathbf{r}(t) = t\mathbf{i} + 3t\mathbf{j}, 0 \le t \le 2$

    $\mathbf{r}'(t) = \mathbf{i} + 3\mathbf{j}$

    $\|\mathbf{r}'(t)\| = \sqrt{10}$

    Lateral surface area:

$$\int_{C_1} f(x, y)\, ds = \int_0^2 [5 + \sin(t + 3t)]\sqrt{10}\, dt = \sqrt{10}\int_0^2 (5 + \sin 4t)\, dt = \frac{\sqrt{10}}{4}(41 - \cos 8) \approx 32.528$$

31. $d\mathbf{r} = (2t\mathbf{i} + 3t^2\mathbf{j}) \, dt$

$\mathbf{F} = t^5\mathbf{i} + t^4\mathbf{j}, \, 0 \leq t \leq 1$

$\displaystyle\int_C \mathbf{F} \cdot d\mathbf{r} = \int_0^1 5t^6 \, dt = \frac{5}{7}$

33. $d\mathbf{r} = [(-2 \sin t)\mathbf{i} + (2 \cos t)\mathbf{j} + \mathbf{k}] \, dt$

$\mathbf{F} = (2 \cos t)\mathbf{i} + (2 \sin t)\mathbf{j} + t\mathbf{k}, \, 0 \leq t \leq 2\pi$

$\displaystyle\int_C \mathbf{F} \cdot d\mathbf{r} = \int_0^{2\pi} t \, dt = 2\pi^2$

35. Let $x = t, \, y = -t, \, z = 2t^2, \, -2 \leq t \leq 2, \, d\mathbf{r} = [\mathbf{i} - \mathbf{j} + 4t\mathbf{k}] \, dt$.

$\mathbf{F} = (-t - 2t^2)\mathbf{i} + (2t^2 - t)\mathbf{j} + (2t)\mathbf{k}$

$\displaystyle\int_C \mathbf{F} \cdot d\mathbf{r} = \int_{-2}^2 4t^2 \, dt = \left[\frac{4t^3}{3}\right]_{-2}^2 = \frac{64}{3}$

37. For $y = x^2, \, \mathbf{r}_1(t) = t\mathbf{i} + t^2\mathbf{j}, \, 0 \leq t \leq 2$

For $y = 2x, \, \mathbf{r}_2(t) = (2 - t)\mathbf{i} + (4 - 2t)\mathbf{j}, \, 0 \leq t \leq 2$

$\displaystyle\int_C xy \, dx + (x^2 + y^2) \, dy = \int_{C_1} xy \, dx + (x^2 + y^2) \, dy + \int_{C_2} xy \, dx + (x^2 + y^2) \, dy$

$\displaystyle\qquad\qquad = \frac{100}{3} + (-32) = \frac{4}{3}$

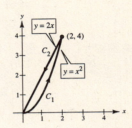

39. $\mathbf{F} = x\mathbf{i} - \sqrt{y}\,\mathbf{j}$ is conservative.

$\text{Work} = \left[\frac{1}{2}x^2 - \frac{2}{3}y^{3/2}\right]_{(0, \, 0)}^{(4, \, 8)} = \frac{1}{2}(16) - \left(\frac{2}{3}\right)8^{3/2} = \frac{8}{3}\left(3 - 4\sqrt{2}\right)$

41. $\displaystyle\int_C 2xyz \, dx + x^2z \, dy + x^2y \, dz = \left[x^2yz\right]_{(0, \, 0, \, 0)}^{(1, \, 4, \, 3)} = 12$

43. (a) $\displaystyle\int_C y^2 \, dx + 2xy \, dy = \int_0^1 \left[(1 + t)^2(3) + 2(1 + 3t)(1 + t)\right] dt$

$\displaystyle\qquad\qquad = \int_0^1 3(t^2 + 2t + 1) + 2(3t^2 + 4t + 1)] \, dt$

$\displaystyle\qquad\qquad = \int_0^1 (9t^2 + 14t + 5) \, dt$

$\displaystyle\qquad\qquad = \left[3t^3 + 7t^2 + 5t\right]_0^1 = 15$

(b) $\displaystyle\int_C y^2 \, dx + 2xy \, dy = \int_1^4 \left[t(1) + 2(t)\left(\sqrt{t}\right)\frac{1}{2\sqrt{t}}\right] dt$

$\displaystyle\qquad\qquad = \int_1^4 (t + t) \, dt$

$\displaystyle\qquad\qquad = \left[t^2\right]_1^4 = 15$

(c) $\mathbf{F}(x, y) = y^2\mathbf{i} + 2xy\,\mathbf{j} = \nabla f$ where $f(x, y) = xy^2$.

Hence,

$\displaystyle\int_C \mathbf{F} \cdot d\mathbf{r} = 4(2)^2 - 1(1)^2 = 15$

45. $\displaystyle\int_C y \, dx + 2x \, dy = \int_0^2 \int_0^2 (2 - 1) \, dy \, dx = \int_0^2 2 \, dx = 4$

47. $\displaystyle\int_C xy^2 \, dx + x^2y \, dy = \iint_R (2xy - 2xy) \, dA = 0$

49. $\int_C xy\, dx + x^2\, dy = \int_0^1 \int_{x^2}^x x\, dy\, dx = \int_0^1 (x^2 - x^3)\, dx = \dfrac{1}{12}$

51. $\mathbf{r}(u, v) = \sec u \cos v\, \mathbf{i} + (1 + 2 \tan u) \sin v\, \mathbf{j} + 2u\, \mathbf{k}$

$0 \le u \le \dfrac{\pi}{3}, \quad 0 \le v \le 2\pi$

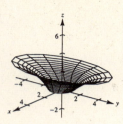

53. (a)

(b)

(c)

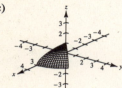

(d)

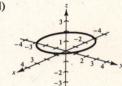

The space curve is a circle:

$$\mathbf{r}\!\left(u, \dfrac{\pi}{4}\right) = \dfrac{3\sqrt{2}}{2} \cos u\, \mathbf{i} + \dfrac{3\sqrt{2}}{2} \sin u\, \mathbf{j} + \dfrac{\sqrt{2}}{2}\, \mathbf{k}$$

(e) $\mathbf{r}_u = -3 \cos v \sin u\, \mathbf{i} + 3 \cos v \cos u\, \mathbf{j}$

$\mathbf{r}_v = -3 \sin v \cos u\, \mathbf{i} - 3 \sin v \sin u\, \mathbf{j} + \cos v\, \mathbf{k}$

$$\mathbf{r}_u \times \mathbf{r}_v = \begin{vmatrix} \mathbf{i} & \mathbf{j} & \mathbf{k} \\ -3 \cos v \sin u & 3 \cos v \cos u & 0 \\ -3 \sin v \cos u & -3 \sin v \sin u & \cos v \end{vmatrix}$$

$$= (3 \cos^2 v \cos u)\mathbf{i} + (3 \cos^2 v \sin u)\mathbf{j} + (9 \cos v \sin v \sin^2 u + 9 \cos v \sin v \cos^2 u)\mathbf{k}$$

$$= (3 \cos^2 v \cos u)\mathbf{i} + (3 \cos^2 v \sin u)\mathbf{j} + (9 \cos v \sin v)\mathbf{k}$$

$$\|\mathbf{r}_u \times \mathbf{r}_v\| = \sqrt{9 \cos^4 v \cos^2 u + 9 \cos^4 v \sin^2 u + 81 \cos^2 v \sin^2 v}$$

$$= \sqrt{9 \cos^4 v + 81 \cos^2 v \sin^2 v}$$

Using a Symbolic integration utility,

$$\int_{\pi/4}^{\pi/2} \int_0^{2\pi} \|\mathbf{r}_u \times \mathbf{r}_v\|\, du\, dv \approx 14.44$$

(f) Similarly,

$$\int_0^{\pi/4} \int_0^{\pi/2} \|\mathbf{r}_u \times \mathbf{r}_v\|\, dv\, du \approx 4.27$$

55. $S: \mathbf{r}(u, v) = u \cos v\mathbf{i} + u \sin v\mathbf{j} + (u - 1)(2 - u)\mathbf{k}, \quad 0 \le u \le 2, 0 \le v \le 2\pi$

$\mathbf{r}_u(u, v) = \cos v\mathbf{i} + \sin v\mathbf{j} + (3 - 2u)\mathbf{k}$

$\mathbf{r}_v(u, v) = -u \sin v\mathbf{i} + u \cos v\mathbf{j}$

$$\mathbf{r}_u \times \mathbf{r}_v = \begin{vmatrix} \mathbf{i} & \mathbf{j} & \mathbf{k} \\ \cos v & \sin v & 3 - 2u \\ -u \sin v & u \cos v & 0 \end{vmatrix} = (2u - 3)u \cos v\mathbf{i} + (2u - 3)u \sin v\mathbf{j} + u\mathbf{k}$$

$\|\mathbf{r}_u \times \mathbf{r}_v\| = u\sqrt{(2u - 3)^2 + 1}$

$$\int\!\!\int_S (x + y) \, dS = \int_0^{2\pi} \int_0^2 (u \cos v + u \sin v) \, u\sqrt{(2u - 3)^2 + 1} \, du \, dv$$

$$= \int_0^2 \int_0^{2\pi} (\cos v + \sin v)u^2\sqrt{(2u - 3)^2 + 1} \, dv \, du = 0$$

57. $\mathbf{F}(x, y, z) = x^2\mathbf{i} + xy\mathbf{j} + z\mathbf{k}$

Q: solid region bounded by the coordinates planes and the plane $2x + 3y + 4z = 12$

Surface Integral: There are four surfaces for this solid.

$z = 0 \quad \mathbf{N} = -\mathbf{k}, \quad \mathbf{F} \cdot \mathbf{N} = -z, \quad \int\!\!\int_{S_1} 0 \, dS = 0$

$y = 0, \quad \mathbf{N} = -\mathbf{j}, \quad \mathbf{F} \cdot \mathbf{N} = -xy, \quad \int\!\!\int_{S_2} 0 \, dS = 0$

$x = 0, \quad \mathbf{N} = -\mathbf{i}, \quad \mathbf{F} \cdot \mathbf{N} = -x^2, \quad \int\!\!\int_{S_3} 0 \, dS = 0$

$2x + 3y + 4z = 12, \mathbf{N} = \dfrac{2\mathbf{i} + 3\mathbf{j} + 4\mathbf{k}}{\sqrt{29}}, dS = \sqrt{1 + \left(\dfrac{1}{4}\right) + \left(\dfrac{9}{16}\right)} \, dA = \dfrac{\sqrt{29}}{4} \, dA$

$$\int\!\!\int_{S_4} \mathbf{F} \cdot \mathbf{N} \, dS = \frac{1}{4}\int\!\!\int_R (2x^2 + 3xy + 4z) \, dA$$

$$= \frac{1}{4}\int_0^6 \int_0^{4 - (2x/3)} (2x^2 + 3xy + 12 - 2x - 3y) \, dy \, dx$$

$$= \frac{1}{4}\int_0^6 \left[2x^2\left(\frac{12 - 2x}{3}\right) + \frac{3x}{2}\left(\frac{12 - 2x}{3}\right)^2 + 12\left(\frac{12 - 2x}{3}\right) - 2x\left(\frac{12 - 2x}{3}\right) - \frac{3}{2}\left(\frac{12 - 2x}{3}\right)^2 \right] dx$$

$$= \frac{1}{6}\int_0^6 (-x^3 + x^2 + 24x + 36) \, dx = \frac{1}{6}\left[-\frac{x^4}{4} + \frac{x^3}{3} + 12x^2 + 36x \right]_0^6 = 66$$

Divergence Theorem: Since div $\mathbf{F} = 2x + x + 1 = 3x + 1$, Divergence Theorem yields

$$\int\!\!\int\!\!\int_Q \text{div } \mathbf{F} \, dV = \int_0^6 \int_0^{(12-2x)/3} \int_0^{(12-2x-3y)/4} (3x + 1) \, dz \, dy \, dx$$

$$= \int_0^6 \int_0^{(12-2x)/3} (3x + 1)\left(\frac{12 - 2x - 3y}{4}\right) dy \, dx$$

$$= \frac{1}{4}\int_0^6 (3x + 1)\left[12y - 2xy - \frac{3}{2}y^2 \right]_0^{(12-2x)/3} dx$$

$$= \frac{1}{4}\int_0^6 (3x + 1)\left[4(12 - 2x) - 2x\left(\frac{12 - 2x}{3}\right) - \frac{3}{2}\left(\frac{12 - 2x}{3}\right)^2 \right] dx$$

$$= \frac{1}{4}\int_0^6 \frac{2}{3}(3x^3 - 35x^2 + 96x + 36) \, dx = \frac{1}{6}\left[\frac{3x^4}{4} - \frac{35x^3}{3} + 48x^2 + 36x \right]_0^6 = 66.$$

59. $\mathbf{F}(x, y, z) = (\cos y + y \cos x)\mathbf{i} + (\sin x - x \sin y)\mathbf{j} + xyz\mathbf{k}$

S: portion of $z = y^2$ over the square in the xy-plane with vertices $(0, 0)$, $(a, 0)$, (a, a), $(0, a)$

Line Integral: Using the line integral we have:

C_1: $y = 0$, $dy = 0$

C_2: $x = 0$, $dx = 0$, $z = y^2$, $dz = 2y\,dy$

C_3: $y = a$, $dy = 0$, $z = a^2$, $dz = 0$

C_4: $x = a$, $dx = 0$, $z = y^2$, $dz = 2y\,dy$

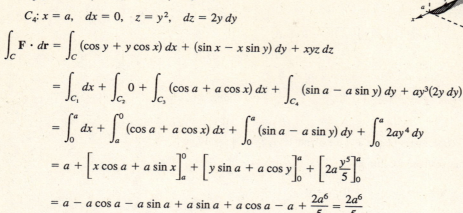

$$\int_C \mathbf{F} \cdot d\mathbf{r} = \int_C (\cos y + y \cos x)\,dx + (\sin x - x \sin y)\,dy + xyz\,dz$$

$$= \int_{C_1} dx + \int_{C_2} 0 + \int_{C_3} (\cos a + a \cos x)\,dx + \int_{C_4} (\sin a - a \sin y)\,dy + ay^3(2y\,dy)$$

$$= \int_0^a dx + \int_a^0 (\cos a + a \cos x)\,dx + \int_0^a (\sin a - a \sin y)\,dy + \int_0^a 2ay^4\,dy$$

$$= a + \left[x \cos a + a \sin x \right]_a^0 + \left[y \sin a + a \cos y \right]_0^a + \left[2a\frac{y^5}{5} \right]_0^a$$

$$= a - a \cos a - a \sin a + a \sin a + a \cos a - a + \frac{2a^6}{5} = \frac{2a^6}{5}$$

Double Integral: Considering $f(x, y, z) = z - y^2$, we have:

$$\mathbf{N} = \frac{\nabla f}{\|\nabla f\|} = \frac{-2y\mathbf{j} + \mathbf{k}}{\sqrt{1 + 4y^2}}, \quad dS = \sqrt{1 + 4y^2}\,dA, \text{ and } \mathbf{curl\,F} = xz\mathbf{i} - yz\mathbf{j}.$$

Hence,

$$\iint_S (\mathbf{curl\,F}) \cdot \mathbf{N}\,dS = \int_0^a \int_0^a 2y^2z\,dy\,dx = \int_0^a \int_0^a 2y^4\,dy\,dx = \int_0^a \frac{2a^5}{5}\,dx = \frac{2a^6}{5}.$$

Problem Solving for Chapter 14

1. (a) $\nabla T = \dfrac{-25}{(x^2 + y^2 + z^2)^{3/2}}[x\mathbf{i} + y\mathbf{i} + z\mathbf{k}]$

$\mathbf{N} = x\mathbf{i} + \sqrt{1 - x^2}\,\mathbf{k}$

$dS = \dfrac{1}{\sqrt{1 - x^2}}\,dA$

$\text{Flux} = \iint_S -k\nabla T \cdot \mathbf{N}\,dS$

$$= 25k \iint_R \left[\frac{x^2}{(x^2 + y^2 + z^2)^{3/2}(1 - x^2)^{1/2}} + \frac{z}{(x^2 + y^2 + z^2)^{3/2}} \right] dA$$

$$= 25k \int_{-1/2}^{1/2} \int_0^1 \left[\frac{x^2}{(x^2 + y^2 + z^2)^{3/2}(1 - x^2)^{1/2}} + \frac{1 - x^2}{(x^2 + y^2 + z^2)^{3/2}(1 - x^2)^{1/2}} \right] dy\,dx$$

$$= 25k \int_{-1/2}^{1/2} \int_0^1 \frac{1}{(1 + y^2)^{3/2}(1 - x^2)^{1/2}}\,dy\,dx$$

$$= 25k \int_0^1 \frac{1}{(1 + y^2)^{3/2}}\,dy \int_{-1/2}^{1/2} \frac{1}{(1 - x^2)^{1/2}}\,dx$$

$$= 25k \left(\frac{\sqrt{2}}{2} \right) \left(\frac{\pi}{3} \right) = 25k\frac{\sqrt{2}\,\pi}{6}$$

—CONTINUED—

1. —CONTINUED—

(b) $\mathbf{r}(u, v) = \langle \cos u, v, \sin u \rangle$

$\mathbf{r}_u = \langle -\sin u, 0, \cos u \rangle, \mathbf{r}_v = \langle 0, 1, 0 \rangle$

$\mathbf{r}_u \times \mathbf{r}_v = \langle -\cos u, 0, -\sin u \rangle$

$\nabla T = \dfrac{-25}{(x^2 + y^2 + z^2)^{3/2}}[x\mathbf{i} + y\mathbf{j} + z\mathbf{k}]$

$\qquad = \dfrac{-25}{(v^2 + 1)^{3/2}}[\cos u\mathbf{i} + v\mathbf{j} + \sin u\mathbf{k}]$

$\nabla T \cdot (\mathbf{r}_u \times \mathbf{r}_v) = \dfrac{-25}{(v^2 + 1)^{3/2}}(-\cos^2 u - \sin^2 u) = \dfrac{25}{(v^2 + 1)^{3/2}}$

$\text{Flux} = \displaystyle\int_0^1 \int_{\pi/3}^{2\pi/3} \dfrac{25k}{(v^2 + 1)^{3/2}}\, du\, dv = 25k\dfrac{\sqrt{2}\pi}{6}$

3. $\mathbf{r}(t) = \langle 3\cos t, 3\sin t, 2t \rangle$

$\mathbf{r}'(t) = \langle -3\sin t, 3\cos t, 2 \rangle, \|\mathbf{r}'(t)\| = \sqrt{13}$

$I_x = \displaystyle\int_C (y^2 + z^2)\rho\, ds = \int_0^{2\pi} (9\sin^2 t + 4t^2)\sqrt{13}\, dt = \dfrac{1}{3}\sqrt{13}\pi(32\pi^2 + 27)$

$I_y = \displaystyle\int_C (x^2 + z^2)\rho\, ds = \int_0^{2\pi} (9\cos^2 t + 4t^2)\sqrt{13}\, dt = \dfrac{1}{3}\sqrt{13}\pi(32\pi^2 + 27)$

$I_z = \displaystyle\int_C (x^2 + y^2)\rho\, ds = \int_0^{2\pi} (9\cos^2 t + 9\sin^2 t)\sqrt{13}\, dt = 18\pi\sqrt{13}$

5. $\dfrac{1}{2}\displaystyle\int_C x\, dy - y\, dx = \dfrac{1}{2}\int_0^{2\pi} [a(\theta - \sin\theta)(a\sin\theta)\, d\theta - a(1 - \cos\theta)(a(1 - \cos\theta))\, d\theta]$

$\qquad = \dfrac{1}{2}a^2 \displaystyle\int_0^{2\pi} [\theta\sin\theta - \sin^2\theta - 1 + 2\cos\theta - \cos^2\theta]\, d\theta$

$\qquad = \dfrac{1}{2}a^2 \displaystyle\int_0^{2\pi} (\theta\sin\theta + 2\cos\theta - 2)\, d\theta$

$\qquad = -3\pi a^2$

Hence, the area is $3\pi a^2$.

7. (a) $\mathbf{r}(t) = t\mathbf{j}, 0 \le t \le 1$

$\mathbf{r}'(t) = \mathbf{j}$

$W = \displaystyle\int_C \mathbf{F} \cdot d\mathbf{r} = \int_0^1 (t\mathbf{i} + \mathbf{j}) \cdot \mathbf{j}\, dt = \int_0^1 dt = 1$

(b) $\mathbf{r}(t) = (t - t^2)\mathbf{i} + t\mathbf{j}, 0 \le t \le 1$

$\mathbf{r}'(t) = (1 - 2t)\mathbf{i} + \mathbf{j}$

$W = \mathbf{F} \cdot d\mathbf{r} = \displaystyle\int_0^1 ((2t - t^2)\mathbf{i} + [(t - t^2)^2 + 1]\mathbf{j}) \cdot ((1 - 2t)\mathbf{i} + \mathbf{j})\, dt$

$\qquad = \displaystyle\int_0^1 [(1 - 2t)(2t - t^2) + (t^4 - 2t^3 + t^2 + 1)]\, dt$

$\qquad = \displaystyle\int_0^1 (t^4 - 4t^2 + 2t + 1)\, dt = \dfrac{13}{15}$

—CONTINUED—

7. —CONTINUED—

(c) $\mathbf{r}(t) = c(t - t^2)\mathbf{i} + t\mathbf{j}, 0 \le t \le 1$

$\mathbf{r}'(t) = c(1 - 2t)\mathbf{i} + \mathbf{j}$

$\mathbf{F} \cdot d\mathbf{r} = (c(t - t^2) + t)(c(1 - 2t)) + (c^2(t - t^2)^2 + 1)(1)$

$\qquad = c^2t^4 - 2c^2t^2 + c^2t - 2ct^2 + ct + 1$

$W = \displaystyle\int_C \mathbf{F} \cdot d\mathbf{r} = \frac{1}{30}c^2 - \frac{1}{6}c + 1$

$\dfrac{dW}{dc} = \dfrac{1}{15}c - \dfrac{1}{6} = 0 \implies c = \dfrac{5}{2}$

$\dfrac{d^2W}{dc^2} = \dfrac{1}{15} > 0 \quad c = \dfrac{5}{2} \text{ minimum.}$

9. $\mathbf{v} \times \mathbf{r} = \langle a_1, a_2, a_3 \rangle \times \langle x, y, z \rangle$

$\qquad = \langle a_2 z - a_3 y, -a_1 z + a_3 x, a_1 y - a_2 x \rangle$

$\mathbf{curl}(\mathbf{v} \times \mathbf{r}) = \langle 2a_1, 2a_2, 2a_3 \rangle = 2\mathbf{v}$

By Stokes's Theorem,

$\displaystyle\int_C (\mathbf{v} \times \mathbf{r}) \, d\mathbf{r} = \iint_S \mathbf{curl}(\mathbf{v} \times \mathbf{r}) \cdot \mathbf{N} \, dS$

$\qquad\qquad = \iint_S 2\mathbf{v} \cdot \mathbf{N} \, dS.$

11. $\mathbf{F}(x, y) = M(x, y)\mathbf{i} + N(x, y)\mathbf{j} = \dfrac{m}{(x^2 + y^2)^{5/2}}[3xy\mathbf{i} + (2y^2 - x^2)\mathbf{j}]$

$M = \dfrac{3mxy}{(x^2 + y^2)^{5/2}} = 3mxy(x^2 + y^2)^{-5/2}$

$\dfrac{\partial M}{\partial y} = 3mxy\left[-\dfrac{5}{2}(x^2 + y^2)^{-7/2}(2y) \right] + (x^2 + y^2)^{-5/2}(3mx)$

$\qquad = 3mx(x^2 + y^2)^{-7/2}[-5y^2 + (x^2 + y^2)] = \dfrac{3mx(x^2 - 4y^2)}{(x^2 + y^2)^{7/2}}$

$N = \dfrac{m(2y^2 - x^2)}{(x^2 + y^2)^{5/2}} = m(2y^2 - x^2)(x^2 + y^2)^{-5/2}$

$\dfrac{\partial N}{\partial x} = m(2y^2 - x^2)\left[-\dfrac{5}{2}(x^2 + y^2)^{-7/2}(2x) \right] + (x^2 + y^2)^{-5/2}(-2mx)$

$\qquad = mx(x^2 + y^2)^{-7/2}[(2y^2 - x^2)(-5) + (x^2 + y^2)(-2)]$

$\qquad = mx(x^2 + y^2)^{-7/2}(3x^2 - 12y^2) = \dfrac{3mx(x^2 - 4y^2)}{(x^2 + y^2)^{7/2}}$

Therefore, $\dfrac{\partial N}{\partial x} = \dfrac{\partial M}{\partial y}$ and $\mathbf{F}$ is conservative.

APPENDIX A

Appendix A.1   Additional Topics in Differential Equations

Solutions to Odd-Numbered Exercises

1.

x	-4	-2	0	2	4	8
y	2	0	4	4	6	8
dy/dx	-2	Undef.	0	$\frac{1}{2}$	$\frac{2}{3}$	1

$\dfrac{dy}{dx} = \dfrac{x}{y}$. For $(x, y) = (-4, 2)$, $\dfrac{dy}{dx} = \dfrac{-4}{2} = -2$.

3. (a), (c)

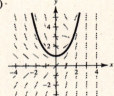

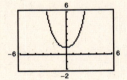

(b) $\dfrac{dy}{dx} = e^x - y$

$\dfrac{dy}{dx} + y = e^x$ Integrating factor: $e^{\int dx} = e^x$

$e^x y' + e^x y = e^{2x}$

$(ye^x) = \displaystyle\int e^{2x}\, dx$

$ye^x = \dfrac{1}{2}e^{2x} + C$

$y(0) = 1 \Rightarrow 1 = \dfrac{1}{2} + C \Rightarrow C = \dfrac{1}{2}$

$ye^x = \dfrac{1}{2}e^{2x} + \dfrac{1}{2}$

$y = \dfrac{1}{2}e^x + \dfrac{1}{2}e^{-x} = \dfrac{1}{2}(e^x + e^{-x})$

5. (a), (c)

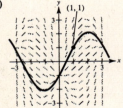

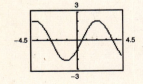

(b) $\dfrac{dy}{dx} = \csc x + y \cot x$

$\dfrac{dy}{dx} - (\cot x)y = \csc x$

Integrating factor: $e^{\int -\cot x\, dx} = e^{-\ln|\sin x|} = \csc x$

$\csc x \cdot y' - \csc x \cot x \cdot y = \csc^2 x$

$(y \csc x)' = \csc^2 x$

$y \csc x = \displaystyle\int \csc^2 x\, dx = -\cot x + C$

$y = -\cos x + C \sin x$

$y(1) = 1 \Rightarrow 1 = -\cos 1 + C \sin 1 \Rightarrow C = \dfrac{1 + \cos 1}{\sin 1}$

≈ 1.83

7.

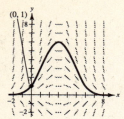

9. $y' = x + y$, $y(0) = 2$, $n = 10$, $h = 0.1$

$y_1 = y_0 + hF(x_0, y_0) = 2 + (0.1)(0 + 2) = 2.2$

$y_2 = y_1 + hF(x_1, y_1) = 2.2 + (0.1)(0.1 + 2.2) = 2.43$, etc.

n	0	1	2	3	4	5	6	7	8	9	10
x_n	0	0.1	0.2	0.3	0.4	0.5	0.6	0.7	0.8	0.9	1.0
y_n	2	2.2	2.43	2.693	2.992	3.332	3.715	4.146	4.631	5.174	5.781

11. $y' = 3x - 2y$, $y(0) = 3$, $n = 10$, $h = 0.05$

$y_1 = y_0 + hF(x_0, y_0) = 3 + (0.05)(3(0) - 2(3)) = 2.7$

$y_2 = y_1 + hF(x_1, y_1) = 2.7 + (0.05)(3(0.05) - 2(2.7)) = 2.4375$, etc.

n	0	1	2	3	4	5	6	7	8	9	10
x_n	0	0.05	0.1	0.15	0.2	0.25	0.3	0.35	0.4	0.45	0.5
y_n	3	2.7	2.438	2.209	2.010	1.839	1.693	1.569	1.464	1.378	1.308

13. $y' = e^{xy}$, $y(0) = 1$, $n = 10$, $h = 0.1$

$y_1 = y_0 + hF(x_0, y_0) = 1 + (0.1)e^{0(1)} = 1.1$

$y_2 = y_1 + hF(x_1, y_1) = 1.1 + (0.1)e^{(0.1)(1.1)} \approx 1.2116$, etc.

n	0	1	2	3	4	5	6	7	8	9	10
x_n	0	0.1	0.2	0.3	0.4	0.5	0.6	0.7	0.8	0.9	1.0
y_n	1	1.1	1.212	1.339	1.488	1.670	1.900	2.213	2.684	3.540	5.958

15. False

$y' + xy = x^2$ is first-order linear.

17. $\dfrac{dy}{dx} + \left(\dfrac{1}{x}\right)y = 3x + 4$

Integrating factor: $e^{\int (1/x)\, dx} = e^{\ln x} = x$

$xy = \int x(3x + 4)\, dx = x^3 + 2x^2 + C$

$y = x^2 + 2x + \dfrac{C}{x}$

19. $\dfrac{dy}{dx} - 3x^2 y = e^{x^3}$

Integrating factor: $e^{-\int 3x^2\, dx} = e^{-x^3}$

$ye^{-x^3} = \int dx$

$ye^{-x^3} = x + C$

$y = (x + C)e^{x^3}$

21. $y' - y = \cos x$

Integrating factor: $e^{\int -1\,dx} = e^{-x}$

$ye^{-x} = \int e^{-x} \cos x \, dx$

$\qquad = \dfrac{1}{2}e^{-x}(-\cos x + \sin x) + C$

$\qquad y = \dfrac{1}{2}(\sin x - \cos x) + Ce^x$

23. $\dfrac{dy}{dx} = \dfrac{x + y}{x} = \dfrac{y}{x} + 1$

$y' - \dfrac{1}{x}y = 1$

Integrating factor: $e^{-\int 1/x\,dx} = e^{-\ln x} = 1/x$

$\dfrac{1}{x}y' - \dfrac{1}{x^2}y = 1/x$

$\left(\dfrac{1}{x}y\right)' = 1/x$

$\dfrac{1}{x}y = \int \dfrac{1}{x}\,dx = \ln|x| + C$

$y = x\ln|x| + Cx$

25. $(3y + \sin 2x)\,dx - dy = 0$

$y' - 3y = \sin 2x$

Integrating factor: $e^{\int -3\,dx} = e^{-3x}$

$ye^{-3x} = \int e^{-3x} \sin 2x \, dx$

$\qquad = \dfrac{1}{13}e^{-3x}(-3\sin 2x - 2\cos 2x) + C$

$\qquad y = -\dfrac{1}{13}(3\sin 2x + 2\cos 2x) + Ce^{3x}$

27. $(x - 1)y' + y = x^2 - 1$

$y' + \left(\dfrac{1}{x - 1}\right)y = x + 1$

Integrating factor: $e^{\int [1/(x-1)]\,dx} = e^{\ln|x-1|} = x - 1$

$y(x - 1) = \int (x^2 - 1)\,dx = \dfrac{1}{3}x^3 - x + C_1$

$\qquad y = \dfrac{x^3 - 3x + C}{3(x - 1)}$

29. $\qquad dy = (y\tan x + 2e^x)\,dx$

$\dfrac{dy}{dx} - (\tan x)y = 2e^x$

Integrating factor: $e^{-\int \tan x\,dx} = e^{\ln|\cos x|} = \cos x$

$y\cos x = \int 2e^x \cos x \, dx = e^x(\cos x + \sin x) + C$

$\qquad y = e^x(1 + \tan x) + C\sec x$

31. $y' - \left(\dfrac{a}{x}\right)y = bx^3$

Integrating factor: $e^{-\int (a/x)\,dx} = e^{-a\ln x} = x^{-a}$

$yx^{-a} = \int bx^3(x^{-a})\,dx = \dfrac{b}{4 - a}x^{4-a} + C$

$\qquad y = \dfrac{bx^4}{4 - a} + Cx^a$

33. $y'\cos^2 x + y - 1 = 0$

$y' + (\sec^2 x)y = \sec^2 x$

Integrating factor: $e^{\int \sec^2 x\,dx} = e^{\tan x}$

$ye^{\tan x} = \int \sec^2 x e^{\tan x}\,dx = e^{\tan x} + C$

$\qquad y = 1 + Ce^{-\tan x}$

Initial condition: $y(0) = 5, C = 4$

Particular solution: $y = 1 + 4e^{-\tan x}$

35. $y' + y\tan x = \sec x + \cos x$

Integrating factor: $e^{\int \tan x\,dx} = e^{\ln|\sec x|} = \sec x$

$y\sec x = \int \sec x(\sec x + \cos x)\,dx = \tan x + x + C$

$\qquad y = \sin x + x\cos x + C\cos x$

Initial condition: $y(0) = 1, 1 = C$

Particular solution: $y = \sin x + (x + 1)\cos x$

37. $y' + \left(\dfrac{1}{x}\right)y = 0$

Integrating factor: $e^{\int (1/x)\,dx} = e^{\ln|x|} = x$

Separation of variables:

$$\dfrac{dy}{dx} = -\dfrac{y}{x}$$

$$\int \dfrac{1}{y}\,dy = \int -\dfrac{1}{x}\,dx$$

$$\ln y = -\ln x + \ln C$$

$$\ln xy = \ln C$$

$$xy = C$$

Initial condition: $y(2) = 2,\ C = 4$

Particular solution: $xy = 4$

39. $x\,dy = (x + y + 2)\,dx$

$$\dfrac{dy}{dx} - \left(\dfrac{1}{x}\right)y = \dfrac{x + 2}{x}$$

Integrating factor: $e^{\int -(1/x)\,dx} = e^{-\ln|x|} = \dfrac{1}{x}$

$$y\left(\dfrac{1}{x}\right) = \int \dfrac{x + 2}{x^2}\,dx = \ln|x| - \dfrac{2}{x} + C$$

$$y = x\ln|x| - 2 + Cx$$

Initial condition: $y(1) = 10$

$$10 = -2 + C \implies C = 12$$

Particular solution: $y = x\ln|x| - 2 + 12x$

41. (a)

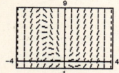

(b) $\dfrac{dy}{dx} - \dfrac{1}{x}y = x^2$

Integrating factor $e^{\int -(1/x)\,dx} = e^{-\ln x} = \dfrac{1}{x}$

$$\dfrac{1}{x}y' - \dfrac{1}{x^2}y = x$$

$$\left(\dfrac{1}{x}y\right) = \int x\,dx = \dfrac{x^2}{2} + C$$

$$y = \dfrac{x^3}{2} + Cx$$

$(-2, 4)\!: 4 = \dfrac{-8}{2} - 2C \implies C = -4 \implies y = \dfrac{x^3}{2} - 4x = \dfrac{1}{2}x(x^2 - 8)$

$(2, 8)\!: 8 = \dfrac{8}{2} + 2C \implies C = 2 \implies y = \dfrac{x^3}{2} + 2x = \dfrac{1}{2}x(x^2 + 4)$

(c)

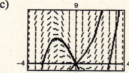

43. $L\dfrac{dI}{dt} + RI = E_0,\ I' + \dfrac{R}{L}I = \dfrac{E_0}{L}$

Integrating factor: $e^{\int (R/L)\,dt} = e^{Rt/L}$

$$I\,e^{Rt/L} = \int \dfrac{E_0}{L}e^{Rt/L}\,dt = \dfrac{E_0}{R}e^{Rt/L} + C$$

$$I = \dfrac{E_0}{R} + Ce^{-Rt/L}$$

45. $L\dfrac{dI}{dt} + RI = E_0 \sin \omega t$

$\dfrac{dI}{dt} + \dfrac{R}{L}I = \dfrac{E_0}{L}\sin \omega t$

Integrating factor: $e^{\int (R/L)\,dt} = e^{Rt/L}$

$Ie^{Rt/L} = \displaystyle\int \dfrac{E_0}{L}e^{Rt/L}\sin \omega t\,dt$

$\qquad = \dfrac{E_0}{L}\left[\dfrac{L^2 e^{Rt/L}}{R^2 + L^2\omega^2}\left(\dfrac{R}{L}\sin \omega t - \omega \cos \omega t\right)\right] + C = \dfrac{E_0 e^{Rt/L}}{R^2 + \omega^2 L^2}(R \sin \omega t - \omega L \cos \omega t) + C$

$\qquad I = \dfrac{E_0}{R^2 + \omega^2 L^2}(R \sin \omega t - \omega L \cos \omega t) + Ce^{-Rt/L}$

47. $\qquad\qquad \dfrac{dP}{dt} = kP + N, N \text{ constant}$

$\qquad\qquad \dfrac{dP}{kP + N} = dt$

$\displaystyle\int \dfrac{1}{kP + N}\,dP = \int dt$

$\dfrac{1}{k}\ln(kP + N) = t + C_1$

$\quad \ln(kP + N) = kt + C_2$

$\qquad kP + N = e^{kt + C_2}$

$\qquad\qquad P = \dfrac{C_3 e^{kt} - N}{k}$

$\qquad\qquad P = Ce^{kt} - \dfrac{N}{k}$

When $t = 0$: $P = P_0$

$\quad P_0 = C - \dfrac{N}{k} \Rightarrow C = P_0 + \dfrac{N}{k}$

$\quad P = \left(P_0 + \dfrac{N}{k}\right)e^{kt} - \dfrac{N}{k}$

49. (a) $A = \dfrac{P}{r}(e^{rt} - 1)$

$\qquad A = \dfrac{100{,}000}{0.06}(e^{0.06(5)} - 1) \approx 583{,}098.01$

(b) $A = \dfrac{250{,}000}{0.05}(e^{0.05(10)} - 1) \approx 3{,}243{,}606.35$

51. $\dfrac{dA}{dt} - rA = -P$

For this linear differential equation, we have $P(t) = -r$ and $Q(t) = -P$. Therefore, the integrating factor is

$u(x) = e^{\int -r\,dt} = e^{-rt}$ and the solution is

$\qquad A = e^{rt}\displaystyle\int -Pe^{-rt}\,dt = e^{rt}\left(\dfrac{P}{r}e^{-rt} + C\right) = \dfrac{P}{r} + Ce^{rt}.$

Since $A = A_0$ when $t = 0$, we have $C = A_0 - (P/r)$ which implies that

$\qquad A = \dfrac{P}{r} + \left(A_0 - \dfrac{P}{r}\right)e^{rt}.$

53. (a) $\dfrac{dQ}{dt} = q - kQ$, q constant

(b) $Q' + kQ = q$

Let $P(t) = k$, $Q(t) = q$, then the integrating factor is $u(t) = e^{kt}$.

$$Q = e^{-kt} \int q e^{kt}\, dt = e^{-kt}\left(\frac{q}{k}e^{kt} + C\right) = \frac{q}{k} + Ce^{-kt}$$

When $t = 0$: $Q = Q_0$

$$Q_0 = \frac{q}{k} + C \implies C = Q_0 - \frac{q}{k}$$

$$Q = \frac{q}{k} + \left(Q_0 - \frac{q}{k}\right)e^{-kt}$$

(c) $\displaystyle\lim_{t\to\infty} Q = \frac{q}{k}$

55. $y' - 2x = 0$

$$\int dy = \int 2x\, dx$$

$$y = x^2 + C$$

Matches c.

57. $y' - 2xy = 0$

$$\int \frac{dy}{y} = \int 2x\, dx$$

$$\ln y = x^2 + C_1$$

$$y = Ce^{x^2}$$

Matches a.